EXPERIMENTS ON SIMPLE MAGNETIC MODEL SYSTEMS

A survey of their experimental status in the light of current theories

L. J. de JONGH
Kamerlingh Onnes Laboratory, Leiden

and

A. R. MIEDEMA
Philips Research Laboratories, Eindhoven

TAYLOR & FRANCIS LTD
10–14 Macklin Street, London WC2B 5NF
1974

First published 1974 *by Taylor & Francis Ltd.*, 10–14 *Macklin Street, London WC2B 5NF*

Reprinted from ADVANCES IN PHYSICS, *Volume* 23, *No.* 1, *January* 1974

Printed and bound in Great Britain by Taylor & Francis Ltd., 10–14 *Macklin Street, London WC2B 5NF*

ISBN 0 85066 085 8

Distributed in the United States of America and its territories by Barnes & Noble Books (a division of Harper & Row Publishers Inc.), 10 *East* 53*rd Street, New York, N.Y.* 10022.

". . . . For the truth of the conclusions of Physical science, observation is the supreme Court of Appeal. . . ." (Sir Arthur Eddington, *The Philosophy of Physical Science.*)

Outline and Aims

In this monograph we review the theoretical and experimental results obtained on simple magnetic model systems. We consider the Heisenberg, XY and Ising types of interaction (ferro and antiferro-magnetic), on magnetic lattices of dimensionality 1, 2 and 3.

Particular attention is paid to the approximation of these model systems in real crystals, *viz.* how they can be realized or be expected to exist in nature. A large number of magnetic compounds which, according to the available experimental information, meet the requirements set by one or the other of the various models are considered and their properties discussed. Many examples will be given that demonstrate to what extent experiments on simple magnetic systems support theoretical descriptions of magnetic ordering phenomena and contribute to their understanding. We also aim to indicate those directions in which we feel there is either a strong need for or the possibility of useful developments in future work.

Contents

EXPERIMENTS ON SIMPLE

MAGNETIC MODEL SYSTEMS

EXPERIMENTS ON SIMPLE MAGNETIC MODEL SYSTEMS

§ 1. Introduction

1.1. *Magnetic model systems*

Most of the progress in theoretical and experimental investigations on critical phenomena has resulted from the introduction of a large variety of lattice models. In that way theoretical physicists have been able to obtain exact or approximate solutions for the behaviour of thermodynamic quantities near phase transitions, calculations that would otherwise not have been possible, due in most cases to the insurmountable mathematical problems associated with cooperative phenomena. In these model systems the magnetic interaction is more or less simplified, while the dimensionality of the lattice may be varied. As an illustration, consider the interaction Hamiltonian

$$\mathscr{H} = -2J \sum_{i>j} [aS_i^zS_j^z + b(S_i^xS_j^x + S_i^yS_j^y)] \tag{1.1}$$

where summation is taken over nearest neighbouring spins and J is the exchange constant. If we put $a = b = 1$ we obtain the Heisenberg model, in which the interaction is wholly isotropic. The other extreme, the anisotropic Ising interaction, is obtained by setting $a = 1$ and $b = 0$. The third case $a = 0$, $b = 1$, is called the XY model, or the planar Heisenberg model if one puts the additional requirement that the spins are constrained to lie within the xy plane.

A useful concept in this connection is that of an order parameter, that is a quantity which is a measure of the amount of ordering present in the system below the critical point†. In magnetic systems the spontaneous magnetization may be taken as the order parameter and it is readily seen that in the Ising model this is a scalar quantity (one-dimensional vector), since the magnetization can only point up or down. In the Heisenberg and the planar Heisenberg model, the order parameter is a three and a two-dimensional vector, respectively.

† Throughout this book we will denote by the symbol T_c the (critical) temperature at which the system undergoes a transition to long-range order. We will not use different symbols for ferro and antiferromagnets (Curie and Néel point).

Apart from the dimension of the order parameter (spin-dimensionality) one may choose a lattice of arbitrary dimensionality, the one, two and three-dimensional lattices being the most frequently studied. In addition the spin value may be varied ; the quantities S_i in (1.1) denote spin operators for quantum-mechanical systems with $S=\frac{1}{2}$, 1, $\frac{3}{2}$, . . ., or they represent classical vectors in the case of the classical models. The latter are treated by replacing J in (1.1) by J'/S^2 and allowing S to tend to infinity.

Another variable quantity is the sign of the exchange constant J which can be either positive or negative, giving rise to ferro or anti-ferromagnetism, respectively. Also, intermediate models have been studied in which the ratio of the constants a and b in (1.1) can take any value, e.g. Yang and Yang (1966), Dalton and Wood (1967). Furthermore one may extend the range of the interaction by including interactions with next-nearest (or still further) neighbours, in this way obtaining information about the influence of further neighbour interactions upon the critical behaviour. Finally, also vector order parameters of a dimension higher than 3 have been studied, the limit of infinite spin-dimensionality being equivalent to the so-called spherical model (Berlin and Kac 1952, Stanley 1968 c).

The original aim of theoretical physicists in devising these various model systems was to get a better understanding of experimental observations. However, in studying lattice models of lower dimensionality ($d=1$, 2), it appeared that certain features of thermodynamic quantities, which were only minor effects in the behaviour observed at that time, turned into gross features in the lower dimensional systems. At first this was a little distressing, since one had hoped that the models would not be oversimplifications of reality so that the main characteristics of the phase transition would remain preserved. But one has come to understand the origin of these effects and in fact the comparison of results obtained for different models has proven to be most rewarding, as it elucidates the way in which, for instance, the lattice dimensionality or the type or range of the interaction influences the general features of a phase transition. Moreover, during the last decade the situation in some way is reversed in that experimental physicists are now supplying theorists with data which may be compared with models that at first sight are most unorthodox. Partly by accident, but in most cases by carefully choosing magnetic substances from the immense reservoir of compounds offered by chemistry and metallurgy, experimentalists have been able to find materials whose properties resemble quite closely those predicted for various theoretical models. Indeed, to such an extent that one sometimes wonders how artificial and unphysical such a model has to be, in order to prevent the discovery of an approximation in the laboratory !

It is the aim of this paper to describe how the conditions set by particular models have been met experimentally and to survey the information available, the experimental results having for most part been obtained during the past ten years. Theoretical calculations will be mentioned in connection with the experiments. The reader who is interested in the details of the theory is referred to the reviews of Domb (1960), Fisher (1965 b, 1967), Stephenson (1971) and the book by Stanley (1971). In addition there are the series edited by Rado and Suhl (' Magnetism ') and by Domb and Green (' Phase Transitions and Critical Phenomena '). Among the reviews which give more emphasis to the experimental data relevant to the present paper, we mention those of Domb and Miedema (1964), Huiskamp (1966), Heller (1967), Kadanoff *et al.* (1967) and Wielinga (1971). Lastly we point to the work of Keffer (1966), who has given an extensive review of spin waves in theory and experiment.

It must be emphasized that, although we shall be concerned only with magnetic systems in this paper, many of the Hamiltonians described above also allow for an interpretation other than a magnetic one. For instance, the Ising Hamiltonian may be used to describe the liquid–gas transition in a lattice gas, as well as the spontaneous phase separation observed in binary fluids and the ordering in alloys. The XY model is the magnetic analogue of a quantum fluid and is therefore of relevance to the theory of superfluidity and superconductivity.

The layout of this paper is as follows. In the remainder of this introduction we shall discuss briefly the effects of lattice dimensionality and the type of interaction. We shall recall how these may be understood in a qualitative way by considering spin–spin correlations. In § 2 we will indicate how general rules can be given for finding compounds that approximate a particular model system. In the next section a collection of the hitherto discovered examples are tabulated with a short account of their individual properties, e.g. quantitative information (if available) about the deviations from the ideal model which evidently will be met in any experimental system. To give a full account of all the publications of the many workers in the field of magnetic transitions clearly is an impossible task and therefore we have aimed merely to offer the interested theoretician a guide to the available data, supplying simultaneously to the experimentalist a survey of the existing theoretical work and of the extensive range of compounds already discovered. Accordingly, we want to apologize beforehand to those investigators whose work has not been included in the present review, through oversight or lack of space.

As concerns the fundamental thermodynamic properties of 1, and 2 and 3-d (d = dimensional) lattices (e.g. specific heat, susceptibility, etc.), comparisons between theory and experiment will be found in § 3. The last section is devoted to a number of special subjects, namely results obtained from various theoretical approaches and their experimental verification. In reading §§ 3 and 4, we think both theorists and experimentalists will share the joy of seeing how extremely well theory and experiment have been found to fit in many cases. Although there certainly remain a number of problems to be solved, one may say that a considerable amount of understanding of the physics in these magnetic model systems has already been reached.

1.2. *Effects of dimensionality and type of interaction*

Changing the dimensionality of a magnetic lattice has a dramatic effect upon the thermodynamic properties. This can clearly be illustrated by considering the specific heat behaviour. In fig. 1 are compared the theoretical specific heats of the Ising model for a 1, 2 and 3-d lattice, together with the molecular field (MF) prediction. We first remark that the MF theory fails to account even for the behaviour of the 3-d Ising model, which corresponds rather closely to what is mostly observed experimentally. The MF prediction for the transition point is too high, the specific heat shows a finite discontinuity instead of diverging, and is furthermore characterized by the absence of the characteristic 'high-temperature tail'. The latter is encountered in all the more sophisticated models as well as experimentally, and is due to the presence of short-range interactions above T_c that are not taken into account in the MF theory. Neglect of the short-range order is in fact the reason why the

Fig. 1

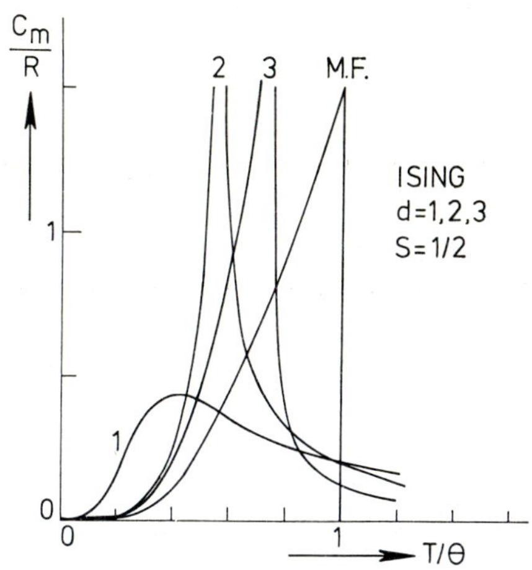

Theoretical magnetic specific heats C_m of the $S=\frac{1}{2}$ Ising model for a 1, 2 and 3-d lattice. The chain curve has been obtained by Ising (1925), who first performed calculations on the model that bears his name. The 2-d curve is also an exact result, derived by Onsager (1944) for the quadratic lattice. The 3-d curve has been calculated by Blöte and Huiskamp (1969) and Blöte (1972) for the simple cubic lattice from the high and low-temperature series expansions of C_m given by Baker *et al.* (1963) and Sykes *et al.* (1972). For comparison, the molecular field prediction (MF) has been included. R denotes the gas constant and θ is the Curie–Weiss temperature ($\theta=\frac{2}{3}zS(S+1)J/k$), which is the transition temperature according to the MF theory.

MF model breaks down if we approach the transition point closely enough, although it has been highly successful in describing the overall properties of (3-d) magnetic substances.

Secondly, one may see by comparing, e.g., the T_c/θ and the high-temperature tails that the relative importance of the short-range order is greatly enhanced by lowering the lattice dimensionality. For the 2-d Ising model there still occurs a transition to long-range order at a finite temperature, reflected as a divergence in the specific heat, although T_c is lowered further with respect to the MF prediction. However for the chain model T_c has moved down to zero, and all of the entropy has to be removed by short-range interactions, resulting in a broad Schottky-type anomaly.

Clearly, the deficiency of the MF theory is more crudely exposed the lower the dimensionality, and this is one of the reasons that makes these 1-d and 2-d systems such interesting objects of study. Moreover, since we know that the origin of the deficiency lies in the introduction of an effective field, which replaces the interactions of a magnetic moment with its neighbours by an average taken over the entire system, it follows that we may obtain a better understanding of the thermodynamic behaviour of magnetic substances by studying the correlations between a given reference spin and its neighbours at a varying distance r. In an elucidating discussion, Fisher (1965 b) considered the static pair correlation functions

$$\Gamma_{\mathbf{r}}(T) = \langle S_0^z S_{\mathbf{r}}^z \rangle / \tfrac{1}{3} S(S+1) \quad (\mathbf{r} = 0, 1, 2, \ldots \infty), \tag{1.2}$$

where the brackets denote the expectation value and $\frac{1}{3}S(S+1)$ is just a normalization factor. The qualitative behaviour of the $\Gamma_{\mathbf{r}}$ as a function of temperature for models possessing a finite transition point is sketched in fig. 2 (*a*). Although exact results have been obtained only for the 2-d Ising model (Kaufman and Onsager 1949, Fisher 1960 b), it can be argued that most of the arguments given below will also have a more general validity.

It is seen from fig. 2 (*a*) that, with the exception of the infinite range correlation Γ_∞ which vanishes at the transition point†, all curves display an inflexion point at T_c. For the 2-d Ising lattice, singularities occur of the form

$$\Gamma_{\mathbf{r}}(T) = A + B|T - T_c| \ln |T - T_c|. \tag{1.3}$$

As A and B are constants, it follows that the curves have tangents with an infinite slope at T_c.

Now it turns out that many thermodynamic quantities may be related to the $\Gamma_{\mathbf{r}}$ in a simple way. For instance, we may take Γ_∞ as a long-range-order parameter and in fact it can be shown that $\langle S_0^z S_\infty^z \rangle$ is proportional

† The self-correlation function Γ_0 evidently forms another exception. This is identical to unity in the models of interest in the present context.

to the square of the spontaneous magnetization (M_s)†. Furthermore, in the case of nearest-neighbour interactions only, we may write for the magnetic energy

$$U_m(T) = -N_0 z |J| \langle S_0^z S_1^z \rangle, \tag{1.4}$$

from which we see that the magnetic energy is simply proportional to $\Gamma_1(T)$. Here N_0 is the total number of magnetic spins, z is the number of nearest neighbours and J is the exchange constant.

Fig. 2

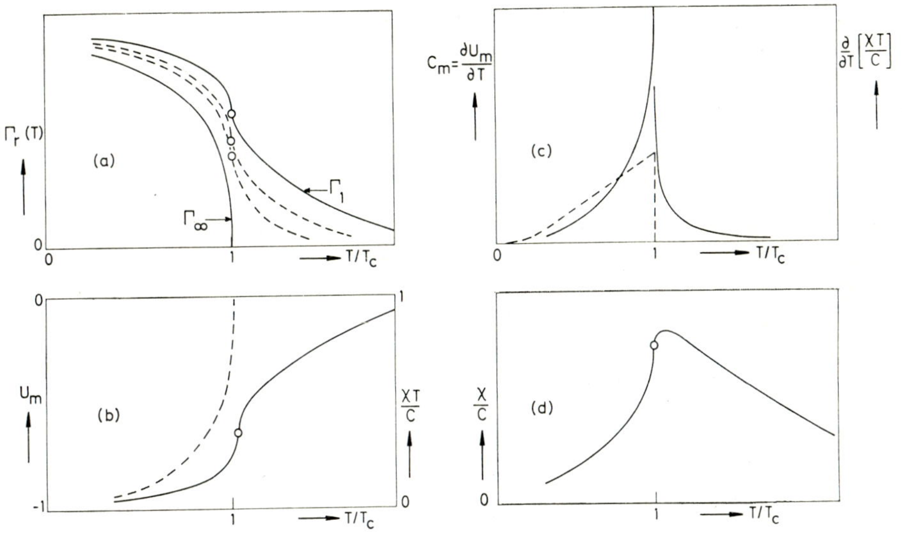

(*a*) Qualitative temperature dependence of the static correlation functions Γ_r (eqn. (1.2)). Plotted are Γ_1 and Γ_∞ (solid curves) and two functions with intermediate r (broken curves). (*b*) Behaviour of the magnetic energy, $U_m \simeq -|\Gamma_1|$, as well as of the product of the antiferromagnetic parallel susceptibility and the temperature (C denotes Curie's constant), $\chi T/C \simeq 1 - |\Gamma_1|$. The broken curve gives the behaviour of the square of the spontaneous magnetization ($M_s^2 \sim \Gamma_\infty$), which should be equal to U_m according to MF theory. (*c*) Temperature dependence of the magnetic specific heat, $C_m = \partial U_m/\partial T$, as well as of the temperature derivative of $\chi T/C$. The broken curve again denotes the MF result. (*d*) Antiferromagnetic parallel susceptibility χ/C versus the relative temperature T/T_c (in (*a*), (*b*) and (*d*) the position of T_c has been indicated by the open circles). (After Fisher 1965 b.)

† Throughout this paper we shall adhere to this statement and shall not go into the theoretical problems involved in establishing this relationship. Likewise, we shall disregard the subtle difference between long–long-range order and short–long-range order. The interested reader may find a discussion of these questions in the paper of Fisher and Jasnow (1971) as well as references to earlier papers bearing on this subject.

Suitably normalized, the behaviour of $U_m(T)$ is that given in fig. 2 (*b*) (solid curve), bearing in mind that U_m is a negative quantity. By taking the temperature derivative of this curve, the specific heat is obtained (fig. 2 (*c*)) and obviously this will rise logarithmically to infinity at T_c if U_m possesses a singularity of the form given by (1.3). Furthermore, we observe that the presence of short-range correlations above T_c accounts indeed for the appearance of a high-temperature tail in the specific heat.

The failure of the MF theory is now also readily traced, since in that theory, in the absence of short-range correlations, $U_m \sim M_s^2 \sim \Gamma_\infty$. The corresponding results for the energy and specific heat are given by the dashed curves in fig. 2 (*b*), (*c*).

Another quantity that is closely related to the correlation functions is the magnetic susceptibility. According to the fluctuation theorem of statistical mechanics the susceptibility in the limit of zero field (initial susceptibility) is given by

$$\chi T/C = 1 + \sum_{\mathbf{r} \neq 0} \Gamma_{\mathbf{r}}(T), \tag{1.5}$$

where C denotes the Curie constant. For a paramagnetic system, all $\Gamma_{\mathbf{r}}$ ($\mathbf{r} \neq 0$) being zero, eqn. (1.5) simply states Curie's law. For a purely ferromagnetic system the sum in (1.5) contains only positive terms and diverges at T_c. On the other hand, in the case of an antiferromagnetic interaction the terms in the series are alternating in sign and it has been argued by Fisher (1962) that one may replace the sum in (1.5) by

$$\chi T/C \simeq 1 - f(T)|\Gamma_1(T)|, \tag{1.6}$$

where the function $f(T)$ accounts for all the omitted terms in the series. It turns out that $f(T)$ is of order unity at T_c and moreover is only very slowly varying with temperature, so that in our qualitative picture we may neglect its presence. It is then easily seen that

$$\chi T/C \simeq 1 - |U_m(T)|. \tag{1.7}$$

This relation between the energy and the antiferromagnetic susceptibility can also be derived from general thermodynamic arguments (Sawatzky and Bloom 1962, Skalyo *et al.* 1967) and has been tested for a variety of antiferromagnets (Fisher 1962, Wolf and Wyatt 1964, Skalyo *et al.* 1967). Consequently, the curve for U_m in fig. 2 (*b*) also represents the qualitative behaviour of $\chi T/C$ (explaining the term 'energetic susceptibility' for the quantity $\chi T/C$). Of much importance is the result for the susceptibility itself depicted in fig. 2 (*d*), obtained by dividing $\chi T/C$ (fig. 2 (*b*)) by the temperature. One may observe that the relationship given by (1.7) implies that the maximum of the antiferromagnetic susceptibility must occur somewhat above T_c, whereas at T_c itself the temperature derivative of the susceptibility reaches infinity, at least for the Ising model, where the specific heat is predicted to diverge in two and three dimensions. For the Heisenberg model C_m is expected to display a finite cusp at T_c for $d=3$, in which case $\partial\chi/\partial T$ will also remain finite, passing through a maximum at T_c. Although in this discussion the parallel susceptibility is considered, a similar result has been obtained for the perpendicular susceptibility (Fisher 1963). The difference between T_c and the temperature of the maximum T_{max} again reflects the presence of short-range order above T_c and therefore will be enhanced by lowering the dimensionality. Obviously, in the MF theory both temperatures coincide.

Thus we have seen that the short-range-order effects are reflected amongst other things in the specific heat tail and in the difference $T_{max}-T_c$. The important quantity in this respect is the constant A in (1.3), since this gives the amount of magnetic energy that is still present at T_c. It will come as no surprise that for the Ising model Sykes and Fisher (1962) found this amount to be about two to three times larger for the 2-d than for the 3-d lattices.

We shall now illustrate the above arguments with a few examples. In fig. 3 the theoretical specific heats of the Ising model in two and three dimensions are once more compared, this time on a temperature scale relative to T_c, so that the enlargement of the high-temperature tail can be seen more clearly. Quantitatively, the ratio of the areas under the specific heat curve below and above T_c is 3·1 for the f.c.c. and 0·41 for the quadratic lattice (see, e.g., Domb and Miedema 1964). In fig. 4 a similar

Fig. 3

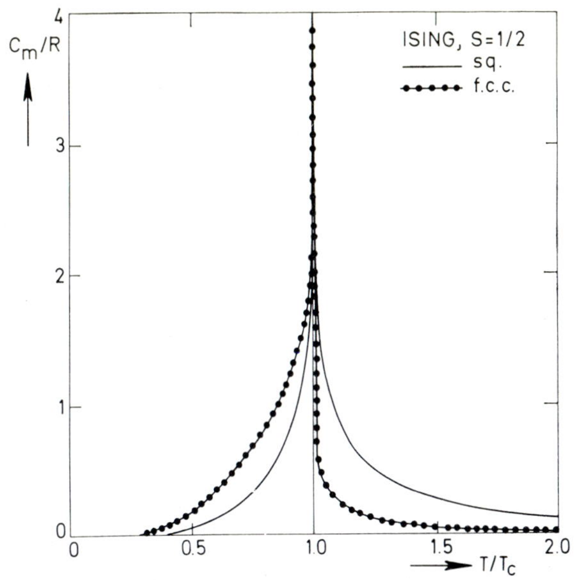

Theoretical specific heats of the $S=\frac{1}{2}$ Ising model. Solid curve: square (simple quadratic) lattice; dotted curve: face-centred cubic lattice. (After Domb 1960.)

comparison is made for the antiferromagnetic susceptibility. Besides the larger reduction of T_c for the 2-d lattice with respect to the MF prediction $\theta = 2zS(S+1)J/3k$, we observe the huge enlargement of the difference $T_{max} - T_c$, and also of the ratio $\chi(T_{max})/\chi(T_c)$, on lowering the dimensionality, in accordance with the qualitative arguments sketched above. As a last example we show in fig. 5 the spontaneous magnetization of a 2-d and a 3-d Ising model. For the triangular lattice it is seen that $M_s(T)$ retains near saturation values up to much higher relative temperatures as compared with the f.c.c. lattice. This is associated with a lower value of the critical exponent β in the power law

$$M_s(T)/M_s(0) \sim (1 - T/T_c)^{\beta}, \tag{1.8}$$

which describes the vanishing of the long-range order as T_c is approached (for a definition of the critical exponents associated with the thermodynamic functions see, e.g., Fisher 1967). For 3-d lattices β is about $\frac{1}{3}$, whereas the 2-d Ising lattices have $\beta = \frac{1}{8}$ (compare with the MF prediction $\beta = \frac{1}{2}$). We may once more use our qualitative picture by saying that, since below T_c the correlation functions will very rapidly approach the limiting Γ_∞ when studied as a function of r (with the exception of the region very close to T_c), a high value of the constant A in the expression of Γ_1 (eqn. (1.3)) will imply a steeper Γ_1 as well as a steeper Γ_∞ curve.

Concerning the behaviour in magnetic chains one can conclude from the absence of long-range ordering at any finite temperature and from

Fig. 4

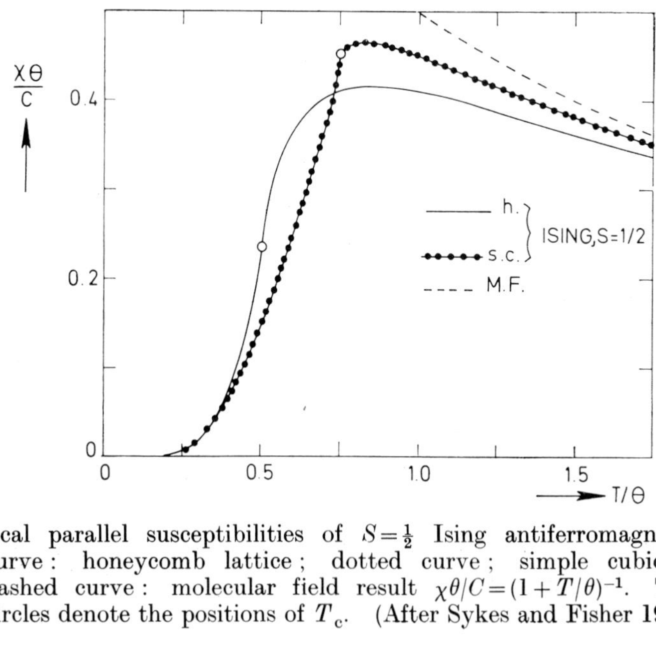

Theoretical parallel susceptibilities of $S = \frac{1}{2}$ Ising antiferromagnets. Full curve: honeycomb lattice; dotted curve; simple cubic lattice; dashed curve: molecular field result $\chi\theta/C = (1 + T/\theta)^{-1}$. The open circles denote the positions of T_c. (After Sykes and Fisher 1962.)

the specific heat curve shown in fig. 1 that the energy must appear as a smooth curve, possessing no singularities of the form given by eqn. (1.3). Likewise for the antiferromagnetic susceptibility, although it will show a broad maximum, it will not have a diverging temperature derivative as in the case of a transition.

Fig. 5

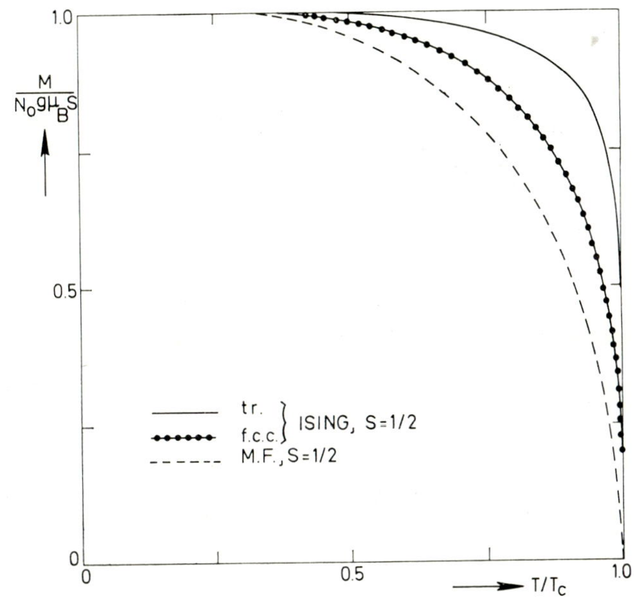

Theoretical spontaneous magnetizations of the $S=\frac{1}{2}$ Ising model. Full curve: triangular lattice; dotted curve: face-centred cubic lattice; dashed curve: molecular field prediction. (After Guttman *et al.* 1970.)

By comparing the specific heat behaviour of the Heisenberg model for $d=1$, 2, 3 as shown in fig. 6 with the corresponding Ising curves (fig. 1), it can be inferred that changing the type of interaction from the anisotropic Ising to the isotropic Heisenberg form has the effect of enhancing the short-range-order contributions. For the 3-d Heisenberg models, which do possess a phase transition, this can be seen by comparing the critical parameters (e.g. T_c/θ) with the corresponding values for the Ising model (Domb and Miedema 1964). A similar qualitative picture for the thermodynamic behaviour as given above will apply to the 3-d Heisenberg model, except for the already mentioned indications obtained from the analysis of series expansions (Baker *et al.* 1967 b) that the specific heat remains finite at T_c, although its derivative is still infinite on both sides of T_c so that the curve displays a sharp cusp. Accordingly the temperature derivative of the correlation functions will not diverge at T_c, so that the Γ_r must be of a different form than that given by eqn. (1.3).

Fig. 6

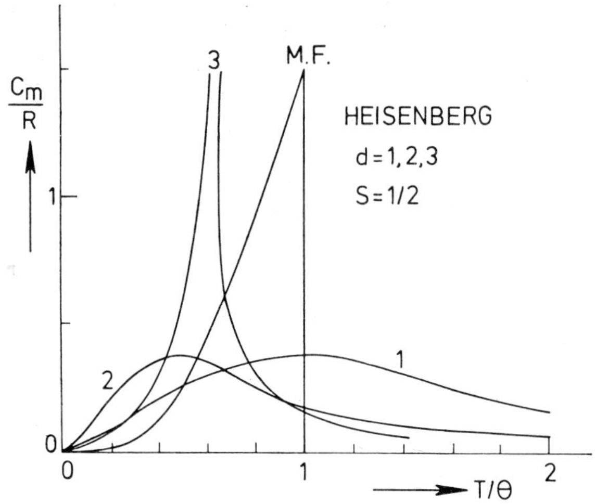

Specific heats of the $S=\frac{1}{2}$ Heisenberg model in 1, 2 and 3 dimensions. The 1-d curve is the result for the antiferromagnetic chain obtained by Bonner and Fisher (1964), from approximate solutions. The 2-d curve applies to the ferromagnetic quadratic lattice and has been constructed by Bloembergen (1971) from the predictions of spin–wave theory ($T/\theta<0{\cdot}1$), from the high-temperature series expansion ($T/\theta>1$), and from the experimental data on approximants of this model ($0{\cdot}1<T/\theta<1$), to be discussed below. The 3-d curve follows from series expansions for the b.c.c. ferromagnet given by Baker *et al.* (1967 b). Also included is the molecular field prediction.

The enhancement of the importance of the short-range-order effects also follows from the fact that in the case of the Heisenberg model a lowering of the dimensionality to 2 is already sufficient to prevent the onset of long-range order at a non-zero temperature (Mermin and Wagner 1966). The thermodynamics of the 2-d Heisenberg model will therefore to a certain extent resemble the behaviour found in the chain models; to a certain extent because there is a possible difference following from the analysis of series expansions of the susceptibility (Stanley and Kaplan 1966), in which indications were found for the existence of non-zero transition points at which the ferromagnetic susceptibility diverges. Thus, although the chain models as well as the 2-d Heisenberg model cannot sustain a spontaneous magnetization at any finite temperature, the latter would distinguish itself by possessing a transition to a phase with an infinite susceptibility. We will return to this intriguing problem later. At this point we merely remark that since the 2-d XY models have been found to possess similar properties as the 2-d Heisenberg model, the anisotropy evidently must be of the Ising form to enable a transition to long-range order to occur in a 2-d lattice.

Finally we mention the influence of the spin value and the interaction range on the properties of a magnetic system. By studying the critical parameters as a function of S one finds that the short-range-order effects are enhanced by lowering S (e.g. Domb and Miedema 1964, Fisher 1967). Evidently the fact that a system is quantum-mechanical, in the sense that it has a finite spin value, also increases the deviations from the MF theory. As concerns the range of the interaction, it has been proven by various workers (see, e.g., Fisher 1967) that the MF theory becomes exact in the limit of an infinite interaction range. Intuitively this may be understood by considering the failure of the effective field concept in the case of short-range interactions, as outlined above. Indeed, the validity of the MF theory in the infinite-range limit can easily be inferred from the qualitative picture obtained with the aid of the correlation functions. From the fact that the MF approximation amounts to replacing the interaction of a given spin with its neighbours by an average taken over the whole magnetic system, one may understand why the MF theory becomes exact in the limit $z \to \infty$ (Brout 1965), that is if one has an infinite number of equivalently interacting magnetic neighbours.

§ 2. Approximation of magnetic lattice models in real crystals

2.1. *General remarks*

In the following sections we shall classify the various model systems considered according to the dimension of the magnetic lattice and the type and sign of the exchange interaction. Concerning the magnetic lattice dimensionality, d, it is clear that experimentally only $d \leqslant 3$ can possibly be achieved. In real crystals a low magnetic dimensionality is approximated when the magnetic atoms interact predominantly with neighbours that are arranged in clusters ($d=0$), in chains ($d=1$), or in planes ($d=2$). In this paper we shall restrict ourselves mainly to $d=1$, 2, 3, because in our opinion the paper by Smart (1965) is still quite representative as concerns the properties of isolated clusters of magnetic atoms. As regards the type of interaction, we shall mainly be concerned with the Ising and Heisenberg models, although a few examples of the planar Heisenberg model have already been found (for a recent review of the XY and planar models see Betts in " Phase Transitions and Critical Phenomena ", edited by Domb and Green, vol. 3, 1973).

The range of the interaction is in most experimental cases not confined to nearest neighbours only, even when the presence of the long-range dipolar forces is neglected. It is in this respect reassuring to recall a well-established result of theoretical studies that, within a given dimension, finer details as, e.g. critical indices, will probably not depend on the number of nearest or further neighbours (Domb and Miedema 1964, Domb and Dalton 1966, Dalton and Wood 1969, Griffiths 1970 b, Paul and Stanley 1971 a, b), in any case much less than on the dimension itself. In particular, when the number of further neighbours is not large and the interactions with these are much weaker than with the nearest, one may expect the properties to be not essentially different from the nearest-neighbour model. Both conditions seem to be fulfilled in most of the experimental systems that will be discussed. Consequently, we will adhere to a classification scheme of $3\times 3\times 2=18$ types of model systems, where the first factor 3 stands for the lattice dimensionality, the second for the type (Ising, XY or Heisenberg) of the interaction, whereas the factor 2 enters because the interaction can be ferro or antiferromagnetic. The compounds considered have, furthermore, various spin values.

We finally remark that the data presented in this paper have all been taken on *insulating* magnetic compounds. The application of the Heisenberg model, with its fully localized moments, to a magnetic metal like Fe or Ni is unjustifiable for fundamental reasons. The agreement that has nevertheless been found in some cases in our opinion does not say anything about the applicability of the model, but rather it points to the fact that certain features of the phase transition are common to all the various order–disorder phenomena (magnetic and liquid–gas transition, binary mixtures ; see, e.g., Kadanoff 1970).

2.2. *Type of interaction*

In theoretical work one arrives at the Ising Hamiltonian simply by putting $b=0$ in eqn. (1.1), thus assuming an anisotropy in the exchange interaction. In practice, however, anisotropic properties often arise not so much from an anisotropy in the interaction mechanism (which may even be wholly isotropic) but from other sources, such as the presence of a crystal field or a magnetic dipolar field that couples the moments to a certain direction in the crystal. It is well known that the former acts via the orbital momentum. Quite generally the effect of the crystal field is to produce a set of orbital levels for the single magnetic ion. At a given temperature only the ground state and the excited states lower than kT will be occupied. For the magnetic properties one need therefore only consider the levels with energies not much larger than kT_c. For instance a rare-earth ion in an axial crystalline field may possess strongly anisotropic properties. It is in this case essential that the crystal field potential is not too large as compared with the spin orbit interaction, which couples L and S to the total moment $\mathscr{J}$, while, on the other hand, the crystal field splittings must be relatively large as compared to kT_c. As an example fig. 7 (*a*) shows the situation for Dy^{3+} for which $L=5$, $S=\frac{5}{2}$ and $\mathscr{J}=\frac{15}{2}$. A purely axially symmetric crystal field would split the 16-fold degenerate $^6H_{15/2}$ ground state into eight doublets, in such a way as to produce a doublet with strongly anisotropic properties lying lowest. At temperatures low compared to the separation of the two lowest doublets, the magnetic moment can, in small fields, be directed only parallel to the symmetry axis of the crystal field (z axis). Writing the exchange constant between two Dy^{3+} ions, which is assumed to be isotropic

Fig. 7

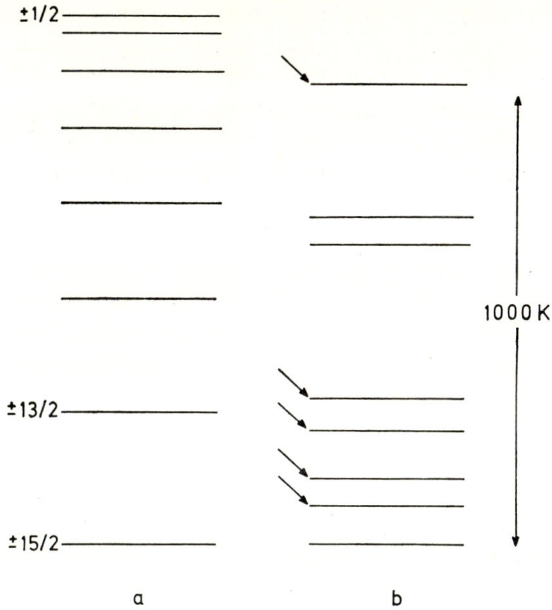

Crystal field splitting of the $^6H_{15/2}$ lowest multiplet of Dy^{3+}. (*a*) The situation that would arise from a purely axially symmetric crystal field. (*b*) Level scheme as calculated by Grünberg *et al.* (1969) for Dy^{3+} in dysprosium aluminium garnet. The arrows indicate levels that have been observed experimentally.

in the true spin Hamiltonian, in terms of an effective spin $\frac{1}{2}$ formalism, one has to introduce an anisotropic interaction which is proportional to the length of the real spin vector S. Meaningful approximate values for the exchange constant, in comparing interactions of different ions or different compounds, may be obtained by writing

$$\left.\begin{aligned} g_{\parallel} &= g_{\parallel}^{S} + g_{\parallel}^{L} \quad J_{\parallel} = J(\tfrac{1}{2}g_{\parallel}^{S})^2 \\ g_{\perp} &= g_{\perp}^{S} + g_{\perp}^{L} \quad J_{\perp} = J(\tfrac{1}{2}g_{\perp}^{S})^2 \end{aligned}\right\} \tag{2.1}$$

where the indices S and L denote the relation of the g components to the contributions of the spin and the orbital angular momentum to the magnetic moment, respectively, and J is the isotropic exchange. In our example $g_{\parallel}^{S} \simeq 20$, $g_{\perp}^{S} \simeq 0$, from which it follows that $J_{\parallel} \simeq 100J$ while $J_{\perp} \simeq 0$.

Of course, in theoretical treatments one may introduce the anisotropy in a similar way; namely, starting from the Hamiltonian of eqn. (1.1) with $a = b = 1$ one may introduce additional terms to account for the effects of the crystal field or the magnetic dipolar interactions, for instance a term of the form DS_z^2, with $D < 0$ in the case of an uniaxial anisotropy of the Ising type, as discussed above. In this way one obtains what may be called an anisotropic Heisenberg model and one may hope that for large values of the anisotropy its behaviour will resemble that of the Ising model. In fact, in comparing experiments on Ising-like substances with theory it has proven in most cases to be unimportant whether the anisotropic properties have been realized by an anisotropy in the exchange or by single-ion anisotropy (note that in the latter case one can approximate closely only the Ising models with $S=\frac{1}{2}$). Nevertheless, one should bear in mind that there is a fundamental difference between the two methods of approach, leading in some problems to contrasting results, and we shall have occasion to point this out below in connection with an example. From the present discussion of the anisotropy, it may be gathered that most of the anisotropic compounds that will be presented in the next section are essentially of the anisotropic Heisenberg rather than of the Ising type.

Evidently, practical cases are more complicated than the idealized picture in fig. 7 (*a*). A more realistic situation is shown in fig. 7 (*b*), which gives the energy levels for the lowest multiplet of Dy^{3+} in dysprosium aluminium garnet as calculated by Grünberg *et al.* (1969). What remains, however, are the strongly anisotropic properties of the lowest doublet ($g_{\parallel}\simeq 18$, $g_{\perp}\simeq 0{\cdot}5$, Faulhaber and Hufner (1969) and Ball *et al.* (1962)), so that at temperatures well below the separation of the two lowest levels which is about 80 K ($T_c\simeq 2{\cdot}5$ K, Keen *et al.* 1967), the Ising requirement for the effective spin $\frac{1}{2}$ is quite accurately fulfilled.

As regards the transition metal ions, many Co^{2+} compounds have also been found to be strongly anisotropic. The situation differs from that in the rare earths in that the crystal field is now stronger than the spin–orbit coupling and the orbital contribution to the magnetic moment may be quenched. Figure 8 (*a*) shows the energy levels of the cobalt ion in $CoCs_3Cl_5$ in which compound the crystal field has a small axial distortion from tetrahedral cubic symmetry. The cubic crystal field splits the sevenfold degenerate $L=3$ multiplet into two triplets and an orbital singlet, the latter lying lowest. Since the spin degeneracy is fourfold, this results in a quartet with g values slightly different from 2·0 ($g^S=2{\cdot}0$, $g^L=0{\cdot}4$). In addition, the axial component of the crystal field produces a splitting of about 10 K, the doublet $S=\pm\frac{3}{2}$ being lowest (fig. 8 (*b*)). Since the magnetic ordering occurs below 1 K ($T_c=0{\cdot}52$ K, Wielinga *et al.* 1967), it can be described within the Ising model with effective spin $\frac{1}{2}$ and $g_{\parallel}=7{\cdot}2$, $g_{\perp}\simeq 0$, $g_{\parallel}{}^S=6$, $J_{\parallel}=9J$.

For many other cobalt salts, e.g. for Co^{2+} in octahedral cubic symmetry, the anisotropy of the lowest doublet is not as complete as in the two examples treated. The reason is that in an octahedral cubic field the orbital levels are reversed so that the lowest-lying level is an orbital triplet. However, what is reassuring in this respect is that theoretical investigations (see, for instance, Dalton and Wood 1967, Griffiths 1970 b) have shown that the intermediate cases, even for say $J_{\perp}/J_{\parallel}\approx 0{\cdot}5$, will still show a critical behaviour that is nearly identical to the fully anisotropic case. This means that compounds like K_2CoF_4, where the lowest doublet according to Folen *et al.* (1968) is described by $g_\parallel = 6{\cdot}3$, $g_\parallel{}^S = 4{\cdot}9$, $g_\perp = 3{\cdot}1$, $g_\perp{}^S = 2{\cdot}4$, and in agreement with (2.1) $J_\perp / J_\parallel = 0{\cdot}24$, may still be considered as good approximations of the Ising model (Breed *et al.* 1969), at least for temperatures not too far above T_c.

Fig. 8

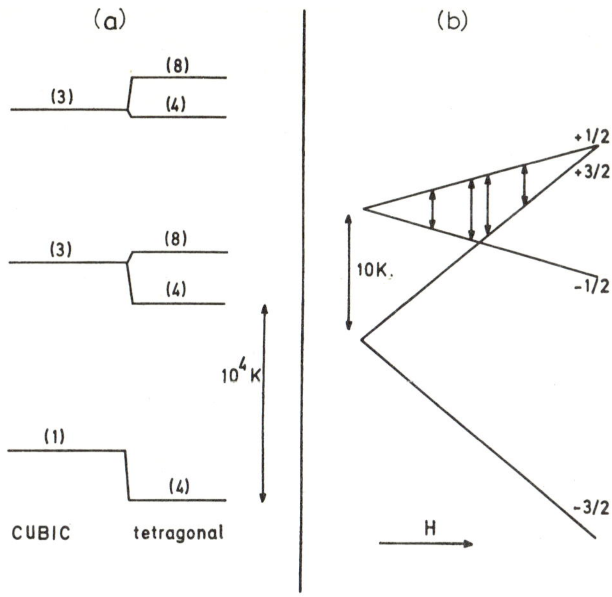

Crystal-field splitting of the lowest multiplet ($L=3$, $S=\frac{3}{2}$) of Co^{2+} in $CoCs_3Cl_5$. (*a*) Energy level scheme in a cubic and in a tetragonal crystal field. The numbers indicate the orbital multiplicity for the cubic and the spin degeneracy for the tetragonal case. (*b*) Splitting of the lowest level (quartet) by the axial component of the crystal-field and field-dependent splitting of the resulting doublets. The arrows are actually observed transitions, as reported by Beljers *et al.* (1964).

In the literature one also finds a large number of Fe^{2+} compounds ($FeCl_2$, $FeCl_2 \,.\, 2H_2O$) and Ni^{2+} compounds ($Ni(CN)_2NH_3C_6H_6$) mentioned as Ising-like materials. However, a word of caution is needed here because in these materials the crystalline field anisotropy and the magnetic interaction are often of the same order of magnitude. As a consequence, at high temperatures these Fe^{2+} and Ni^{2+} compounds will behave like Heisenberg compounds with $S=2$ and $S=1$, respectively, whereas at low temperatures they become strongly anisotropic. Hence a comparison with theoretical models is made difficult, since, for instance, part of the heat capacity anomaly is due simply to the crystalline field splittings

(Schottky anomaly is the absence of interaction). Clearly one meets the complication that the effective spin quantum number must be considered as being temperature dependent. Therefore, many of these compounds may be described by the Ising model as far as their low temperature properties ($T < T_c$) are concerned and to a considerable extent, the critical behaviour may also be well described within the Ising model. However, as regards their magnetic and thermal properties over the full temperature range, they will differ strongly from the predictions of the Ising model with the appropriate spin value. A beautiful example of such a situation is given in fig. 9 which shows the specific heat of $FeCl_2 . 4H_2O$, drawn from data of Friedberg *et al.* (1961) and Raquet and Friedberg (1973). Here the Schottky anomaly, having its maximum at 3 K, could be clearly resolved from the lattice specific heat as well as from the sharp peak at about 1 K, which is due to a transition to antiferromagnetic order.

Fig. 9

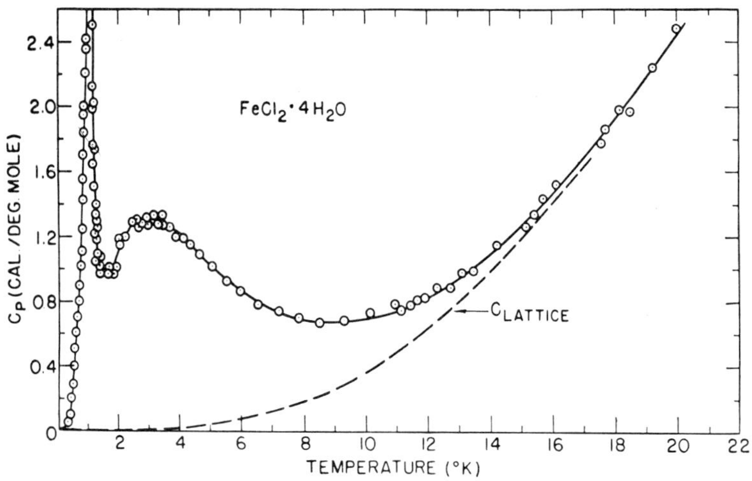

Specific heat of $FeCl_2 . 4H_2O$ drawn from data obtained by Friedberg *et al.* (1961) and Raquet and Friedberg (1973). The peak near 1 K is only partly shown, the highest value of the specific heat measured being above 6 cal/mol K.

The above discussion of the anisotropy has of necessity been very brief. The dissatisfied reader will find more thorough treatments in the review papers of Kanamori (1963), Wolf (1970) and Baker (1971 a), or in the extensive literature on paramagnetic resonance.

Next we will try to indicate how materials possessing highly isotropic interactions as required by the Heisenberg model can be found in Nature. Obviously, this condition is more difficult to be met, in view of the many possible sources of anisotropy existing in real crystals. It is evident that one must start with ions possessing a very small single-ion anisotropy.

In principle it would then be sufficient to find compounds in which the magnetic ions occupy sites of cubic symmetry. Unfortunately, however, this does not work out in practice whenever the crystal field energy can be sufficiently lowered by a distortion from cubic symmetry, which is quite often the case (Jahn–Teller effect). It is for this reason in particular that S-state ions are preferable, for which crystal field splittings play only a minor role (Mn^{2+}, Fe^{3+}, Gd^{3+}, Eu^{2+}). Of course it always remains advantageous to have a cubic crystal structure.

We would like to point out that in the literature it is mostly forgotten to incorporate in the above group of ' spin only ' ions the large assembly of free radicals. In these materials one (or more) of the electrons does not partake in the bonding, so that they have ' spin only ' magnetic moments with $S = \frac{1}{2}$ and a fully isotropic g value. Since S-state ions with $S > \frac{1}{2}$ generally will have some crystal field splittings through mixing with other states, free radical compounds are clearly rather attractive. A disadvantage, however, is the difficulty of preparing samples, in particular single crystals, while also the structure and concentration of the magnetic moments in radical compounds is not always known sufficiently well.

Another widely exploited possibility of realizing an isotropic interaction is offered by Cu^{2+} compounds. A free Cu^{2+} ion has $L = 2$ and $S = \frac{1}{2}$, but, as in the case of Co^{2+} in Cs_3CoCl_5, the orbital moment may be nearly quenched by the crystal field. At the temperatures of interest, generally only two levels are populated (Kramers' doublet) with g values not much different from 2.

Taking $CuK_2Cl_4 \cdot 2H_2O$ as an example we have $g_{\parallel} = 2{\cdot}38$, $g_{\perp} = 2{\cdot}06$ (Ono and Ohtsuka 1958), where the anisotropy in g is predominantly an orbital contribution. Consequently, the anisotropy in the exchange arising from single-ion Cu^{2+} properties and corresponding to $(g^S)^2$ is very small.

Highly isotropic single-ion properties also exist in a number of Ni^{2+} and Cr^{3+} compounds. Again the orbital moment is nearly completely quenched, resulting in a triplet ($S = 1$) and a quartet ($S = \frac{3}{2}$), respectively, as the only levels populated at the relevant temperatures and g values which are approximately equal to the spin-only value 2. Axial crystal-field splittings that may nevertheless occur are generally small, so that if the transition temperature happens to be well above 1 K, also Ni^{2+} and Cr^{3+} compounds may be considered to be quasi-Heisenberg magnets. Note that the same arguments would have made the interaction in $CoCs_3Cl_5$ of the Heisenberg type, if the transition temperature would have been much larger than 10 K instead of being well below 1 K.

In conclusion we would like to sum up the possible sources of anisotropy that may ' spoil ' the Heisenberg interaction. We have already mentioned the single-ion, or crystal-field anisotropy, and also the anisotropy arising from dipolar interactions. Apart from contributing to the anisotropy the latter also influences the range of the interaction (see below). Other sources are: biquadratic exchange interactions, electric multipole interactions, virtual phonon exchange and the superexchange mechanism itself. Of these the dipolar contribution can be simply calculated; its relative magnitude depends on the crystal structure and on the concentration and magnitude of the magnetic moments. Electric multipole interactions, as produced by aspherical charge distributions, and virtual phonon exchange thus far have only been shown to be large for some of the rare earths and for actinide compounds (see Wolf 1970, Baker 1971 b). The anisotropy occurring from the superexchange mechanism, which sensitively depends on the type of intervening nonmagnetic anion and on the relative positions of the atoms in the crystal (the overlap of wave-functions), can be very large and difficult to predict, as for example in magnetic complexes such as $[Fe(CN)_6]^{3-}$ in $K_3Fe(CN)_6$. For this complex the exchange does not at all correlate with the g values; Ohtsuka (1961 a, b) found $J_{\parallel}/J_{\perp} \approx 1{\cdot}25$, whereas $g_{\parallel}/g_{\perp} \approx 0{\cdot}4$. However, in the case of fairly ionic compounds, to which class most of the materials discussed below belong, an anisotropy in the exchange other than simply derived from single-ion anisotropy has been observed in a few cases only.

Finally, as concerns the planar Heisenberg model, one expects this to be realized whenever the anisotropy is such as to produce a strong preference for an alignment within an easy plane. An example is $CsNiF_3$, in which, in the absence of an exchange interaction, the crystal-field splittings result in a singlet and a doublet, the former lying lowest. Since the exchange energy is comparable in magnitude with the level separation, a magnetic ordering with effective $S=1$ is found, in combination with a large uniaxial anisotropy of the form DS_z^2 (with $D>0$), restricting the spins in a planar configuration (Steiner 1971). More generally, the planar model will be approximated whenever the anisotropy is of orthorhombic symmetry, with the distinction between the hardest and the easy axis much greater than between the next preferred and the easy axis†.

As we have mentioned in the introduction the difference between the XY and the planar model is that in the former only an anisotropic exchange ($\alpha=0$ in eqn. (1.1)) is required, whereas in the latter one puts the additional restriction that the spins lie within a plane by adding the term DS_z^2 ($D>0$) to the XY Hamiltonian. In view of the above discussion, it is clear, that the experimental examples will approximate the planar, rather than the XY model. Note however that, in theoretical treatments, for $S=\frac{1}{2}$ a term of the form DS_z^2 is simply an additive constant, since S_z is then a multiple of the unit operator.

---

† In what follows we will often use the term anisotropy field, which is the effective field H_A associated with the anisotropy ($g\mu_B H_A = 2DS$). In the case of an anisotropy of orthorhombic symmetry we will differentiate between the anisotropy H_A^{I} within the easy plane formed by the preferred and next preferred axes (in-plane anisotropy), and the anisotropy H_A^{II} between the easy axis and the hardest direction (out-of-plane anisotropy).

2.3. *Interaction range*

Within a given dimension one prefers to have a well-defined number of interacting magnetic neighbours. We have already mentioned this problem in § 2.1. As far as exchange interactions are concerned, the ratio of the interaction between next nearest (J_2) and nearest neighbours (J_1) may in principle be arbitrarily small, because the superexchange interaction depends critically on the mutual separation r of the magnetic atoms, viz. like r^{-10} or even more rapidly (see, for example, Bloch 1966, and Hutchings *et al.* 1968). As the interaction is of such a short range, this leads to the rule of thumb that each additional intervening anion reduces the exchange interaction with at least a factor 10^2. However, in practice the presence of additional long-range dipole–dipole interactions (varying only as r^{-3}) will make the above argument of limited value. Considering the experimental data for the exchange constants (for a given ion and varying exchange path) and estimating the dipolar contribution, one must conclude that apart from accidental cancellations the ratio J_2/J_1 will in most cases be of the order of 10^{-2}, in particular if the number of anions along the two paths is different. Examples are $KMnF_3$, for which Pickart *et al.* (1966) report a ratio of 0·03, and $KNiF_3$ and K_2NiF_4, for which Yamaguchi and Sakamoto (1969) found 0·005 and 0·01, respectively. It is also obvious from the above that if one prefers to have the short-ranged exchange forces to be the predominant interaction, only those crystals are of interest in which the exchange path between the nearest neighbours involves not more than one or two intervening atoms since otherwise the superexchange may become comparable in strength with the dipolar interaction. We also point out that in the dipolar interaction the total magnetic moment enters. Hence, ions with a large orbital contribution to the moment, like most of the rare earths, will have relatively large dipolar contributions. On these grounds the pure S-state ions, the free radicals and those transition elements in which the orbital contribution is quenched are preferable if one wants to diminish the effects of long-range forces. That these may markedly influence the characteristics of the phase transition has already been mentioned at the end of § 1.2. Perhaps superfluously it is noted that the dipolar interactions cancel in the case of a cubic structure.

The reader will find, in what follows, that with a few exceptions we have restricted ourselves in this paper to the nearest-neighbour-only models, treating the interactions of a longer range—if perceivable—as unwanted by-effects.

2.4. *Dimension of the magnetic lattice*

The experiments on low-dimensional magnets described in this paper have all been performed on 3-d crystals. Obviously, Nature can provide only approximations of the ideal low-dimensional magnetic structure; it is even more striking how extremely good these approximations can be made in practice.

There are a number of reasons which may cause the lack of an appreciable magnetic interaction between neighbours along one or more spatial directions in a crystal, the most obvious being the extremely short range of the superexchange interaction. Since the magnetic and the crystallographic lattice need not be identical, this property may be utilized by choosing a lattice in which the distance between the magnetic ions along a given direction is much longer than along the other ones. The magnetic ions may, for instance, be largely separated along certain axes by putting non-magnetic atoms in between them. In addition a lower dimensionality may be a consequence of the fact that both signs of the superexchange interaction do occur, depending on path lengths and bond angles. This offers the possibility of an accidental cancellation of the interaction in a given direction, if there exist different bonds in that direction.

Fig. 10

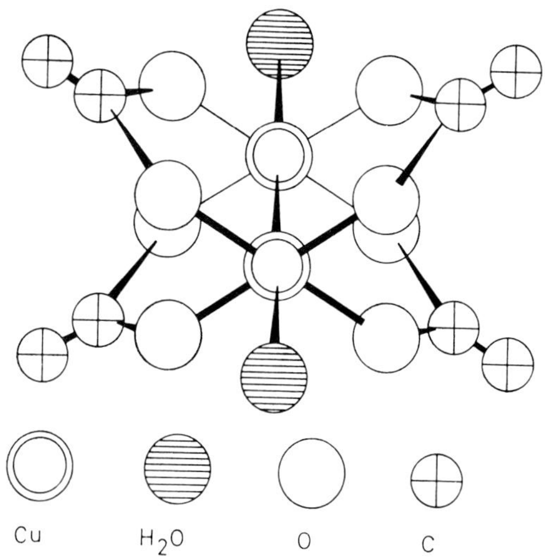

Structure of the cupric acetate monohydrate molecule, $Cu(CH_3COO)_2 \cdot H_2O$. (From Van Niekerk and Schoening 1953.)

The structural low dimensionality is best illustrated with a few examples. A zero-dimensional magnetic system, or in practice an assembly of nearly isolated clusters of magnetic atoms, is for instance approximated in copper acetate. Figure 10 gives the structure, as reported by Van Niekerk and Schoening (1953). The two copper ions are very close together, their separation (2·64 Å) being only slightly larger than in the copper metal. Following Smart (1965), one may describe the structure by saying that each pair of Cu^{2+} ions is enclosed in a cylindrical cage, with four acetate molecules along the sides and a water molecule at each end. The exchange will in this case be due mainly to direct overlap, in agreement with the observed antiferromagnetic sign of the interaction. The number of magnetic atoms forming a cluster may of course vary. For example, a tri-nuclear cluster is found in the compound $Cr_3(CH_3COO)_6OCl \cdot 5H_2O$ (Figgis and Robertson 1965, Uryû and Friedberg 1965).

Similar geometrical arguments explain the 2-d character of the copper compounds with general formula $(C_nH_{2n+1}NH_3)_2CuCl_4$ ($n=0$, 1, 2, 3 . . .). These salts may be looked upon as being derived from $(NH_4)_2CuCl_4$ and

Fig. 11

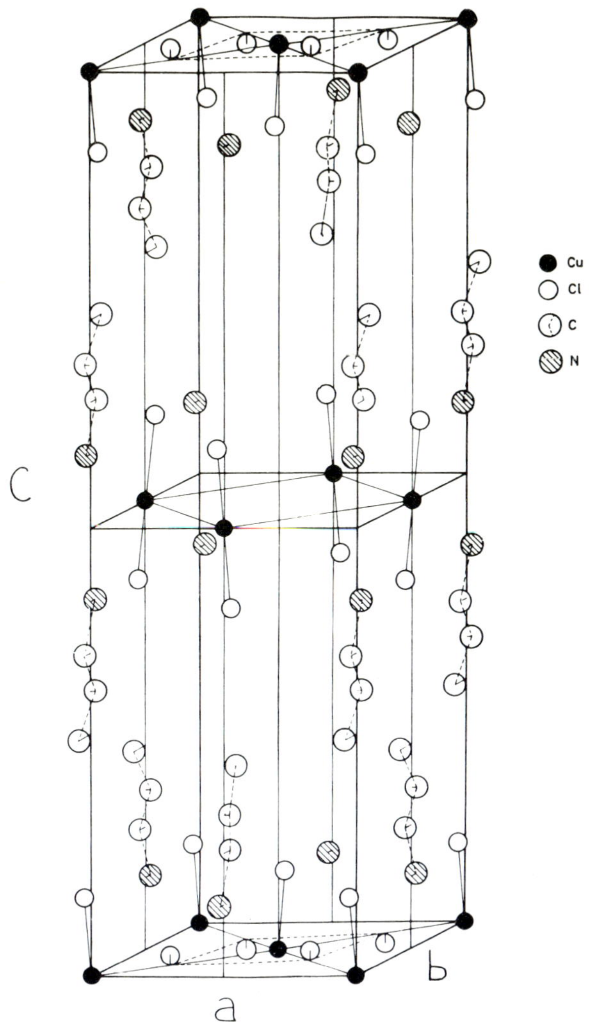

Crystal structure of $(C_3H_7NH_3)_2CuCl_4$ (from De Jongh *et al.*, 1969). Part of the propyl ammonium groups and H atoms have been omitted for the sake of clarity.

consist of ferromagnetic Cu^{2+} layers, separated by two layers of non-magnetic alkyl ammonium groups. This may be seen in fig. 11 where the structure of the compound with $n=3$ is given, drawn after crystallographic data of Barendregt and Schenk (1970). By varying n, the distance between Cu^{2+} ions from neighbouring layers is increased from 9·97 Å ($n=1$) to 25·8 Å ($n=10$), while the configuration within the copper layers is not appreciably changed (the Cu–Cu distance within the layer being about 5·25 Å). A nice feature of such a series of compounds is that one may study the various properties of interest as a function of the inter-layer distance and thereby make extrapolations to the ideal 2-d system (Bloembergen *et al.* 1970, De Jongh and Van Amstel 1970). Concerning the magnetic interaction J' between the layers we may distinguish between the dipolar coupling and the superexchange interaction. The former has been calculated by Colpa (1972 b) to be smaller than 10^{-5} of the exchange J within the layer, for all values of n larger than 3. The superexchange between the layers may only be estimated ; using the rule of thumb just mentioned (§ 2.3) one obtains $J' \approx 10^{-3}$–$10^{-20}J$ for $n=1$–10. For comparison, the measured interlayer coupling J' for the cases $n=0$ and 2, was found to be $|J'/J| = 3{\cdot}2\times10^{-3}$ and $8{\cdot}5\times10^{-4}$, respectively (De Jongh *et al.* 1972, Bloembergen and Franse 1972, Lécuyer *et al.* 1972).

Unequal magnetic lattice parameters assisted by a symmetry argument lead to 2-d antiferromagnetism in the K_2NiF_4 structure. Since the Ni ion can be replaced by Mn, Fe, Co and likewise K by Rb or Cs and F by Cl, quite a lot of examples have become known in recent years. As shown in fig. 12, the tetragonal K_2NiF_4 structure can be looked upon as being derived from the cubic (perovskite) $KNiF_3$ structure by adding an extra layer of KF between the NiF_2 sheets. By this simple fact a 3-d antiferromagnetic lattice is transformed into a magnetic layer structure. It is of importance that the interaction within the layer is antiferromagnetic, since this causes a cancellation of the interaction between neighbouring layers in the ordered state, as was first pointed out by Legrand and Plumier (1962 a, b). This may be understood by observing that the neighbouring planes are shifted over $a_0/2$, $b_0/2$ ($a_0=b_0$) with respect to each other. In the case of antiferromagnetic order within the layers the spin in the centre has in the adjacent planes an equal number of neighbours with spin up as with spin down ; there will be no net interaction, at least as far as the static properties (at $T=0$) are concerned. In that case the interaction in the third dimension is with the next-nearest layer and the interplanar superexchange interaction has to take place via four intervening anions, so that it may be expected to be 10^{-6} of the intralayer exchange. The dipolar coupling between next-nearest layers has been calculated by Colpa (see Colpa 1972 a) to be of the order of 10^{-7}–10^{-8} of J for the various compounds. As is shown by the Ba_2ZnF_6 structure (Von Schnering 1967) the interlayer distance can be increased further by the addition of still more non-magnetic layers. In this compound the Zn atom may also be replaced by Fe, Co, Ni or Cu (with the exception of the copper compound, they are also tetragonal).

Fig. 12

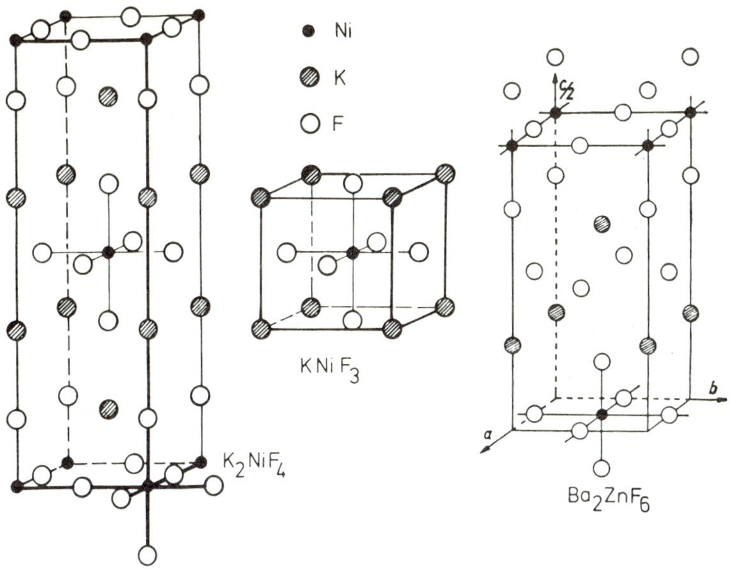

Comparison of three related crystal structures, two of which are 2-d in magnetic respect. In the middle the cubic perovskite structure of $KNiF_3$, on the left the tetragonal K_2NiF_4 unit cell. On the right the structure of Ba_2ZnF_6 (Von Schnering 1967). These crystal structures offer the possibility of comparing the 2-d and 3-d properties of compounds which are quite similar in other respects.

We remark at this point that the above-mentioned symmetry argument, that has been repeated by many authors including ourselves, is of itself insufficient to explain the fact that experimentally 2-d behaviour has been observed over nearly the whole temperature range, in particular also in the paramagnetic regime where there is no long-range antiferromagnetic order†. It would indeed fail completely if the superexchange interaction J' between central spins and corner spins itself was not much smaller than J, regardless of any symmetry considerations. For instance, if $J' \simeq J$, the behaviour would certainly be 3-d. As we see it now the correct argument would go as follows. Since the interaction J' involves three ligands it will likewise be much smaller than J (at least a factor 10^4 by the reasoning given above). Accordingly at high temperature the only correlations that come into play are those within the layer. As a

† The authors acknowledge stimulating correspondence and discussion with D. D. Betts and R. P. van Stapele on this subject.

consequence of the pronounced 2-d character the amount of short-range order established at a given temperature $kT/|J|$ will be relatively much larger than in a 3-d system. As the temperature is lowered the correlation length becomes sufficiently large that substantial clusters of antiferromagnetically correlated spins exist within the layer. The symmetry argument will then tend to reduce the effect of the coupling J'. This in turn enhances the two-dimensionality, etc., leaving ultimately in the case of complete order the interaction between next-nearest neighbouring layers (along the c axis) as the only interlayer coupling. By this argument one may understand why these systems show a very nearly pure 2-d behaviour over the *whole* range of temperatures (see below), except for an extremely small region around the transition temperature ($|T-T_c|/T_c < 10^{-4}$!). It is pointed out in conclusion that similar (partial) cancellations because of symmetry can also be found in other structures, for instance in compounds that consist of nearly isolated antiferromagnetic chains.

A lowering of the dimensionality which is not just simply related to the mutual separation of the magnetic ions is illustrated by the following examples. In fig. 13 a projection of the unit cell of $Cu(NH_3)_4SO_4 . H_2O$ is shown, as determined by Mazzi (1955). The 1-d properties of this structure originate from the difference in exchange paths connecting the Cu^{2+} ions. Since the superexchange interaction via the oxygen ion is more favourable than that via the two NH_3 groups it is not surprising that the crystal behaves as an assembly of nearly isolated magnetic chains (running along the c axis), with $|J'/J|$ being about 5×10^{-3}.

Fig. 13

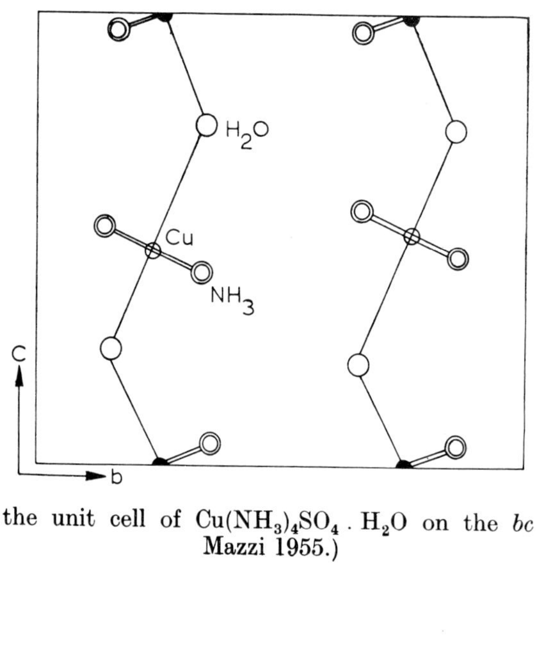

Projection of the unit cell of $Cu(NH_3)_4SO_4 . H_2O$ on the bc plane. (After Mazzi 1955.)

Rather difficult to conceive at first sight is the 1-d magnetic character of $KCuF_3$, since, according to Okazaki and Suemune (1961 a), the crystal structure is very nearly cubic (perovskite : $KNiF_3$ in fig. 12). However, the distortion of the fluorine octahedron surrounding the Cu^{2+} ion has extreme consequences for the magnetic interaction. As shown in fig. 14, taken from Hirakawa *et al.* (1970), a special alignment of the wave-functions of the dγ orbitals of the Cu^{2+} ion is produced, in such a way that there is a strong overlap through the intervening fluorine ions along the *c* axis and practically no overlap along the *a* axes. Accordingly, the interaction in the *c* plane is a factor 10^2 weaker than the antiferromagnetic exchange along the *c* axis.

We remark that quite generally, it is considerably more difficult to realize a quasi 1-d system with a very small interchain interaction, than a quasi 2-d system with a very small interlayer coupling. This

Fig. 14

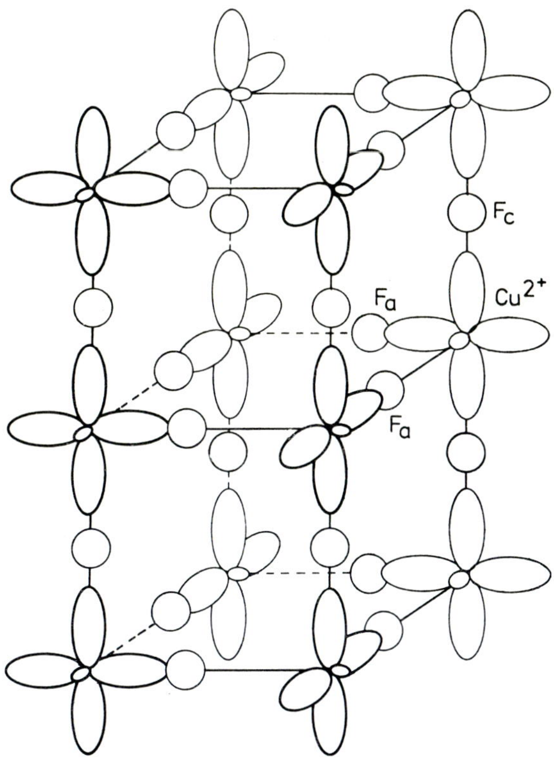

The alignments of the $d\gamma$-orbitals of the Cu^{2+} ion in $KCuF_3$ according to Hirakawa *et al.* (1970). The positions of the fluorine anions $F_{a\ c}$, are also shown. The 1-d behaviour arises because there is hardly any overlap of the wave functions along the a-axes (actually there exist two different types of alignments, of which the one shown has been found to possess the most pronounced 1-d properties).

arises because in the former case the interaction within a crystallographic plane has to be minimized, in the latter only along one crystallographic axis.

As examples of to a certain extent accidental low dimensionality one may consider all those cases in which a lower dimensionality cannot be inferred *a priori* from the crystal structure, but is instead deduced from the observed magnetic and thermal behaviour. The compound $CuNO_3 . 2\frac{1}{2}H_2O$, for instance, has properties that clearly indicate the presence of nearly isolated pairs of Cu^{2+} ions, although such a clustering is not obvious from the crystal structure (Friedberg and Raquet 1968, Bonner *et al.* 1970). As a second example we may mention the pronounced difference in magnetic behaviour of the two isomorphous cobalt compounds $CoCs_3Cl_5$ and $CoCs_3Br_5$. According to Wielinga *et al.* (1967) and Mess *et al.* (1967), the properties of the chloride can be described by a 3-d Ising antiferromagnetic model, whereas those of the bromide are in good agreement with predictions for the quadratic Ising lattice. Apparently, the interaction for the nearest neighbours along the tetragonal axis in the quasi simple cubic lattice of magnetic atoms is cancelled in the case of the bromine compound.

§ 3. Examples of simple magnetic model systems in real crystals

3.1. *Chain structures*

3.1.1. *Introduction*

As has already been mentioned above, the study of low-dimensional magnetic systems has been quite rewarding for theoretical physicists, since they provided the means of obtaining exact solutions of cooperative phenomena. Most of the existing exact results have in fact been acquired on chain models and we have thought it worth while to give a short review, prior to a discussion of the experimental results.

For the Ising chain, calculations of the energy, the specific heat and the susceptibility are available for $S=\frac{1}{2}, 1, \frac{3}{2}$ (Ising 1925, Obokata and Oguchi 1968, Suzuki *et al.* 1967). Katsura (1962) has solved the energy, specific heat and perpendicular susceptibility of the XY chain (transverse coupled chain). The thermodynamic behaviour of the classical ($S=\infty$) Heisenberg model in one dimension has been calculated by Fisher (1964) (see also Stanley 1969 b). In the case of the $S=\frac{1}{2}$ Heisenberg chain an exact result for the magnetization curve at zero temperature was obtained by Griffiths (1964 a). For temperatures $T>0$ approximate solutions for the thermodynamic behaviour have been derived by Bonner and Fisher (1964) and Griffiths (1962), by calculating the properties of closed rings containing an increasing number of spins and subsequently extrapolating to the infinite chain. In the temperature region above $kT/J \approx 0{\cdot}5$ the results found in this way are in close agreement with those obtained by Baker *et al.* (1964), from Padé approximant analyses of high-temperature series expansions. Results for the Heisenberg chain with $S > \frac{1}{2}$ have been reported by Weng in his thesis (1969).

Fig. 15

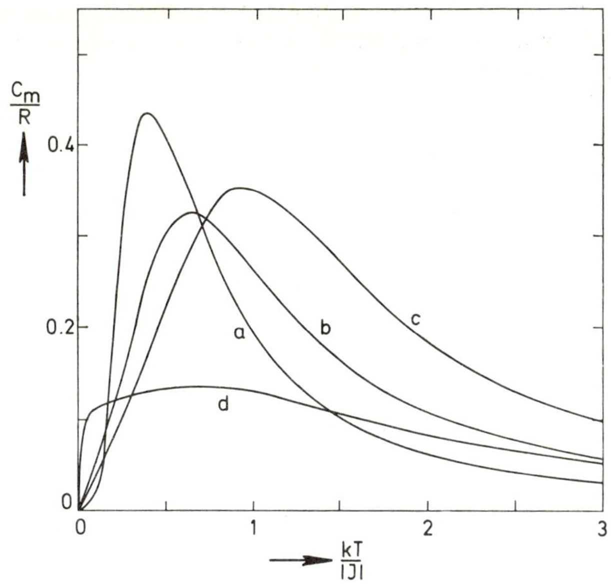

Theoretical heat capacities of a number of magnetic chains with $S = \frac{1}{2}$. (*a*) and (*b*) correspond to the Ising and the XY model, respectively (ferro and antiferromagnetic). Curves (*c*) and (*d*) are for the antiferromagnetic and ferromagnetic Heisenberg chain, respectively. For the references, see the text.

The thermodynamic behaviour of 1-d systems is governed by the intrinsic property common to all of them, namely the absence of long-range order at any non-zero temperature. That the Heisenberg and XY chains cannot sustain a spontaneous magnetization for $T > 0$ has been rigorously proven by Mermin and Wagner (1966). For the ferromagnetic Ising chain there is a simple argument due to Landau (Landau and Lifschitz 1958), which we reproduce here because of its elucidating nature (see also Fisher 1973). The argument may be put in the following way. Suppose we have a line of N spins that are ordered in parallel. Consider the change in the free energy $F = U - TS$, where U is the internal energy and S is the entropy, when this alignment is broken by reversing the direction of the first L spins. If we are dealing with short-range forces, the change in energy will be simply the amount ΔU lost across the interface between the up domain and the down domain and ΔU will be independent of N. On the other hand, there will be an entropy change since there are N possible choices of L, so that $\Delta S = kT \ln N$. We have therefore

$$\Delta F \simeq \Delta U - kT \ln N, \qquad (3.1)$$

from which it follows that for any $T>0$ and sufficiently large N the change in free energy will be negative, so that the system will break up spontaneously into oppositely aligned segments. Moreover it is seen that a finite system may become completely aligned if only T is made small enough, and also that the argument will no longer work when long-range interactions are considered (Thouless 1969), because in that case ΔU becomes dependent of N.

Fig. 16

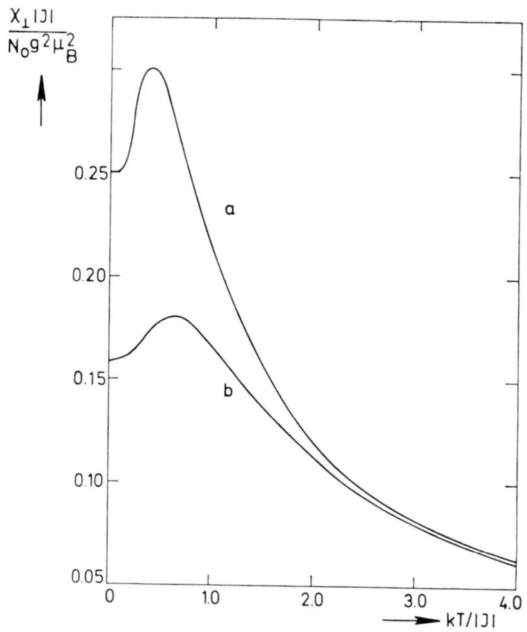

Theoretical curves for the perpendicular susceptibility of the $S=\frac{1}{2}$ Ising (a) and XY (b) chain model. The values for $\chi_{\perp}(0)$ at $T=0$ are $N_0 g^2 \mu_B{}^2/2z|J|$ and $N_0 g^2 \mu_B{}^2/\pi z|J|$, respectively. For the references, see the text.

As a consequence of the absence of long-range ordering in the ideal infinite 1-d system, the entropy has to be removed in short-range order processes. This is reflected in the specific heat and the susceptibility, both of which display broad maxima, occurring at temperatures of the order of the exchange interaction along the chain. In figs. 15–18 we have reproduced the specific heat and susceptibility curves of the Heisenberg, XY and Ising models, which all show this characteristic feature. These curves have been taken from the references cited above. Only antiferromagnetic susceptibilities have been shown. For the ferromagnetic models the susceptibility diverges as T approaches zero. The divergence is exponentially fast for the Ising models (Suzuki *et al.*

Fig. 17

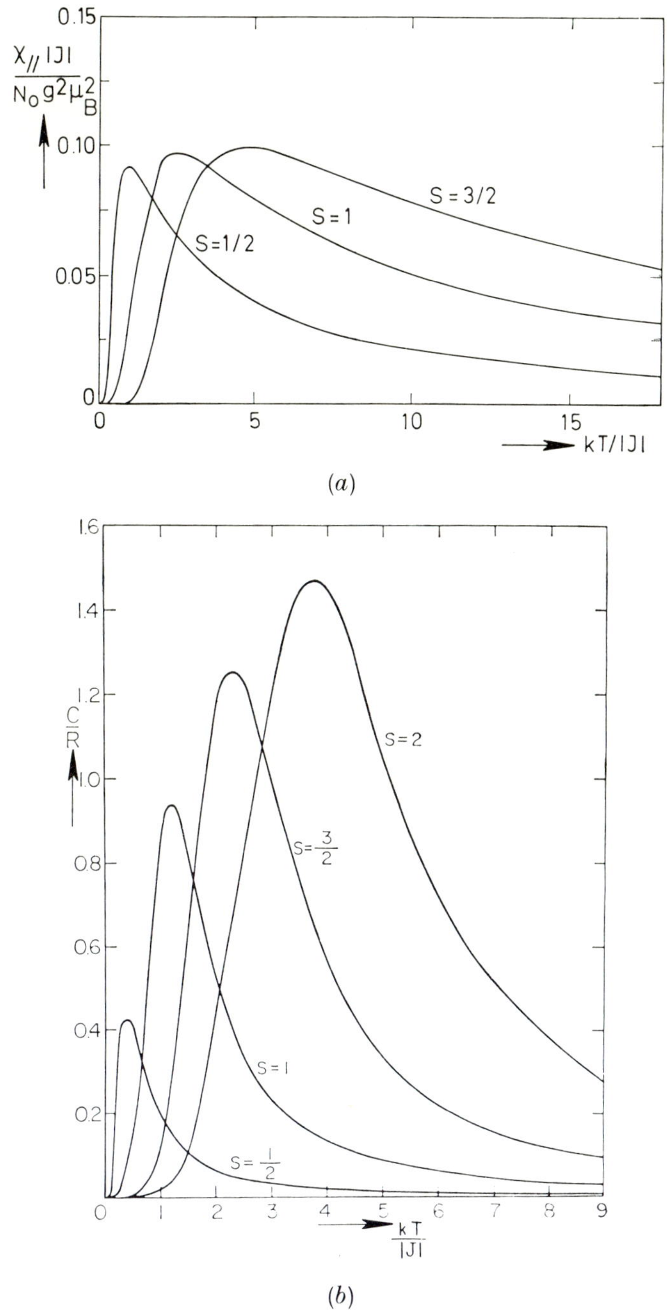

Dependence on the spin value of the parallel susceptibility (*a*) and the specific heat (*b*) of the Ising chain. (After Suzuki *et al.* 1967 and Obokata and Oguchi 1968.)

Fig. 18

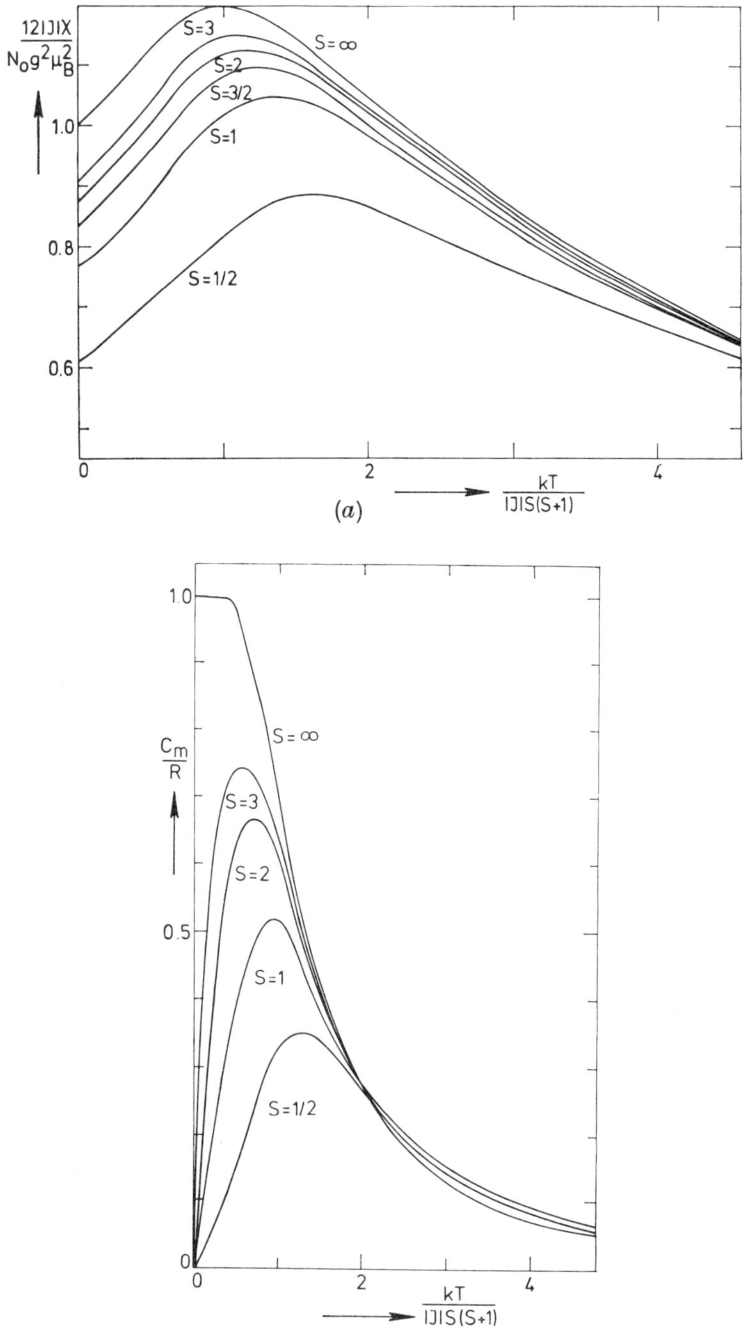

Dependence on the spin value of the susceptibility (*a*) and the specific heat (*b*) of the Heisenberg chain. (After Weng 1969.)

1967) and of the power-law form T^{-n} for the Heisenberg models, with n equal or nearly equal to 2 (Fisher 1964).

The comparison of the curves obtained for the various chain models can also illustrate the difficulties that may arise when the experimenter, on the basis of a limited amount of data, must make a choice between them to fit his experiment (with J/k as the adjustable parameter). In some cases the choice may be an obvious one, for instance on the basis of the considerations as sketched in § 2. But more often than not a choice will be not wholly justifiable unless measurements of more than one thermodynamic quantity are available. As a warning we mention here that the specific heat and the susceptibility of isolated pairs of magnetic atoms exhibit similar broad maxima (Smart 1965) as those of figs. 15–18, and also that the result of a weak antiferromagnetic inter-chain coupling on the (diverging) susceptibility of an assembly of ferromagnetic chains may well yield a curve similar to those of figs. 16 and 18 (*a*), the maximum in χ now occurring at the transition temperature T_c at which the inter-chain coupling drives the system into (long-range) 3-d ordering.

Obviously a determination of the magnetic structure with neutron diffraction is a most valuable tool. On the other hand, one can go a long way by simply combining specific heat and susceptibility measurements. This may be inferred from table 1, in which numerical results for a number of models have been collected. The height of the specific heat maximum C_{max}, the ratio of the temperatures at which the maxima in C_m and χ occur and the quantity $(1/g)^2\chi_{max}T(\chi_{max})$ together provide a handy set of criteria for the determination of the model appropriate to the investigated compound (the values for the Ising model with $S > \frac{3}{2}$ have been provided by H. T. Witteveen, private communication). We shall have the opportunity to use these below.

Let us now focus attention on the question of what the thermodynamic behaviour of the approximations of the chain models studied in the laboratory will look like. In the 1-d systems the only possible deviation from ideality that can have the effect of establishing a long-range (3-d) ordering is the presence of a weak but finite interchain coupling J'. Although rigorous calculations for a 3-d array of loosely coupled chains are not available, we may turn to the work of Onsager (1944) which provides us with a 2-d analogue. In fig. 19 (*a*) the specific heat of the quadratic Ising lattice, with different interactions J and J' along the two axes, is plotted for three values of J'/J. For $J' = J$ (dot-dash curve), we retrieve the heat capacity of the quadratic lattice, already shown in the introduction. In the case $J' = 0$, $J \neq 0$ (solid curve), we have a system of completely isolated Ising chains and the result is the same as curve *a* in fig. 15. Of most interest in the present context, however, is the dashed curve that is obtained for $J'/J = 0{\cdot}01$, since this corresponds to the specific heat of a 2-d assembly of loosely coupled chains. It is seen that at high temperatures there is no appreciable difference with the isolated chain,

Table 1. Numerical results for the 1-d Heisenberg, XY and Ising models. Listed are the spin value S; the temperatures at which the maxima in the specific heat and the susceptibility occur, divided by the exchange constant $|J|/k$; the heights of the specific heat and susceptibility maxima in reduced units; the ratio of $T(\chi_{max})$ and $T(C_{max})$; the quantity $(1/g)^2\chi_{max}T(\chi_{max})$. The values have been obtained from the references cited in the text. In the case that they have been derived from figures, or when they are a bit uncertain for other reasons, a $\simeq$ sign has been attached. The plus and minus signs in brackets denote whether the interaction is ferro or antiferromagnetic, respectively.

		Specific heat		Antiferromagnetic susceptibility			
Interaction	S	$\frac{kT(C_{max})}{\lvert J\rvert}$	$\frac{C_{max}}{R}$	$\frac{kT(\chi_{max})}{\lvert J\rvert}$	$\frac{\chi_{max}\lvert J\rvert}{N_0 g^2\mu_\beta^2}$	$\frac{T(\chi_{max})}{T(C_{max})}$	$\frac{\chi_{max}T(\chi_{max})}{g^2}$
Heisenberg	$\frac{1}{2}$	0·962 (−) 0·70 (+)	0·350 (−) 0·134 (+)	1·282 (−) —	0·07346 (−) —	1·33 (−) —	0·0353 (−) —
Heisenberg	1	1·8 (−) 1·6 (+)	0·52 (−) 0·28 (+)	2·70 (−) —	0·088 (−) —	1·5 (−) —	0·089 (−) —
Heisenberg	$\frac{3}{2}$			≃ 4·75 (−) —	≃0·091 (−) —		≃0·16 (−) —
Heisenberg	2	≃4·25 (−) ≃3·5 (+)	≃0·67 (−) ≃0·46 (+)	≃ 7·1 (−) —	≃0·094 (−) —	≃1·7 (−) —	≃0·25 (−) —
Heisenberg	$\frac{5}{2}$			≃10·6 (−) —	≃0·095 (−) —		≃0·38 (−) —
Heisenberg	3	≃6·6 (−) ≃5·7 (+)	≃0·74 (−) ≃0·56 (+)	≃13·1 (−) —	≃0·096 (−) —	≃2·0 (−) —	≃0·47 (−) —
XY	$\frac{1}{2}$	≃0·64	≃0·326	≃0·64 $(\chi_\perp)$	≃0·174 $(\chi_\perp)$	≃1·0	≃0·0417
Ising	$\frac{1}{2}$	0·416	0·445	1 $(\chi_\parallel)$ 0·4168 $(\chi_\perp)$	0·09197 $(\chi_\parallel)$ 0·2999 $(\chi_\perp)$	2·40 1·0	0·0345 0·0469
Ising	1	≃1·22	≃0·94	≃ 2·55 $(\chi_\parallel)$	≃0·098 $(\chi_\parallel)$	≃2·09	≃0·094
Ising	$\frac{3}{2}$	≃2·32	≃1·26	≃ 4·70 $(\chi_\parallel)$	≃0·100 $(\chi_\parallel)$	≃2·03	≃0·18
Ising	2	≃3·72	≃1·47	≃ 7·46 $(\chi_\parallel)$	≃0·101 $(\chi_\parallel)$	≃2·01	≃0·28
Ising	$\frac{5}{2}$	≃5·41	≃1·61	≃10·8 $(\chi_\parallel)$	≃0·1015 $(\chi_\parallel)$	≃2·00	≃0·41
Ising	3	≃7·41	≃1·70	≃14·8 $(\chi_\parallel)$	≃0·102 $(\chi_\parallel)$	≃2·00	≃0·57

but as the temperature is lowered the weak interchain coupling begins to take effect and finally causes a transition to long-range order that is reflected in C_m as a sharp spike, sitting as it were upon the broad maximum due to the short-range-order effects along the chain. As J'/J is made smaller and smaller the peak moves to the left, finally vanishing at zero temperature as $J' \to 0$. As may be inferred from fig. 19 (*a*), the position of the spike depends logarithmically on J'/J; for $J'/J = 10^{-2}$ the transition temperature is lowered by a factor 2 only with respect to that of the quadratic lattice.

Another interesting point that may be learned from these exact calculations on the 2-d Ising model is that the critical properties of the 2-d array of chains are essentially the same as those of the quadratic lattice with $J' = J$. In fig. 19 (*b*), taken from Chang (1952), the magnetization is plotted as a function of J'/J. The important feature here is that, if one comes close enough to T_c, for all values of J'/J the critical index β in the power law describing the vanishing of the magnetization at T_c is exactly the same 2-d value $\beta = \frac{1}{8}$. Likewise, for every value of J'/J, however small, the specific heat spike of fig. 19 (*a*) displays, close enough to T_c, the same critical behaviour as found for the quadratic lattice (logarithmic divergence at both sides of T_c). With this in mind it will no longer come as a surprise when we shall find below that the critical behaviour of 3-d arrays of loosely coupled chains is the same as that of the 'usual' 3-d systems (for instance the observed β values are all near $\frac{1}{3}$).

At this point we would like to stress the fact that in the experimental examples of magnetic chains it is only the interchain coupling J' that can be held responsible for the occurrence of long-range order. This contrasts with the situation in the 2-d Heisenberg model, where the anisotropy constitutes another mechanism that may yield a finite T_c, as will be discussed below. In the magnetic chains, since both the isotropic and anisotropic models ideally do not possess a transition point, the influence of the anisotropy will only consist of a shift in the position of the T_c brought about by the interchain coupling. This arises from the fact that the 3-d Ising and Heisenberg models have a different T_c, with respect to the MF value θ. Also, the way in which the T_c of a 3-d assembly of loosely coupled chains depends on the interchain coupling may be expected to be different for the anisotropic as for the isotropic case.

Finally, we point out that in the case of an example of the Ising chain, it will be much more difficult to resolve the broad chain maximum in the specific heat from the superimposed anomaly due to the 3-d ordering caused by J'. In fig. 1 it can be seen that in the Ising model the chain maximum occurs below the T_c of the 3-d lattice. Assuming the same logarithmic dependence of T_c on J'/J just found for the quadratic lattice to hold also for the 3-d analogue, it follows that a value of J'/J as low as 10^{-2} brings down the position of T_c to the temperature of the chain maximum only. In the Heisenberg model the situation is quite different, as can be inferred from fig. 6. In this case the T_c of the 3-d model occurs

Fig. 19

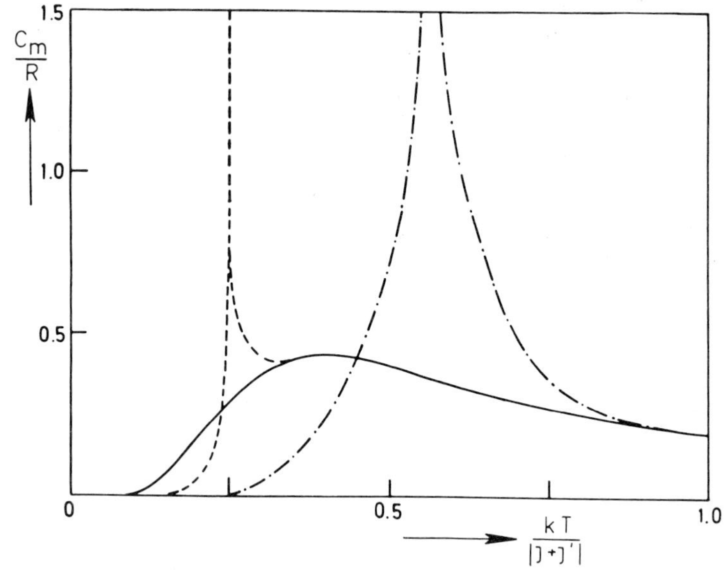

(*a*)

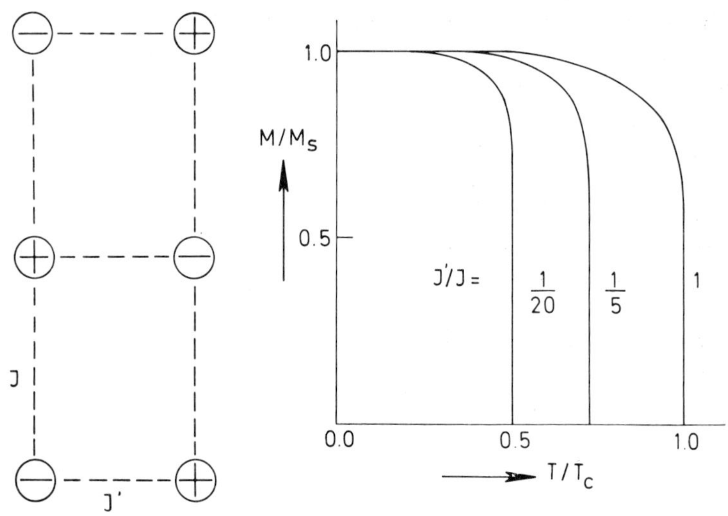

(*b*)

Specific heat and magnetization of the 2-d quadratic Ising lattice with different exchange interactions J and J' along the two crystallographic axes. (*a*) Specific heat for the cases $J'/J=1$ (dot-dash curve), $J'/J=0{\cdot}01$ (dashed curve) and $J'=0$, $J\neq 0$ (solid curve). (*b*) Magnetization for the cases $J'/J=1$, $0{\cdot}2$ and $0{\cdot}05$. References are mentioned in the text.

well below the temperature of the chain maximum, so that even for relatively high values of J'/J, the T_c of the experimental chain system will be found at a much lower temperature than that of the chain maximum.

Armed with this extensive amount of theoretical information, we now turn to what the experimenters have to offer us.

3.1.2. *Survey of experimental results*

Most of the experimental work on magnetic chain structures has been performed on Heisenberg systems. This may have to do with the above-mentioned difficulty of observing the 1-d properties of Ising chains. In any case, the oldest examples of pronounced chain-like behaviour that have been found belong to the Heisenberg class and we will therefore start this review with a discussion of the 1-d Heisenberg antiferromagnets.

In table 2 we have collected most of the examples available in the literature, grouped according to spin value, together with those properties that are of interest in the present context. Listed are some of the quantities of table 1, with, in addition, the exchange constant as determined from the data, the transition temperature T_c at which 3-d ordering sets in and the ratio of T_c and the Curie–Weiss θ, the latter being calculated from the exchange constant ($\theta = \frac{2}{3}zS(S+1)|J|/k$). The last column gives estimates of the ratio $|J'/J|$ of inter and intrachain exchange, obtained from a relation between $|J'/J|$ and $kT_c/|J|$ derived by Oguchi (1964) on the basis of a Green function method. The only check as to the correctness of these estimates is provided by the result $|J'/J| = 3{\cdot}5 \times 10^{-3}$ obtained by Skalyo *et al.* (1970) from the spin-wave dispersion curve of $CsMnCl_3 \cdot 2H_2O$, as measured with neutron diffraction. Comparing this with the value listed in table 2, it is found that the prediction from the Oguchi relation is in close agreement. Moreover these estimates are useful in comparing the various compounds, which we shall now discuss successively.

$CuSO_4 \cdot 5H_2O$ *and* $CuSeO_4 \cdot 5H_2O$

These copper salts belong to the earliest examples of chain-like behaviour. As was pointed out by Geballe and Giauque (1953) they have the peculiar property of consisting of two different magnetic systems, which is a consequence of the two inequivalent positions of the copper ions in the unit cell. This was verified by Miedema *et al.* (1962) by magnetic and caloric experiments, from which it also became clear that the coppers in one of the subsystems have a much larger exchange interaction and form nearly isolated linear chains, whereas the other subsystem remains paramagnetic down to at least 0·1 K. Additional evidence has subsequently been obtained from N.M.R. experiments ; see Wittekoek *et al.* (1968), in which paper references to the earlier work may be found. To complete the list of references, we mention the work of Giauque *et al.* (1970), a paper that is number V of a series in which extensive magnetothermodynamical studies on $CuSO_4 \cdot 5H_2O$ are reported.

Table 2. Properties of examples of the antiferromagnetic Heisenberg chain. The compounds have been grouped according to spin value. Listed are some of the quantities from table 1 with, in addition, the exchange constant as derived from the data, the transition temperatures T_c at which the 3-d order between the chains is found to set in, the ratio of T_c to the Curie–Weiss θ as calculated from J/k and lastly the estimate of the ratio of interchain (J') to intrachain (J) exchange obtained from the value of $kT_c/|J|$ with the aid of the Oguchi relation (1964).

	S	$T(\chi_{max})$ (K)	$T(C_{max})$ (K)	$\frac{C_{max}}{R}$	$\frac{T(\chi_{max})}{T(C_{max})}$	$\frac{\chi_{max}T(\chi_{max})}{g^2}$	J/k (K)	T_c (K)	T_c/θ	$\|J'/J\|$ (Oguchi)
$CuSO_4 . 5H_2O$	$\frac{1}{2}$	1·71	1·41	0·355	1·21	0·032	−1·45	?(<0·03)	<0·02	$\simeq 10^{-4}$
$CuSeO_4 . 5H_2O$	$\frac{1}{2}$	0·94	0·8	0·32	1·18	0·031	−0·8	0·045	0·056	$1·3 \times 10^{-3}$
$Cu(NH_3)_4SO_4 . H_2O$	$\frac{1}{2}$	3·4	3·0	0·359	1·13	0·029	−3·15	0·37	0·12	6×10^{-3}
$Cu(NH_3)_4SeO_4 . H_2O$	$\frac{1}{2}$		2·45				−2·36	?(<1·2)	?	?
$Cu(NH_3)_4(NO_3)_2$	$\frac{1}{2}$		3·56				−3·70	?(<1·2)	?	?
$CuCl_2 . 2NC_5H_5$	$\frac{1}{2}$	17·5				0·035	−13	1·13	0·087	4×10^{-3}
$KCuF_3$	$\frac{1}{2}$	243				0·036	−190	{38 22	{0·20 0·12	{$1·6 \times 10^{-2}$ 6×10^{-3}
$CuCl_2$	$\frac{1}{2}$	70	40	0·61	1·75	0·035	≃ −60	23·91	≃0·4	$\simeq 6 \times 10^{-2}$
Iminoxyl radical	$\frac{1}{2}$	≃6	4	0·328	≃1·5	0·035	−4·2	?(<1·3)	?	?
$CsNiCl_3$	1	35				0·085	−13	4·65	0·13	7×10^{-3}
$RbNiCl_3$	1	45				0·079	−17	11	0·24	2×10^{-2}
VF_2	$\frac{3}{2}$	42·5	27	0·69	1·57	0·13	−9·0	7·0	0·16	1×10^{-2}
$CrCl_2$	2	40	35	0·88	1·14	0·17	−5·6	16·06	0·36	5×10^{-2}
$CsMnCl_3 . 2H_2O$	$\frac{5}{2}$	32				0·38	−3·57	4·89	0·12	6×10^{-3}
$[(CH_3)_4N][MnCl_3]$	$\frac{5}{2}$	55				0·34	−6·5	0·84	0·011	$< 10^{-4}$

Fig. 20

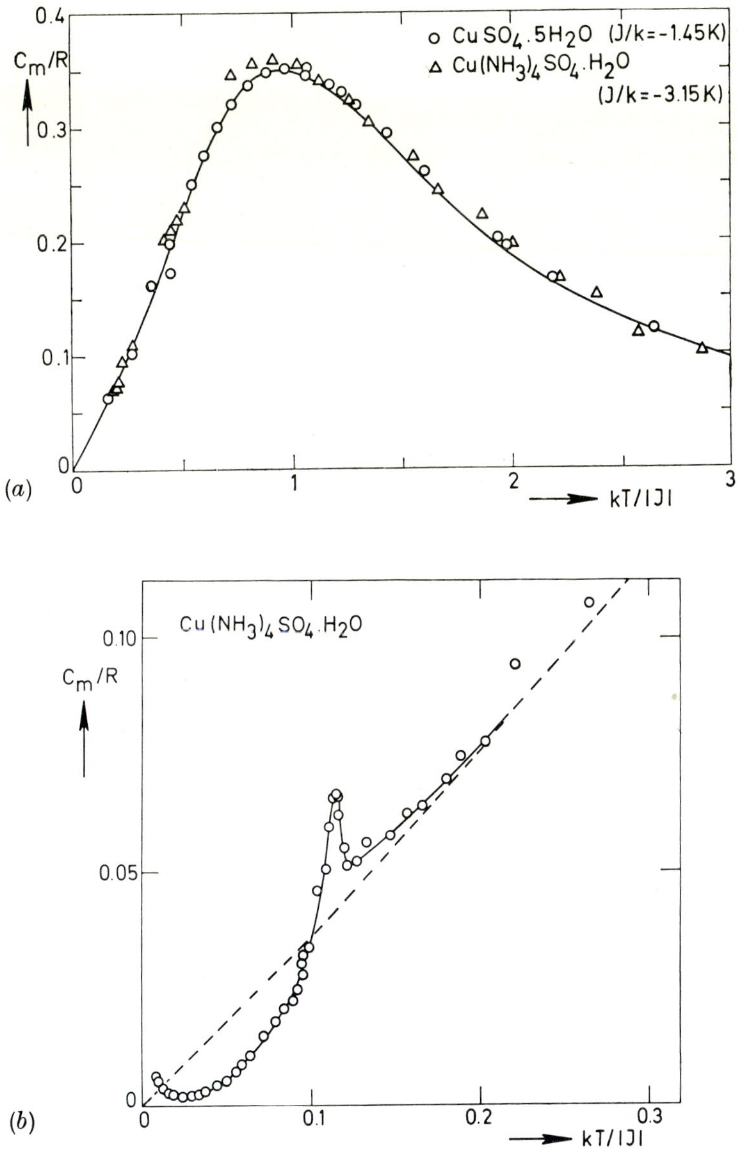

Specific heats of two examples of the antiferromagnetic, $S=\frac{1}{2}$, Heisenberg chain (for the references, see the text). (*a*) Fit of the data to the theoretical curve of Bonner and Fisher (1964) (J/k is the only adjustable parameter). (*b*) The low-temperature region, in which in the case of $Cu(NH_3)_4SO_4 . H_2O$ the transition to long-range 3-d ordering has been observed. The dashed curve is the theoretical specific heat, which is linearly dependent on temperature at the lowest temperatures. The increase in C_m observed below $kT/|J|\simeq 0{\cdot}03$ is due to the hyperfine contribution.

The excellent fit of the specific heat to the theoretical curve of Bonner and Fisher (1964) is shown in fig. 20 (*a*), in which we have combined the heat capacity data of Miedema *et al.* (1962), Giauque *et al.* (1970) and Duijckaerts (1951). In the case of the susceptibility the fit to theory is slightly less, as may be seen in fig. 21. Wittekoek *et al.* (1968) have attributed the discrepancy for $kT/|J| < 1\cdot0$ to the presence of anisotropy. The fact that the agreement for $KCuF_3$ is better, although it has a far larger value of $|J'/J|$, seems indeed to exclude the inter-chain coupling as a possible source. On the other hand, the influence of anisotropy is not appreciable in the heat capacity, so that one may not wholly discard the effect of the paramagnetic subsystem, the susceptibility contribution of which had to be subtracted in order to obtain the data of fig. 21. Unfortunately no estimate of the amount of anisotropy has as yet been obtained.

Fig. 21

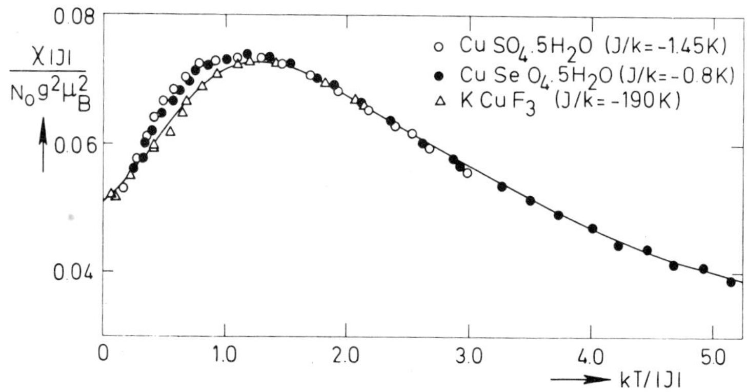

Reduced susceptibilities of three examples of the antiferromagnetic, $S=\frac{1}{2}$, Heisenberg chain. The fit to theory (full curve, obtained by Bonner and Fisher 1964) is again brought about by choosing the right J/k. It is reassuring to observe how compounds with exchange constants differing by a factor 200 may be similarly well fitted (for the references, see the text).

$Cu(NH_3)_4SO_4 . H_2O$ *and* $Cu(NH_3)_4SeO_4 . H_2O$

Copper tetrammine sulphate monohydrate seems to be the first magnetic linear chain compound recognised as such in Nature. The earlier χ and C_m measurements of Watanabe and Haseda (1958) and Fritz and Pinch (1959) were extended and reviewed by Haseda and Miedema (1961), who arrived at a linear chain arrangement. The crystallographic argument supporting this view has already been exposed in §2.4 (see fig. 13).

The heat capacity data of Haseda and Miedema (1961) and Fritz and Pinch (1959) have also been included in fig. 20 (*a*). As a consequence of the larger value of J' the transition to 3-d ordering could also be observed,

This is depicted in fig. 20 (*b*), which shows the specific heat at the lowest temperatures ($kT/|J|<0{\cdot}3$). It may be concluded from figs. 20 (*a*), (*b*) that the observed thermodynamic behaviour of the non-ideal chain is indeed in accord with that predicted in the preceding section. This is also corroborated by the χ measurements of Haseda and Miedema (1961) in which the onset of long-range 3-d antiferromagnetic ordering below $T_c=0{\cdot}37$ K could be deduced from the anisotropy in χ below that temperature. These authors compared their experiments with the Ising chain, since at that time it was the only available theoretical prediction for a 1-d system. Afterwards it was shown by Griffiths (1964 b) that the calculations for the Heisenberg chain gave a much better fit to the experimental susceptibility and specific heat data.

A nice example of the magnetization curve of the antiferromagnetic Heisenberg chain is provided by the measurements of Haseda and Kobayashi (1964) on this compound. Their result is in excellent agreement with the curve obtained by Griffiths (1964 a) and Bonner and Fisher (1964), as shown in fig. 22. The departure from the theoretical curve at the highest fields may be understood by considering the influence of an antiferromagnetic interchain coupling (the effect of the finite $kT/|J|=0{\cdot}35$ on the theoretical curve shown, which applies to $T=0$, is much less than the observed discrepancy between theory and experiment). We remark that Kaseda and Kobayashi compared their measurements with the

Fig. 22

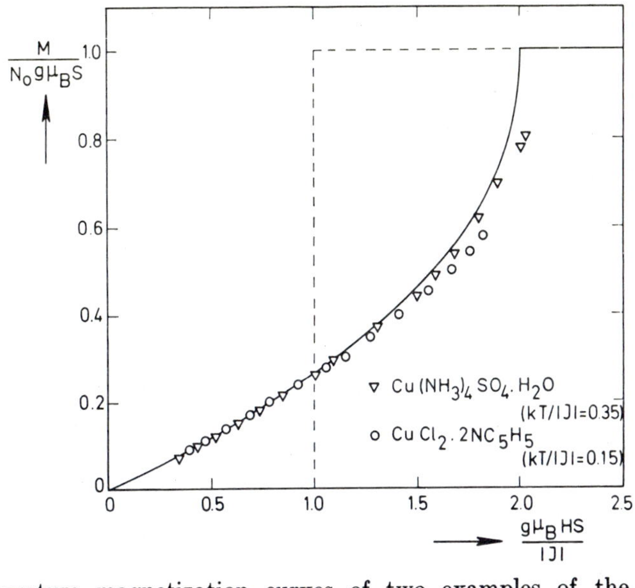

Low-temperature magnetization curves of two examples of the antiferromagnetic, $S=\frac{1}{2}$, Heisenberg chain, compared with the theoretical result obtained by Griffiths (1964 a), applying to $T=0$ K. The dashed curve shows the behaviour to be expected for the Ising chain.

theoretical curve obtained by Inawashiro and Katsura (1965), which lies somewhat above the Griffiths curve in most of the field region of interest.

Considering the isomorphous selenate compound, only specific heat measurements are available until now, which lie in the region $0{\cdot}85 < kT/|J| < 8{\cdot}5$ (Lowndes *et al.* 1969). The data could likewise be fitted excellently to the Bonner–Fisher curve, taking $J/k = -2{\cdot}36$ K. A disadvantage of the experiment is, however, that the heat capacity was not measured absolutely, so that not only $T(C_{max})$ but also the height of the maximum C_{max} had to be scaled onto the theoretical curve.

$Cu(NH_3)_4(NO_3)_2$

Although from a crystallographic viewpoint there is no direct structural evidence for the occurrence of chains in this compound, E.P.R. studies and specific heat results seem to indicate a 1-d behaviour (Rogers and Dempesey 1967). The heat capacity could be fitted to the Bonner–Fisher curve in the region $0{\cdot}27 < kT/|J| < 4{\cdot}3$, yielding $J/k = -3{\cdot}70$ K.

$CuCl_2 . 2NC_5H_5$

This compound may in some sense be thought of as being derived from $CuCl_2 . 2H_2O$ by replacing the water molecules by the pyridine molecules (Takeda *et al.* 1970). However, in $CuCl_2 . 2H_2O$ the unit cell is orthorhombic, with magnetic ions on the corners and on the centres of the upper and lower faces, whereas $CuCl_2 . 2NC_5H_5$ has a body-centred monoclinic structure ($\gamma = 91° 52'$). Nevertheless, in comparing the chain-like properties of these compounds the above concept is certainly of value and one may anticipate that the 1-d character is greatly enhanced by the substitution of the pyridines, which results in highly inequivalent exchange paths along and between the chains.

The experiments bear out this expectation. The susceptibility, the heat capacity and the magnetization data can all be fitted to the Heisenberg chain predictions with $J/k \simeq -13$ K (Takeda *et al.* 1970, 1971 b, Matsuura 1971). In the region $2 < T < 4$ K the specific heat varies linearly with temperature, as expected from theory. Takeda's data did not extend below 2 K. Recently, Duffy (private communication, to be published) has found the sharp anomaly due to the 3-d ordering to be at $T_c = 1{\cdot}13$ K. With the aid of the Oguchi relation the estimate $|J'/J| \simeq 4 \times 10^{-3}$ is obtained on the basis of the $kT_c/|J|$ value. This is an order of magnitude smaller than the value $|J'/J| \simeq 4 \times 10^{-2}$ needed by Matsuura (1971) to explain the deviation of the experimental magnetization curve at high field values from the Griffiths prediction (fig. 22). The explanation may well be the existence of a symmetry argument which reduces the inter-chain coupling, similar to that present in the K_2NiF_4 structure. A necessary prerequisite for such a cancellation is, as we have seen, an antiferromagnetic alignment within the layer or along the chain. Since, in fields near to the (ferromagnetic) saturation value, the antiferromagnetic orientation will have been very nearly broken up, one will in that case measure the *true* interaction between the chains. In zero field, on the other hand, the symmetry argument can take effect, reducing the true interchain coupling to a lower effective J'.

$KCuF_3$

The rather exceptional origin of the occurrence of 1-d magnetism in $KCuF_3$ has already been discussed in § 2.4. There exist two polymorphisms (Okazaki 1969 a), according to the particular alignment of the wave-functions of the Cu^{2+} ions. The two forms have different values for the interchain coupling. In table 1 it may be seen from comparison of the T_c/θ values that in one of them J' is about 60% smaller than in the other. The transition temperatures have been obtained from neutron diffraction and E.S.R. experiments (Hutchings *et al.* 1969, Ikebe and Date 1971).

Although one would expect 1-d correlations over long distances to persist to temperatures well above T_c, they could not be detected in the neutron diffraction experiment. However, their presence could also not be ruled out†. Furthermore, a large reduction (55%) of the expected moment $g\mu_B S \simeq 1{\cdot}1\ \mu_B$ was observed. Although 10% of this reduction was attributed by the authors to covalency effects, there remains a large value of about 45%, close to the value of about 50% found in the N.M.R. experiments of Hirakawa *et al.* (1970) to be explained. More will be said about the spin reduction in antiferromagnets in § 4.2. This effect arises from the existence of deviations from the fully aligned Néel state even at zero temperature, and it is another example of a property of which the relative importance is greatly enhanced by lowering the lattice dimensionality (see § 4.2).

Additional evidence for the 1-d behaviour is found from the susceptibility curve (Kadota *et al.* 1967, Hirakawa *et al.* 1971), the absence of a specific heat singularity (Kadota *et al.* 1967) and the N.M.R. experiments (Hirakawa and Kadota 1967, Hirakawa *et al.* 1970). From the fit of the susceptibility to theory, shown in fig. 21, the large value $J/k = -190$ K is obtained. As a consequence, the maximum of the magnetic chain specific heat is expected to occur at a temperature at which the lattice contribution is about 30 times larger, so that one can scarcely hope to be able to separate the magnetic part. The transition to long-range order at T_c is likewise not appreciable in the specific heat. The anisotropy in the susceptibility can, within the uncertainties, be accounted for by the anisotropy in the g tensor. Since only a fraction of this g anisotropy

† Recently an additional neutron diffraction study has been reported by Ikeda and Hirakawa (1973), which may be summarized as follows. Evidence for the 1-d correlations above T_c ($=39{\cdot}51$ K) was obtained. The inter-chain coupling was estimated as $|J'/J| \simeq 2{\cdot}7 \times 10^{-2}$ (compare the Oguchi result $1{\cdot}6 \times 10^{-2}$). Below T_c the sublattice magnetization was found to be 3-d in character, with $\beta = 0{\cdot}355 \pm 0{\cdot}010$, $B = 1{\cdot}53 \pm 0{\cdot}05$ in the region $1 \times 10^{-3} < 1 - T/T_c < 0{\cdot}1$, and with a T^2 decrease in the low-temperature (spin-wave) region.

will be due to the spin contribution to the magnetic moment, the applicability of the Heisenberg model seems to be justified. One may object that from the resonance experiments (Hirakawa *et al.* 1971, Ikebe and Date 1971) it follows that the anisotropy in the direction of the c axis (H_A^{II}) is much larger than that in the easy plane perpendicular to it (H_A^{I}), which would make the planar model seem more appropriate. However, considering the fact that the exchange field following from J/k is $2{\cdot}6 \times 10^6$ Oe, while H_A^{I} is only of the order of 5 Oe, it follows that a H_A^{II} which is 500 times as large as H_A^{I} would still be only 0·1% of the exchange field, small enough to choose for the Heisenberg model. Not surprisingly, the susceptibility results (e.g. $(1/g^2)\chi_{max}T(\chi_{max})$) exclude the applicability of the planar model.

$CuCl_2$

This compound is important for historical reasons, since the specific heat measurements on this salt belong to the first that were analysed in terms of a linear chain arrangement (Stout and Chisholm 1962). It is, however, a rather poor example of a chain structure. This may be inferred from the relatively high value of T_c which is 60% of $T(C_{max})$ and the amount of entropy that is already gained below T_c (17% of R ln 2). For $Cu(NH_3)_4SO_4 \, . \, H_2O$ the corresponding numbers are 12% and 4%, respectively.

Also it is not clear whether the interaction is indeed of the Heisenberg type, although it is listed as such in table 2. Stout and Chisholm (1962) used the Ising model to analyse their data, which was the only available chain model at the time. However, for a Cu compound the Heisenberg model is in general more appropriate and the ratio of $T(\chi_{max})/T(C_{max})$ is indeed nearer to the Heisenberg than to the Ising value (for $\chi_{\parallel}$). The fact that the value $C_{max}/R = 0{\cdot}61$ is much too high for both the Ising and Heisenberg model can be explained by considering that a large lattice contribution had to be corrected for in order to obtain C_m. Consequently, the absolute values of C_m are uncertain.

Susceptibility measurements have been performed only on powdered specimens (De Haas and Gorter 1931, Starr *et al.* 1940). It is difficult therefore to derive a value of J/k. Starting from the Ising model, as Stout and Chisholm did, one arrives at $J/k \simeq 70$ K. Applying the Heisenberg model a much lower value of about 50 K would be obtained. The corresponding values of T_c/θ are 0·34 and 0·48, clearly much higher than for the other Cu chains.

On the isomorphous bromine compound $CuBr_2$ only a susceptibility experiment has been performed (Barraclough and Ng 1964). The maximum in χ was located at 226 K.

Free radicals

A considerable number of free radicals have been found to possess chain-like properties (Hamilton and Pake 1963, Edelstein 1964, Duffy and Strandburg 1967, Karimov 1969, Duffy *et al.* 1972). As one of the best examples, we have listed in table 2 the properties of the iminoxyl radical 2,2,6,6-tetramethyl 4-piperidinol 1-oxyl. Both the heat capacity (Lemaire *et al.* 1968) and the magnetic data (Karimov 1969) are in good numerical agreement with the Heisenberg chain predictions. Other iminoxyl radicals with a similar chemical structure, but with sometimes quite different values of the exchange, show a similar agreement (Karimov 1969). For more references to recent experimental work see Yamaguchi *et al.* (1970), Saito *et al.* (1970) and Hone (1971).

$CsNiCl_3$ *and* $RbNiCl_3$

These members of the hexagonal $ABCl_3$-type group of compounds ($A \equiv$ monovalent cation, $B \equiv$ divalent transition metal ion) have been reported to be representatives of the $S = 1$ antiferromagnetic Heisenberg chain. In this structure magnetic chains are formed along the c axis, since the exchange interaction between metal ions of neighbouring chains has to take place via two chlorine ions which are far apart (3·6–4·0 Å), whereas there are short B–Cl–B paths along the c axis (e.g. the B–B distance is only about 3 Å). The interaction can be thought to be reasonably isotropic since the single-ion anisotropy of the Ni^{2+} ion in an octahedral field is usually small. The susceptibility measurements (Achiwa 1969) do show a fairly isotropic behaviour, in contrast to those of the isomorphous cobalt compound.

Instead of the estimates of the exchange of Achiwa (1969) and of Smith *et al.* (1970), we have listed in table 2 J/k values obtained from the theoretical $kT(\chi_{max})/|J|$ listed in table 1. In this way we find $J/k = -13$ and -17 K for $CsNiCl_3$ and $RbNiCl_3$, respectively. Note that the value of $(1/g^2)\chi_{max}T(\chi_{max})$ is in excellent agreement with the theoretical estimate.

A specific heat measurement has up to the present only been performed on $CsNiCl_3$ (Mekata *et al.* 1970). An advantage is the existence of the isomorphous diamagnetic compound $CsMgCl_3$, which can be of help in the evaluation of the lattice contribution to the heat capacity. Although these authors attributed an apparent linear temperature dependence of C_m in the region $6 < T < 12$ K to the 1-d character, the argument is fallacious since the linear dependence is not expected to extend beyond $T \simeq 0{\cdot}2|J/k|$ (Weng 1969), which corresponds to a temperature of about 2·5 K. Accordingly, the linear part between 6 and 12 K does not extrapolate to zero for $T \to 0$. The amount of entropy gained below T_c was found to be only 5% of $R \ln 3$, confirming the 1-d character.

As concerns the precise spin structure below T_c, where the inter-chain coupling has established a 3-d ordering, the various investigators are in disagreement with one another. From their neutron diffraction measurements Minkiewicz *et al.* (1970), Mekata *et al.* (1970) and Cox and Minkiewicz (1971) conclude that the antiferromagnetic chains along the c axis are coupled together in a triangular array, the moments lying in a plane perpendicular to the basal plane. Notwithstanding this, another neutron diffraction experiment of Epstein *et al.* (1971) and the N.M.R. measurements of Clark and Moulton (1972) seem to indicate a collinear structure with the spins along the *c* axis.

Fig. 23

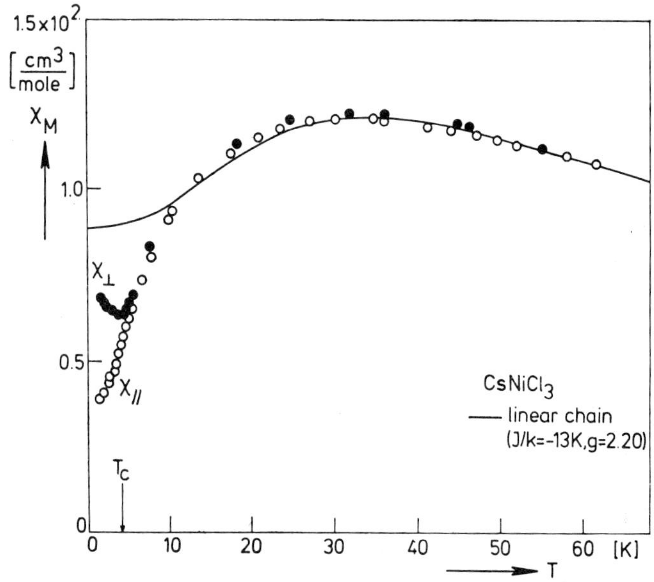

Susceptibility of $CsNiCl_3$ (after Achiwa 1969), which is an example of the anti-ferromagnetic $S=1$ Heisenberg chain. The full curve is Weng's theoretical result (1969), calculated with $J/k=-13$ K and $g=2{\cdot}20$. Note that below T_c the otherwise fairly isotropic susceptibility is split up by the small anisotropy into a parallel ($\|c$ axis) and perpendicular part, as a consequence of the appearance of long-range 3-d ordering below this temperature, which may itself be attributed to the inter-chain coupling.

As a contribution to the solution of this controversy, we have compared in fig. 23 the susceptibility of $CsNiCl_3$ with Weng's prediction (1969) for the $S=1$ Heisenberg chain, taking $J/k=-13$ K. At high temperatures the fit is quite good ; below $T\simeq 3T_c$ deviations occur which we attribute to the influence of the inter-chain coupling. Interestingly, in this region the experimental points lie below the theoretical curve, in contrast with what one would expect. Also the $\chi_{\|}$ curve, measured parallel to the *c* axis does not extrapolate to zero. These features are difficult to explain within the collinear structure, but may be understood quite well from the model proposed by Minkiewicz *et al.* If the spins are arranged in a triangular array the $\chi_{\|}$ obviously is non-zero at $T=0$ (the expected Van Vleck contribution is only 2% of the value attained at the susceptibility maximum, Achiwa 1969). We mention that the $\chi_{\perp}$ as measured by Achiwa (perpendicular to the *c* axis) is actually the susceptibility in the

[110] direction, which is at an angle of 60° with the plane in which the moments lie.

The neutron experiments do agree to a low value for the magnetic moment extrapolated to 0 K, as compared to the high-temperature result of about 2 μ_B found from the χ measurements. Mekata *et al.* (1970) derived $1{\cdot}5 \pm 0{\cdot}1\ \mu_B$ from a powder measurement, while Cox and Minkiewicz (1971) reported $1{\cdot}0 \pm 0{\cdot}1\ \mu_B$ for a single crystal. This considerable reduction is attributed mainly to the effect of zero-point spin deviations. The sublattice magnetization was found to vary like a power law with an exponent $\beta = 0{\cdot}27 \pm 0{\cdot}03$ (Cox and Minkiewicz 1971), $\beta = 0{\cdot}32 \pm 0{\cdot}03$ (Clark and Moulton 1972) and $\beta = 0{\cdot}35 \pm 0{\cdot}05$ (Mekata *et al.* 1970). The '3-dimensional' value of β is in accord with the argument outlined in the preceding section.

As for the nature of the scattering above T_c it was reported by Minkiewicz *et al.* (1970) for $RbNiCl_3$ that the 3-d correlations disappear at $T \simeq 4T_c$, in accordance with the susceptibility behaviour discussed above. Above this temperature only the strong correlations along the c axis are observed. This 1-d scattering will be treated in more detail in § 4.1.

VF_2

The same phenomenon of 1-d correlations persisting up to temperatures $T \gg T_c$, as detected by neutron diffraction techniques, has been reported by Child *et al.* (1970) and Lau *et al.* (1969) for the compound VF_2. This salt seems to order below T_c in a spiral structure around the c axis, with the spins perpendicular to the c axis. At first sight VF_2 does not seem a likely candidate for 1-d magnetism, since it has the tetragonal rutile structure and is therefore isomorphous to MnF_2, FeF_2, CoF_2 and NiF_2. Stout and Boo (1966) proposed the following possible explanation for its chain-like character. In these sister compounds the exchange J_1 between the metal atoms along the c axis, which are nearest neighbours, is smaller than the interaction J_2 between the next-nearest neighbours, which are at the distance $\frac{1}{2}(c^2 + 2a^2)^{1/2}$. If now, on the other hand, J_1 would be much larger than J_2 this would result in an assembly of chains running along the c axis.

A justification for classifying VF_2 as a Heisenberg compound is the very small anisotropy observed in the susceptibility by Stout and Lau (1967), which is of the order of 0·1% only. The susceptibility and specific heat show the usual 1-d characteristics. The entropy gained below T_c was found to be a mere 9% of the expected entropy change $R \ln 4$. Considering the relatively high spin value, this would qualify VF_2 as a good approximation of the isolated chain.

Calculations of the exchange constant from $T(\chi_{max})$ and $T(C_{max})$ with the aid of table 1 agree to $J/k = -9{\cdot}0$ K. The resulting value $T_c/\theta = 0{\cdot}16$, which is about the same as those of the Ni chains in spite of the larger spin value, again points to a pronounced 1-d character. Interpolating between the values of $T(C_{max})$ for $S=1$ and 2 of table 1, we obtain the prediction $T(\chi_{max})/T(C_{max}) \simeq 1{\cdot}6$, in good agreement with the experimental result. The value for $(1/g)^2 \chi_{max} T(\chi_{max})$ is a little lower than the theoretical one. Also the maximum value of the specific heat exceeds the prediction obtained by interpolating once more between $S=1$ and 2 (table 1). However, this may be attributed to the difficulties associated with the subtraction of the lattice contribution.

$CrCl_2$

This compound was investigated by Stout and Chisholm (1962) together with the isomorphous $CuCl_2$. The amount of entropy gained below T_c was found to be 18%, which is fairly low for a chain-like compound with a high spin value. But the discrepancies between the experimental values of $T(\chi_{max})/T(C_{max})$ and $(1/g)^2 \chi_{max} T(\chi_{max})$ and theory point to the fact that it is not too good an example. From the temperatures $T(\chi_{max})$ and $T(C_{max})$ one obtains for the exchange $J/k = -5{\cdot}6$ and $-8{\cdot}2$ K, respectively. We have adhered to the former, since the heat capacity maximum is more difficult to obtain experimentally. The resulting value for $T_c/\theta = 0{\cdot}36$ points to a rather large interchain coupling.

$CsMnCl_3 . 2H_2O$

Of the two manganese salts listed in table 2, $CsMnCl_3 . 2H_2O$ has the largest inter-chain coupling. In this compound Cl–Mn–Cl–Mn chains exist along the *a* axis of the orthorhombic unit cell. The neighbouring chains are linked by exchange paths involving several non-magnetic atoms or H_2O groups.

Susceptibility measurements have been performed by Smith and Friedberg (1968) and by Kobayashi *et al.* (1972), who analysed the data above T_c in terms of the calculations of Weng and Griffiths for $S \geqslant 1$ (Weng 1969), obtaining $J/k = -3{\cdot}0$ K, which is fairly close to the value $-3{\cdot}57$ K derived subsequently from the spin-wave dispersion curve as measured by neutron diffraction by Skalyo *et al.* (1970). A more extensive discussion of this neutron work will be given in § 4.1. From the analysis of the dispersion curve (fig. 24) the interchain coupling could also be obtained. As mentioned before, their value $|J'/J| = 3{\cdot}5 \times 10^{-3}$ is in reasonable agreement with the Oguchi prediction (6×10^{-3}) on the basis of the $kT_c/|J|$ value. The reason why we listed the Oguchi result in table 2 was to enable a comparison of the chain properties with the other compounds, for which a value of J' has not been derived experimentally. The anisotropy was also determined in the neutron work, but using $J/k = -3{\cdot}57$ K the more accurate value $H_A/H_E = 5 \times 10^{-3}$ can be derived from the measured spin-flop field (Butterworth and Woollam 1969, Botterman *et al.* 1969). This is the anisotropy in the easy plane ($H_A{}^{I}$). The out-of-plane anisotropy ($H_A{}^{II}$) has been reported to be about three times larger (Nagata and Tazuke 1970, cf. note on page 20).

Fig. 24

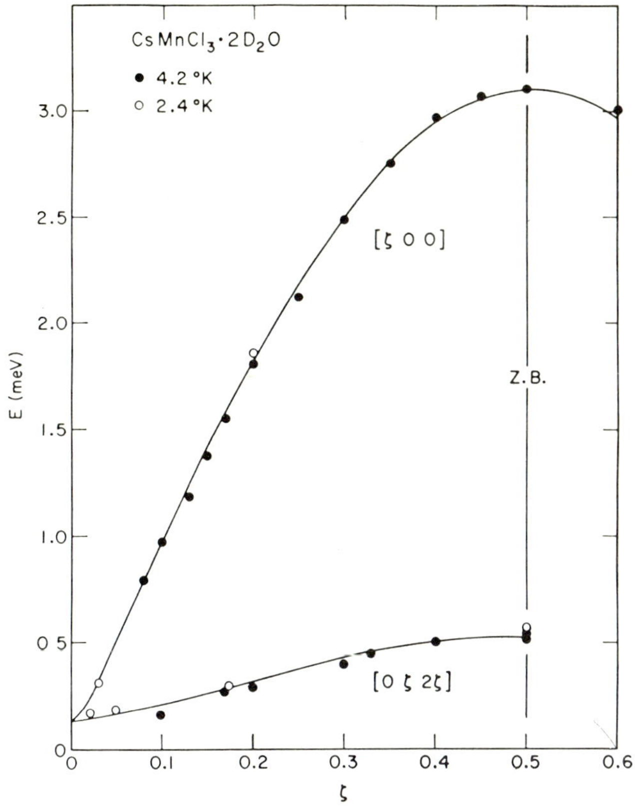

Magnon dispersion curve of $CsMnCl_3 . 2D_2O$, which is the deuterated isomorph of the antiferromagnetic, $S=\frac{5}{2}$, Heisenberg chain $CsMnCl_3 . 2H_2O$. The solid lines are spin-wave theoretical results for a 3-d assembly of weakly coupled chains. From the fit to the data the exchange along the chain, as well as the interchain coupling and the anisotropy, can be derived. The wave-vector is denoted by ζ and the zone boundary is indicated by the vertical line. The 1-d character follows from the lack of an appreciable dispersion perpendicular to the chain direction $[\zeta, 0, 0]$ in reciprocal space. (After Skalyo *et al.* 1970.)

Not surprisingly the critical index β of the sublattice magnetization was found to be $\beta=0{\cdot}30$ (Skalyo *et al.* 1970). In this neutron work evidence for the existence of 1-d scattering was given for the first time. The correlations along the chain were found to persist at temperatures $T \gg T_c$. For instance, at $T=3T_c$ the average number of correlated spins within a chain is about five. The effect of the interchain interactions begins to be felt at $T \simeq 2T_c$.

$[(CH_3)_4N][MnCl_3]$

It may be observed from table 2 that this compound is the best approximation of a 1-d magnet found so far, since the interchain interaction J'

is 10^{-4}–10^{-5} of J, which is one or two orders of magnitude smaller than for the other examples.

In $[(CH_3)_4N][MnCl_3]$ the chain character is once again a consequence of the crystal structure, which consists of chains of chlorine octahedra, surrounding the manganese ions, the adjacent octahedra sharing {111} faces. Thus three Mn–Cl–Mn paths link the magnetic ions along the chain axis, whereas there are no direct bonds between the chains, which are separated from one another by the tetramethyl ammonium groups.

By fitting the susceptibility data to Fisher's classical chain model, scaled to $S=\frac{5}{2}$, the exchange was found to be $J/k=-6{\cdot}3$ K (Dingle *et al.* 1969). At temperatures below the maximum the susceptibility rises again, a phenomenon also observed in $CsMnCl_3 \,.\, 2H_2O$ and $KCuF_3$. The most likely explanation for this rather steep increase is the presence of an impurity. An indication for this is also the fact that the values for J/k derived from the experimental χ_{max} and $T(\chi_{max})$ with the aid of table 1 are considerably lower ($J/k=-5{\cdot}8$ and $-5{\cdot}2$ K, respectively) than the result obtained by the neutron diffraction work (see below). The impurity contribution would result in too high a χ_{max}, and also shift the maximum to a lower temperature. In this respect the better agreement of the J/k value derived by the fit to Fisher's $S=\infty$ model scaled to $S=\frac{5}{2}$ (Dingle *et al.* 1969) will be fortuitous and may be understood by considering that the susceptibility curve of this model lies above the result of Weng (1969) for $S=\frac{5}{2}$, which seems to be a better approximation (see Smith and Friedberg 1968). The value of $T_c=0{\cdot}84$ K, which was derived from the susceptibility behaviour (Hutchings *et al.* 1972 b) was later confirmed by additional χ measurements (Walker *et al.* 1972) and low-temperature neutron diffraction data (Birgeneau *et al.* 1972 a). Earlier neutron experiments, also performed on the deuterated material have been reported by Birgeneau *et al.* (1971 a) and Hutchings *et al.* (1972 b). Apart from establishing the 1-d character of the magnetic structure, it was found that the nature of the scattering could be fully accounted for by Fisher's calculations for the correlations in a classical Heisenberg chain (1964). These measurements extended down to $T\simeq 1{\cdot}1$ K. The dispersion curve could be described within the experimental error by a sine curve, as predicted by simple spin-wave theory (cf. fig. 24). More will be said about these important results in § 4.1. From the additional neutron studies below $T=1$ K reported by Birgeneau *et al.* (1972 a), the critical index β for the magnetization was found to be $\beta=0{\cdot}26$ for $10^{-3}<1-T/T_c<10^{-1}$.

From the susceptibility measurements a small anisotropy of about 1% of the planar type has been deduced (Hutchings *et al.* 1972 b). As discussed by Walker *et al.* (1972) this may be attributed to the dipolar interactions. The same authors compared the χ between 60 and 170 K with the series result of Rushbrooke and Wood (1958), obtaining $J/k=-6{\cdot}5$ K, in good agreement with the value $J/k=-6{\cdot}6$ K derived from the spin-wave dispersion curve. In contrast with this a larger value, $J/k=-7{\cdot}7$ K was found in the quasi-elastic scattering experiments (Birgeneau *et al.*

1971 a), a discrepancy which is not yet understood. For the present we shall adhere to the value $J/k = -6{\cdot}5$ K. Lastly, we mention the interesting E.P.R. study of Dietz *et al.* (1971).

This concludes the discussion of the examples of the antiferromagnetic Heisenberg chain, which is by far the largest group of 1-d compounds. In fact there are only few chain structures left, which have been listed in table 3, the compounds being grouped according to the type of the interaction. Their properties will now be briefly reviewed.

$CsCuCl_3$

This salt, which is very nearly isomorphous with $CsNiCl_3$, is listed here because it is the only example of the $S = \frac{1}{2}$ ferromagnetic Heisenberg chain available at present. Susceptibility experiments have been performed by Achiwa (1969) and by Rioux and Gerstein (1970). If the interaction along the chain is indeed ferromagnetic, as we assume here, it follows that there must be a fairly large antiferromagnetic interchain interaction, since a negative deviation from the Curie law is observed up to $T \simeq 100$ K (Rioux and Gerstein 1970). Our suggestion of a ferromagnetic intrachain exchange is based upon the positive value for the Curie–Weiss θ found by Achiwa from the susceptibility in the region $100 < T < 300$ (the data of Rioux and Gerstein do not extend beyond 150 K), and on the shape of the susceptibility curve, which shows a gradual increase up to the temperature $T_c = 10{\cdot}4$ K at which a transition to 3-d ordering was observed in the heat capacity (Rioux and Gerstein 1969). The absence of a maximum in χ above T_c and the failure of the attempts to analyse the χ curve in terms of an antiferromagnetic intrachain interaction are explained by assuming a model of antiferromagnetically coupled ferromagnetic chains. Neutron experiments to check this would be very welcome. In any case it is clear that $CsCuCl_3$ will be a rather poor example of a ferromagnetic chain, the interchain coupling being relatively large.

$[(CH_3)_4N][NiCl_3]$

$[(CH_3)_4N][NiCl_3]$ seems to be a better candidate for 1-d ferromagnetism in view of the high degree of 1-dimensionality of the isomorphous manganese compound (although the ferromagnetic intrachain exchange will result in a larger dipolar coupling between the chains). It consists of chains quite analogous to those in $CsNiCl_3$ but with a larger interchain distance (9 Å instead of 7 Å) and more unfavourable interchain exchange paths. Unfortunately only susceptibility measurements have hitherto been performed (Gerstein *et al.* 1972 a). In the region $1{\cdot}6 < T < 79$ K these could be very well fitted to a ferromagnetic chain model, based upon Fisher's infinite spin model (1964), scaled to $S = 1$. The value $J/k = 1{\cdot}0$ K was obtained from this fit. No evidence for a transition to 3-d ordering was found in the investigated region, implying that

Table 3. Properties of available examples of the other simple chain models: the ferromagnetic Heisenberg chain, the antiferromagnetic and ferromagnetic Ising chain and the ferromagnetic planar Heisenberg (PH) chain. The examples are grouped according to type and sign of the interaction. Listed are the spin value S, the intra-chain exchange J, the transition temperature T_c and T_c/θ with θ calculated from J.

	Type of interaction	Sign of interaction	S	J/k (K)	T_c (K)	T_c/θ
$CsCuCl_3$	Heisenberg	+	$\frac{1}{2}$	?	10·4	?
$[(CH_3)_4N][NiCl_3]$	Heisenberg	+	1	1·0	? (<1·7)	? (<0·6)
$K_3Fe(CN)_6$	Ising	−	$\frac{1}{2}$	−0·23	0·129	0·56
$CsCoCl_3$	Ising	−	$\frac{1}{2}$	$\simeq -100$ (?)	$\simeq 8$ (?)	$\simeq 0·08$ (?)
$Dy(C_2H_5SO_4)_2 \cdot 9H_2O$	Ising	+	$\frac{1}{2}$	0·2	0·13	0·5
$CoCl_2 \cdot 2NC_5H_5$	Ising	+	$\frac{1}{2}$	9·5	3·5	0·37
$RbFeCl_3$	PH	+	1	?	2·55	?
$CsNiF_3$	PH	+	1	$\simeq -8·3$	2·61	$\simeq 0·12$

$T_c/\theta < 0{\cdot}5$. More experiments on this compound in particular in the region below 1 K would be welcome.

$K_3Fe(CN)_6$

Potassium ferricyanide is also one of the earliest investigated chain-like compounds. Ohtsuka (1961 a, b) concluded to the existence of antiferromagnetic chains, with an anisotropy in the interaction of about 25%. The effective spin is $\frac{1}{2}$, the ground doublet lying several hundred cm^{-1} below the nearest excited doublet. The interchain coupling was estimated by him to be about $0{\cdot}1\ |J|$.

The anisotropy value places the compound in between the Ising and Heisenberg model (see the results of Bonner and Fisher 1964). The specific heat, which was measured by Duffy *et al.* (1962) and Rayl *et al.* (1968) is indeed intermediate between these two extremes. As explained in the preceding section, the rounded maximum is not clearly resolved from the large peak that reflects the 3-d ordering. The amount of entropy already removed above T_c was found to be 65%. The maximum in the susceptibility seems to be masked by the appearance of a weak ferromagnetic moment in the ordered state.

From the specific heat and susceptibility experiments a value for the exchange constant of about $J/k = -0{\cdot}23$ can be deduced. The spin reduction of 18% found by Ôno *et al.* (1970) in Mössbauer experiments is rather low for a 1-d compound with $S = \frac{1}{2}$, which may be attributed to the large anisotropy and to the relatively high value of $|J'/J|$.

$CsCoCl_3$

Another member of the hexagonal $ABCl_3$-type group of compounds (see $CsNiCl_3$). The susceptibility measured by Achiwa (1969) shows extremely anisotropic character and was analysed in terms of the Ising model. Apparently Achiwa could not correct for the Van Vleck contribution to the susceptibility, which makes the analysis dubious. The value of $J/k = -85$ K derived by him from $T(\chi_{\parallel\,max})$ and $\chi_{\parallel\,max}$ is therefore unreliable. If we tentatively take the Van Vleck contribution to be equal to the value of $\chi_\parallel$ extrapolated to $T = 0$, and subtract this from the value attained at the maximum, we obtain $J/k \simeq -115$ K from the χ_{max} so obtained. As concerns the value of T_c, only slight indications of a transition are found at about 8 K in $\chi_\parallel$ and $\chi_\perp$.

Clearly more information is needed to put things on a sure footing. The values $J/k \simeq 100$ K and $T_c \simeq 8$ K listed in table 3 may therefore only be considered as very rough estimates.

The hydrated compound $CsCoCl_3 . 2H_2O$, which is isomorphous to $CsMnCl_3 . 2H_2O$ has been investigated by Herweyer *et al.* (1972). The chain character turns out to be less pronounced than in $CsCoCl_3$, there being relatively large interchain interactions.

$Dy(C_2H_5SO_4)_2 . 9H_2O$

The Ising character of this compound with effective spin $\frac{1}{2}$ is quite well established with $g_{\parallel}=10\cdot8$ and $g_{\perp}\simeq0$ for the lowest doublet, while the next doublet is found at $\Delta E/k=22\cdot5$ K. It has been extensively investigated by Cooke *et al.* (1959, 1968) and also by Wielinga (1968). The magnetic interactions are predominantly of the dipolar kind, effecting a ferromagnetic alignment along the *c* axis. The dipolar interactions along the *c* axis are about one order of magnitude stronger than the others, which accounts for the chain-like character above T_c. Thus, although the compound is a rather poor example of a 1-d magnet, its magnetic and thermal properties at temperatures sufficiently above T_c can be described by a model of loosely coupled Ising chains. Below $T_c=0\cdot115$ K the system is brought into 3-d ferromagnetic ordering.

$CoCl_2 . 2NC_5H_5$

A much better example of the ferromagnetic Ising chain is $CoCl_2 . 2NC_5H_5$, which is isomorphous to $CuCl_2 . 2NC_5H_5$ already discussed earlier, and has been studied by Takeda *et al.* (1970, 1971 b). The specific heat is plotted in fig. 25 (*a*) and it represents the closest approximation to the theoretical Ising chain heat capacity found at present. It clearly illustrates the difficulty mentioned above of resolving the rounded chain maximum from the peak due to the 3-d ordering, in case of Ising-type compounds. In fact the specific heat of $CoCl_2 . 2NC_5H_5$ shows a close resemblance to the dashed curve in fig. 19 (*a*), bearing in mind that here we have to do with a 3-d assembly of Ising chains. The difference observed at $4<T<6$ K may be attributed to the anisotropy not being complete. Below T_c the ferromagnetic chains are ordered antiparallel with respect to one another, as can be deduced from the susceptibility behaviour.

Another illustration of the 1-d character of $CoCl_2 . 2NC_5H_5$ can be found in fig. 25 (*b*), in which the entropy versus relative temperature T/T_c curves of this compound and of $CoCl_2 . 2H_2O$ are compared. As already mentioned in the discussion of $CuCl_2 . 2NC_5H_5$, one may imagine $CoCl_2 . 2NC_5H_5$ to be derived from $CoCl_2 . 2H_2O$ by replacing the water molecules by the larger pyridine molecules, thereby enhancing the 1-d character. This is indeed obvious from fig. 25 (*b*), since one may observe that the amount of entropy gained below T_c is decreased from 60% in the case of $CoCl_2 . 2H_2O$ to a mere 15% in $CoCl_2 . 2NC_5H_5$.

$RbFeCl_3$

Although the susceptibility of this $ABCl_3$-type compound was analysed by Achiwa (1969) in terms of an uniaxial anisotropic antiferromagnetic chain, the neutron diffraction measurements of Davidson *et al.* (1971) showed the magnetic structure below $T_c=2\cdot55$ K to consist of ferromagnetic chains, with the moments in the plane perpendicular to the chain axis in a triangular array. From this it may be expected that the anisotropy will be of the planar Heisenberg form. The 3-d correlations were found to persist up to 10 K, but at 20 K the observed scattering

Fig. 25

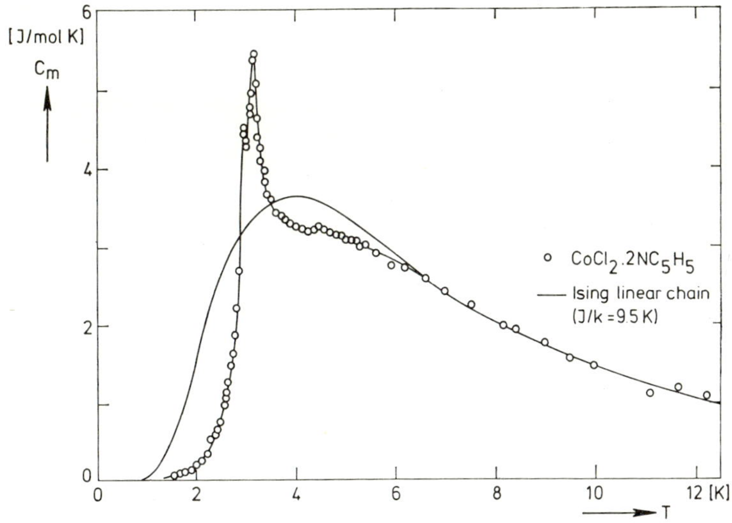

(*a*)

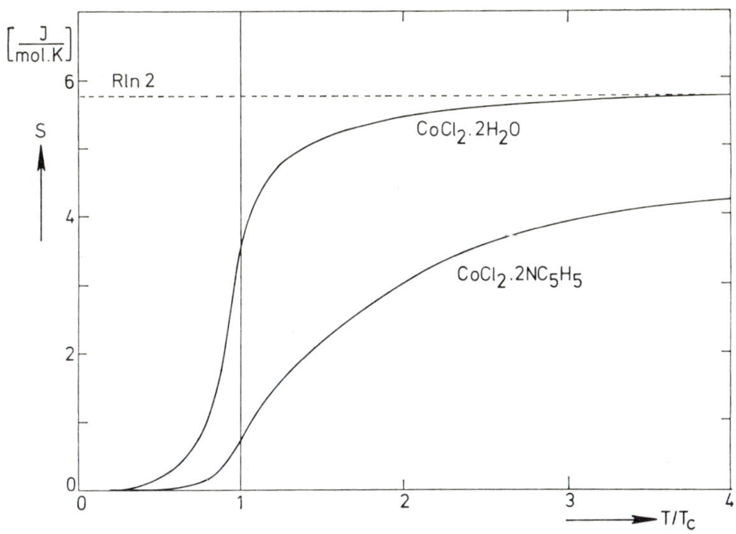

(*b*)

(*a*) Magnetic specific heat versus temperature of $CoCl_2 . 2NC_5H_5$. The solid curve is the theoretical prediction for the $S=\frac{1}{2}$ Ising chain, calculated with $J/k=9{\cdot}5$ K. (After Takeda *et al.* 1971 b.) (*b*) Comparison of the entropy versus temperature curves of two Co salts. The large enhancement of the 1-d character by substituting the NC_5H_5 molecules for the H_2O molecules can be clearly seen from the large reduction in the amount of entropy gained below T_c. (After Takeda *et al.* 1971 b.)

was wholly 1-d in character. From the Mössbauer study the value of the magnetic moment at $T=0$ was estimated to be about 2 μ_B, which again indicates considerable reduction, since the g factor deduced from the χ measurements was 4·49 ($S=1$).

This example clearly illustrates the difficulty of deducing the type of magnetic structure from χ measurements alone. As an additional complication the effective spin value will be temperature dependent due to the contribution of higher excited levels. In view of all the uncertainties involved, we have refrained from listing a value for J/k.

$CsNiF_3$

The other available example of the linear planar Heisenberg model is $CsNiF_3$, which also has the hexagonal $ABCl_3$ structure. Neutron diffraction and magnetization measurements have been performed by Steiner *et al.* (1971). The behaviour of the specific heat and the susceptibility is currently being investigated by Lebesque of our laboratory. Again the easy plane is perpendicular to the chain axis. The transition to long-range order, as found in the magnetization and heat capacity measurements, is $T_c=2{\cdot}61$ K. In fig. 26 the magnetic specific heat as reported by Lebesque *et al.* (1972) is shown as an example of the behaviour found in a magnetic chain with large planar anisotropy. Also in this case the estimate of J/k is still uncertain due to the complications involved. The evaluation of the lattice specific heat is therefore not too certain, since the J/k value was used in the analysis. But the overall appearance is as expected for a $S=1$ chain with large planar anisotropy, in that the curve is intermediate between the $S=1$ Ising and Heisenberg chains. Once again there is a small spike observed at T_c.

In concluding this section on 1-d magnets we make some remarks concerning the need for future work. On the theoretical side we have seen that considerable information is available, although for the isotropic chains discrepancies between the approximate solutions and the spin-wave theory of these short-range ordered systems have been found that need further investigation (see § 4.2). As far as the experiments are concerned, one may state that the existence in some cases of extremely good examples of magnetic chains has been quite well established, at first from measurements of the heat capacity and susceptibility, and lately on quite rigorous grounds by neutron diffraction investigations.

Since a large number of antiferromagnetic Heisenberg chains have already been discovered, there is at present more need for investigations of the other models. In particular examples of the Ising models with $|J'/J|<10^{-3}$ would be welcome. Concerning the XY and the planar Heisenberg model there is also much work left to be done.

Comparing ferro and antiferromagnetic chains, one finds that the first type has scarcely been found. This is not surprising since antiferromagnetism is more commonly found in Nature. The antiferromagnetic chain has the advantage that it is easier to obtain reliable values for J/k

Fig. 26

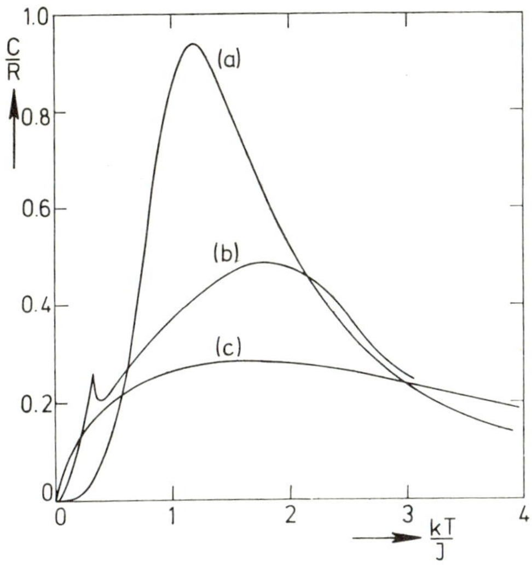

Magnetic specific heat of $CsNiF_3$ (curve *b*), which is an example of a ferromagnetic $S=1$ chain with large planar anisotropy. As expected the curve is intermediate between the predictions for the $S=1$ Ising (*a*) and ferromagnetic Heisenberg (*c*) chain. The small spike observed at low temperatures reflects the transition to long-range 3-d ordering. (After Lebesque *et al.* 1973.)

from susceptibility or specific heat data. Also values for the anisotropy can be obtained readily in the antiferromagnetic case from, for instance, spin-flop measurements. It is a pity that hitherto, only in one case, such an experiment has been performed, since knowledge of the amount of anisotropy is of importance when analysing results in terms of the Heisenberg model.

In conclusion we have seen that, as far as quantitative theoretical results are available for the specific heat and the susceptibility of chain models that could be tested experimentally, the best experimental examples yield excellent fits to theory. This is not trivial, since there is often only one adjustable parameter, the exchange constant J/k, needed to fit the various thermodynamical quantities.

3.2. *Layered structures*

3.2.1. *Introduction*

As discussed in § 1.2, in going from the 1-d to the 2-d lattices there arises a profound difference between the Ising model on the one hand and the Heisenberg and XY models on the other. In the case of the ideal chain model there is no transition to long-range order except at $T=0$ for any type of interaction. But whereas changing the dimension to 2 is sufficient for the Ising model to order at a finite temperature, this is not so for the other models. For the isotropic systems the dimension has to be raised to 3 before a phase transition occurs (cf. fig. 6). As a consequence, a review of 2-d magnetism naturally divides itself into two parts, *viz.* the Ising and the isotropic systems. We will start with the former.

The work of Onsager (1944) on the quadratic Ising lattice has already been mentioned in the preceding pages. In this paper he calculated the partition function in zero field, from which the behaviour of thermodynamic quantities such as the energy and specific heat can be derived. In later work (Onsager 1949, Yang 1952) the spontaneous magnetization and the correlation functions were obtained. The properties of the other planar lattices (triangular, honeycomb) have also been investigated and were found to be qualitatively identical to those of the quadratic network.

As concerns the field-dependent behaviour, however, no exact results have been acquired, the available information having been drawn from numerical studies, which none the less have reached a high degree of accuracy. The only exception is Fisher's solution (1960 b) of a special kind of Ising lattice, *viz.* the ‘decorated superexchange’ Ising model, in which the magnetic spins decorate the bonds of the lattice and interact antiferromagnetically via non-magnetic spins on the lattice nodes. For this particular model the free energy could also be calculated for all values of the applied field.

For an extensive review of the existing theoretical information on the 2-d Ising lattices the reader may consult the review paper of Fisher (1967). From the discussions in the foregoing sections the extreme importance of the results obtained on this model, for the qualitative understanding of phase transitions will have become clear. In fact we have already shown much of the thermodynamic behaviour of the 2-d Ising model in figs. 1–5 and 19 of the preceding sections.

Before leaving the Ising model we would like to mention that in analogy with Landau's proof of the absence of a phase transition in an Ising chain, there exists a similar argument, originally due to Peierls (1936) but put on a rigorous footing by Griffiths (1964 c) and by Dobrushin (1965), which proves the existence of a phase transition in the 2-d case.

Turning now to the isotropic models we will first discuss the theoretical arguments predicting the absence of ordering in the 2-d Heisenberg and XY models. This has been rigorously proven by Mermin and Wagner (1966), using an idea of Hohenberg (see, e.g., Hohenberg 1967). They showed that for sufficiently small applied fields the magnetization in two dimensions is bounded in the following way :

$$M(H, T) \leqslant Q\{T\,|\ln|\,H\,||\}^{-1/2}. \tag{3.2}$$

In this inequality Q is a constant and H denotes an arbitrary field, for instance, the applied magnetic field or an anisotropy field. By letting $H \to 0$ it follows that there will be no spontaneous magnetization at any finite temperature. Note however, that the presence of even a very small field will spoil the argument. Evidently under experimental conditions there will, for example, always be a minute amount of anisotropy or interlayer coupling, which may both be represented in the form of an effective field acting on the magnetic moments.

Recently, Fisher and Jasnow (1971) have extended Mermin and Wagner's argument in the sense that they proved the absence of a spontaneous magnetization without needing the introduction of a symmetry breaking field H (eqn. (3.2)). Moreover they showed that the 2-d system need not consist of a single monolayer but that the arguments also apply to a system contained between two infinitely extending parallel planes of a finite separation. Thus the system may be 3-d in the sense that it may contain a large number of monolayers. But provided this number remains finite (however large) the system will not order spontaneously. It should be noted that the above results apply to short-range interactions and that similar proof has been given for the 1-d isotropic systems (§ 3.1.1).

Earlier arguments concerning the absence of long-range order in 1 and 2-d systems were based upon spin-wave theoretical arguments. To illustrate this we consider the fractional decrease $\Delta M_s(T)/M_s(0)$ of the spontaneous magnetization of an isotropic ferromagnet due to the excitation of spin waves. This quantity is equal to $(N_0S)^{-1}\sum_k n_k$, where

$$\sum_k n_k = \int d\omega\, g(\omega)\langle n(\omega)\rangle \tag{3.3}$$

is the total number of magnons created at a temperature T, $\langle n(\omega)\rangle$ is the average value of the number of magnons of frequency ω, and $g(\omega)$ is the number of magnons of frequency ω per unit frequency range. Since we are dealing with the Bose distribution we can write

$$\sum_k n_k = \int d\omega \frac{g(\omega)}{\exp(\hbar\omega/kT)-1}. \tag{3.4}$$

Substituting $x=\hbar\omega/kT$, it can be shown that

$$\sum_k n_k \simeq T^{d/2}\int_0^\infty \frac{x^{(d/2)-1}}{\exp(x)-1}\,dx \tag{3.5}$$

for dimension $d=1$, 2, 3. Since the integral in (3.5) diverges for $d<3$, it was argued that the magnetization for $d=1$, 2 must be zero at any finite temperature. Furthermore, one also can see how in this case the argument fails if a finite field or a finite sample size is introduced, since both have the effect of changing the lower integration bound from zero to a finite value. In the case of a finite field this occurs via the field term in the spin-wave dispersion relation, while the finite sample size implies a minimum wave-vector and consequently a minimum ω. It should be stressed that although these arguments may serve as a nice illustration, they constitute no rigorous proof of the absence of a transition, if only because spin-wave theory is merely a low-temperature approximation of the system of isotropically interacting spins (Stanley and Kaplan 1966, Fisher 1973).

To sum up, we have seen that just as in the case of magnetic chains, theory predicts the absence of a spontaneous magnetization at a non-zero temperature for the isotropic 2-d systems and that a transition to long-range order may likewise be brought about by the existing deviations from the ideal model. The dependence of the so-obtained T_c on the strength of these deviations has been studied by various authors (Lines 1964, 1970, Dalton and Wood 1967, Obokata *et al.* 1967, Kats 1969, Ishikawa and Oguchi 1971). These results have in common that T_c decays to zero if the strength of the deviations is decreased to zero, or in other words, if extrapolation is made to the ideal system. This is to be expected since in these studies T_c is identified with the temperature of the onset of the spontaneous magnetization.

Consequently, as concerns the occurrence of long-range order the ideal 2-d Heisenberg and XY models behave similarly as the chain models. There is, however, one fundamental difference indicated by the analyses of the high-temperature series expansions for the initial susceptibility ($\chi = (\partial M/\partial H)_T$ in the limit $H \to 0$). It has already been pointed out in 1958 by Rushbrooke and Wood that the series for the 2-d lattices suggested the existence of finite temperatures at which the susceptibility diverges, just as in the case of 3-d lattices, where these temperatures are commonly identified with the transition to long-range order. Since such an identification cannot possibly be made for the isotropic 2-d lattices, there certainly is a problem. Of course one may doubt the results of the series analyses, since only a relatively small number of terms in the series is known. But the method works well in lattices that do show a transition to long-range order, so there is no *a priori* reason to doubt it in the 2-d isotropic case. More recently, Stanley and Kaplan (1966) brought up this matter again, using longer series (see also Stanley 1967, 1968 a, b, Moore 1969, Watson *et al.* 1970, Berezinskii 1970, Lines 1971, Betts *et al.* 1971, Ishikawa and Oguchi 1971, Wood and Dalton 1972, Ritchie and Fisher 1973). Moreover, they made the important remark that the two requirements of a zero spontaneous magnetization and an infinite susceptibility need not be incompatible, if, for example, certain restrictions are put on the dependence of the pair correlation function $\langle S_0 \,.\, S_r \rangle$ on the separation r. It is in this respect noteworthy that in all the proof excluding the existence of long-range order, the possibility of a diverging susceptibility cannot be ruled out (Mermin and Wagner 1966, Fisher and Jasnow 1971). For instance, the bound imposed on the magnetization by the presence of a magnetic field itself contains the property of an infinite initial susceptibility at non-zero T, as can readily be observed from eqn. (3.2). In fig. 27 (*a*) and (*b*) we have compared qualitatively the magnetization curves appropriate to this new magnetic phase with those encountered in the case of a normal ferromagnetic transition. At and above T_c the

Fig. 27

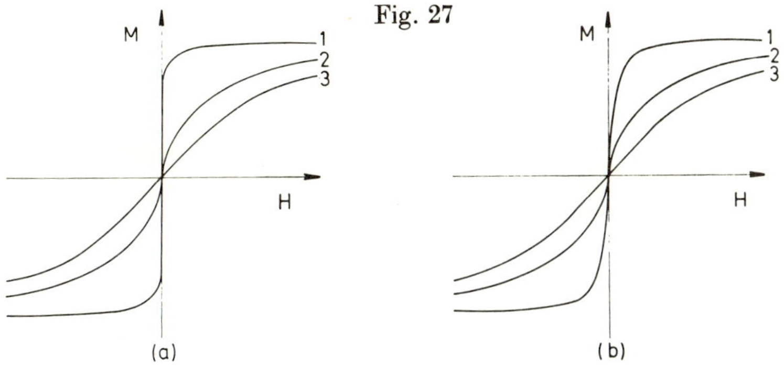

A qualitative comparison of the magnetization curves for the case of a normal ferromagnetic transition (*a*) and for the special type of transition to a state of infinite susceptibility but without spontaneous magnetization (*b*). The isotherms are labelled by the numbers 1, 2 and 3, referring to temperatures below, at and above T_c, respectively.

qualitative behaviour would be similar. Below T_c a finite amount of spontaneous magnetization is found for $H=0$ in the normal ferromagnet (*a*), whereas in case of a Stanley–Kaplan transition all magnetization curves below T_c would start from the origin with an infinite slope (*b*). We finally mention that there seems to be no thermodynamic argument excluding the possible occurrence of this new phase (Stanley 1971).

What would be the consequences on the experimental findings of the existence of a finite transition temperature at which the ferromagnetic susceptibility (and also the staggered† antiferromagnetic susceptibility, Stanley 1969 a) of a 2-d isotropic lattice diverges ? The importance of this prediction lies herein, that it implies that under experimental conditions long-range order will be found below these temperatures. One may understand this by considering that at these temperatures the correlations have become of such a long range that any existing deviation from the ideal system, however small, will ‘ trigger ’ the occurrence of long-range order within the magnetic layers‡. And since we shall not be dealing with isolated mono-layers, it may be understood from simple energy considerations that also an interplanar long-range order will be established throughout

† By staggered it is meant the susceptibility in a staggered field $\tilde{H}$ (likewise in the limit $\tilde{H}\rightarrow 0$), which is a (theoretical) field that is alternating in sign in going from one sublattice of the antiferromagnet to the other. Evidently, the spontaneous magnetization in an antiferromagnet is a staggered quantity.

‡ It is pointed out that, since the interlayer coupling constitutes a deviation from ideality, this argument implies that the new phase with infinite χ and yet no long-range order cannot occur in 3-d lattices. Otherwise one could have conceived of an intermediate temperature range, in between the temperature of the susceptibility divergence (T_{SK}) and that of the onset of spontaneous magnetization (T_c), since the two need not necessarily coincide. Indeed, no indications of such a situation were found in the numerical studies of Baker *et al.* (1970) on the 3-d Heisenberg model (nor in the calculations on the 2-d and 3-d Ising models).

the sample. Depending on the strength of the existing deviations (anisotropy, interlayer coupling) the experimental T_c so obtained will be shifted upwards by a small amount with respect to the temperature at which the susceptibility of the ideal mono-layer would diverge. Thus the crucial difference from the situation found in the magnetic chains is that in the latter the deviations from ideality cause an upward shift of T_c with respect to $T=0$, whereas in the isotropic layers the shift is with respect to the Stanley–Kaplan temperature T_{SK} (it is worth mentioning that, as we have seen in the preceding section, the ferromagnetic susceptibility of isotropic chains is indeed predicted to diverge at $T=0$). If the strength of the deviations can be made sufficiently weak it follows that T_c will very nearly coincide with T_{SK}. We will come back to this point after having reviewed the hitherto discovered examples of 2-d magnetism, and will then present experimental evidence in favour of the existence of such a T_{SK}.

As concerns the behaviour of other thermodynamic quantities of interest, we may expect that in the absence of long-range order the specific heats will exhibit similar broad finite maxima as in the case of magnetic chains (see fig. 6). The transition to long-range order, due to the deviations from ideality, will again be reflected as a sharp spike sitting on the flank of the broad short-range-order maximum. As to the precise form of the latter there is no theory available. At low temperatures, simple spin-wave theoretical arguments predict a temperature dependence that is linear for the ferromagnetic and quadratic for the antiferromagnetic Heisenberg lattices (§ 4.2). In the high-temperature region the series expansion results for the heat capacity provide a reliable prediction. But in the rather large intermediate region there are only experimental results available (see below).

For the antiferromagnetic susceptibility one expects a similar behaviour as found for the isotropic antiferromagnetic chains (figs. 18 (*a*), 21 and 23), since they have in common the absence of long-range order and of anisotropy. Accordingly, a broad maximum due to the short-range-order effects should be found at the higher temperatures, whereas at $T=0$ the susceptibility should attain a finite value. Since there is no closed-form theory available, we must once more rely on the high-temperature series expansions and on spin-wave theory in the high and low-temperature range, respectively.

The series expansion results have been studied by Lines (1970), who used the six terms in the susceptibility series for general lattice and spin given by Rushbrooke and Wood (1958). Calculating the antiferromagnetic susceptibility of the quadratic Heisenberg lattice for various S values, Lines found that the series prediction just covers the temperature range in which the broad maxima occur. At lower temperatures it becomes unreliable because of the finite number of terms known in the series (which is given in ascending powers of J/kT). Lines also produced a formula relating the exchange constant for various S values to the temperature at which the maximum occurs, thus enabling an estimate of J/k of the experimental examples.

De Jongh (1972 c) has extended this work with the aid of additional terms in the series, finding slightly different quantitative results. He points out that more reliable estimates of the exchange may be obtained by comparing with theory the height of the observed susceptibility maximum rather than the temperature at which it occurs, since the former quantity can be established with a greater accuracy from the finite number of terms known in the series and is mostly also better determined experimentally. The values for the temperatures and heights of the maxima for various S are listed in table 4, and we shall make use of them in what follows in determining the exchange constants of the experimental examples. Anticipating the discussion of these examples below, we show in fig. 28 (De Jongh 1972 c) the susceptibilities of six approximations of the quadratic antiferromagnetic Heisenberg layer, with different spin values. In the region below T_c ($kT_c/|J|S(S+1) \simeq 1$) the data represent the measured perpendicular susceptibilities. In the high-temperature region above T_c the susceptibility is mostly found to be fairly isotropic, except for K_2NiF_4 in which there is a discernible difference between $\chi_\perp$ and $\chi_\parallel$ also above T_c, so that for this salt only the $\chi_\parallel$ data have been plotted for $T > T_c$. By fitting the exchange constants, the experimental curves have been scaled upon the high-temperature series predictions, which have been drawn in the temperature region for which the calculations with a varying number of terms do not differ by more than a few per cent. Note the larger deviation from the molecular field theory when S is decreased. Although there are some discrepancies, which may be caused, for example, by a temperature dependence of J/k or by the limited number of terms in the series, there is on the whole very good agreement.

It is also observed from fig. 28 that the experimental perpendicular susceptibilities, extrapolated to $T = 0$, are in close agreement with the $\chi_\perp(0)$ values as predicted by spin-wave theory for a 2-d Heisenberg antiferromagnet. For $S = \frac{5}{2}$ and $S = 1$ these values have been indicated in fig. 28 by the horizontal lines labelled a and b. We point out that the difference between the spin-wave predictions for $\chi_\perp(0)$ and the MF result $\chi_\perp{}^0 = N_0 g\mu_B S/2H_E$† is, apart from an anisotropy effect that is extremely small in the compounds considered, wholly due to the effects of zero-point spin deviations (De Jongh 1972 b, c). Spin-wave theory yields (see, e.g., Keffer 1966)

$$\chi_\perp(0) = \frac{\chi_\perp{}^\circ}{1+\frac{1}{2}\alpha}\left[1 - \frac{\Delta S(\alpha)}{S} - \frac{e(\alpha)}{(2+\alpha)zS}\right]. \tag{3.6}$$

Here $\chi_\perp{}^\circ$ is the MF prediction defined above, $\alpha = H_A/H_E$, S is the spin value, z the number of nearest neighbours, whereas the anisotropy

† H_E is the effective field associated with J, according to the MF relation: $g\mu_B H_E = 2\pi|J|S$.

Table 4. Estimated values for the temperatures at which the maxima occur in the antiferromagnetic susceptibility of the quadratic Heisenberg lattices. The values attained by the susceptibility at these temperatures are also given. Note that the product $\chi_{max}T(\chi_{max})/C$ is only very slowly varying with S. For comparison the result for the parallel susceptibility of the quadratic Ising lattice ($S=\frac{1}{2}$) has been included. Also listed are the number of terms available in the series expansions used in deriving the listed results. After De Jongh (1972 c).

	Quadratic Heisenberg lattice						Ising ($\chi_{\parallel}$)
S	$\frac{1}{2}$	1	$\frac{3}{2}$	2	$\frac{5}{2}$	∞	$\frac{1}{2}$
$\frac{kT(\chi_{max})}{\lvert J\rvert S(S+1)}$	2·53 ±0·05	2·20 ±0·02	2·10 ±0·01	2·07 ±0·01	2·05 ±0·01	2·01 ±0·01	2·325
$\frac{\chi_{max}\lvert J\rvert}{N_0 g^2 \mu_B{}^2}$	0·0469 ±0·0001	0·0521 ±0·0001	0·0539 ±0·0001	0·0547 ±0·0001	0·0551 ±0·0001	0·0561 ±0·0001	0·05370
$\frac{\chi_{max}T(\chi_{max})}{C}$	0·356	0·344	0·340	0·340	0·339	0·338	0·375
Number of terms	10	7	7	7	7	9	

Fig. 28

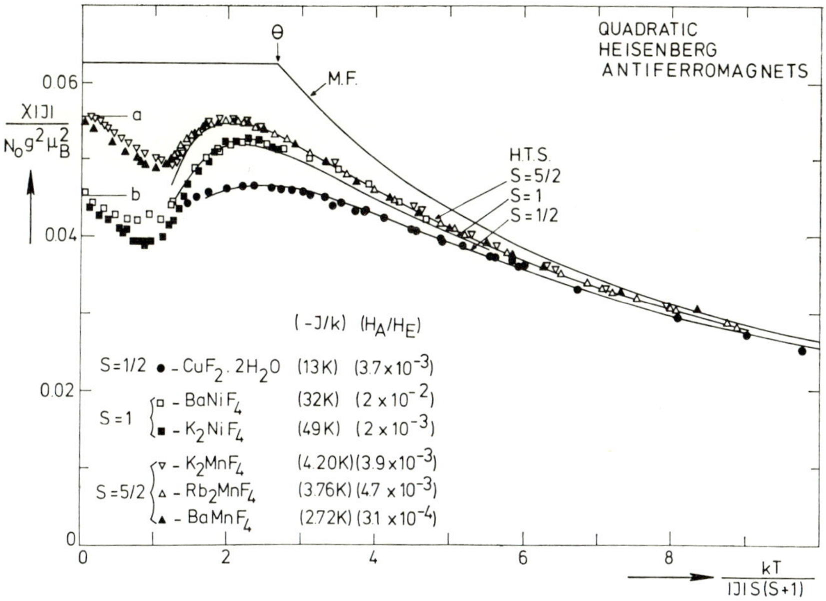

The susceptibilities of six examples of the quadratic Heisenberg antiferromagnet. The experimental data in the high-temperature region ($T > T_c$; $kT_c \simeq S(S+1)|J|$) have been fitted to the theoretical (solid) curves by choosing the right exchange constants J/k. These curves have been calculated from the high-temperature series expansions for the susceptibility (H.T.S.). Note the large deviation from the molecular field result (MF) for the susceptibility in the paramagnetic region and below the transition temperature ($\chi_\perp$). For references to the data, see the text. Below T_c the measured perpendicular susceptibilities of two $S=\frac{5}{2}$ and two $S=1$ compounds have been included. The extrapolated values to $T=0$ may be compared with the $\chi_\perp(0)$ values predicted by spin wave theory (in the limit $H_A=0$), which have been indicated by the horizontal lines *a* and *b* for $S=\frac{5}{2}$ and $S=1$, respectively. The differences between $\chi_\perp$ for compounds with the same S reflect the difference in anisotropy.

dependent quantities $\Delta S(\alpha)$ and $e(\alpha)$ reflect the effects of zero-point spin deviations on the effective length of the magnetization vector and on the ground-state energy, respectively. Consequently, the fact that the experimental $\chi_\perp$ curves agree with the spin-wave predictions at $T=0$, constitutes experimental evidence for the existence of these zero-point deviations, since the parameters α and J/k have been determined independently from other sources. We shall return to this interesting point in § 4.2. Alternatively, the above argument may be reversed: knowing $\Delta S(\alpha)$ and $e(\alpha)$ one may use the experimental $\chi_\perp(0)$ value to determine the exchange constant with the aid of eqn. (3.5). We shall frequently do this below, making use of the $\Delta S(\alpha)$ values tabulated by Lines (1970) and Colpa *et al.* (1971). As shown by Breed (1969), the quantity $e(\alpha)$ can be taken equal to $e(0) = 0{\cdot}632$ (for the quadratic lattice, see Keffer 1966) for anisotropy values $\alpha < 10^{-2}$.

The applicability of spin-wave theory for $T > 0$ and finite α is illustrated in fig. 29 (De Jongh 1972 c). Under experimental conditions the deviations from ideality will cause a transition to long-range order, as discussed above, and once this has been established the finite amount of anisotropy that will always be present will split up the susceptibility into a perpendicular and a parallel part, so that one obtains the picture seen in fig. 29. Knowing J/k and α, both $\chi_{\perp}$ and $\chi_{\parallel}$ may be calculated from spin-wave theory, as has been done by Breed, using the parameters appropriate to K_2MnF_4 of which he measured the susceptibility. Renormalization effects (Oguchi's correction terms, Oguchi 1960) were incorporated in the calculations. As far as the limiting low-temperature behaviour is concerned the agreement is seen to be good, but there is certainly a need for a better theory. Note that the behaviour of the parallel susceptibility is reminiscent of that of the 2-d Ising antiferromagnet (fig. 4). Also in this case T_c coincides with the temperature at which $\partial\chi_{\parallel}/\partial T$ reaches

Fig. 29

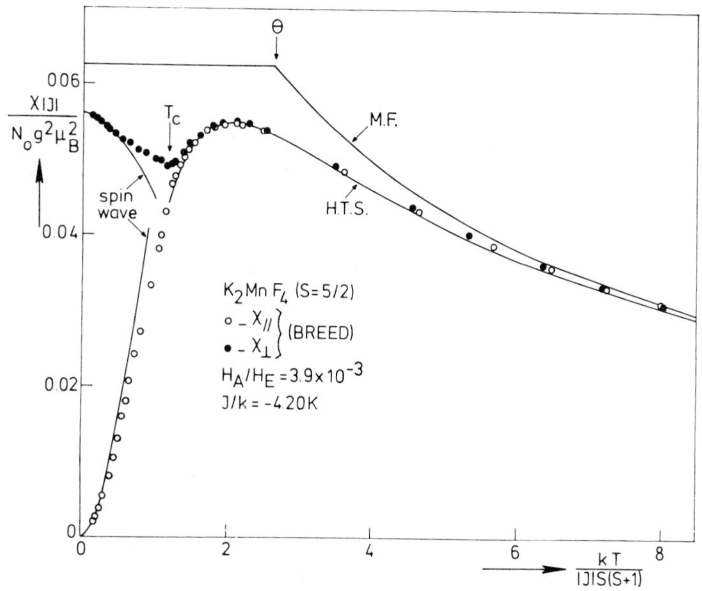

The measured parallel and perpendicular susceptibility of K_2MnF_4, which is an example of the quadratic $S = \frac{5}{2}$ Heisenberg antiferromagnet. The value of J/k has been determined by fitting the high-temperature susceptibility to the series expansion prediction (H.T.S.). The value of the anisotropy parameter H_A/H_E is also indicated. Using these values of J/k and H_A/H_E, Breed (1969) has calculated the spin-wave prediction for $\chi_{\perp}$ and $\chi_{\parallel}$ shown in the figure. The measurements were also performed by Breed. The position of the transition temperature T_c has been indicated.

its maximum, but the derivative remains finite here. The anisotropy in the susceptibility is observed to persist in a small region above T_c, an effect that is commonly met in these 2-d antiferromagnets and is due no doubt to the substantial short-range order that is present above T_c.

Thus, simple spin-wave theory, including a small anisotropy term, seems to give a reasonable description of the limiting low-temperature behaviour of the 2-d antiferromagnetic susceptibility. It is interesting to note that the general appearance of the perpendicular susceptibility is rather similar to that of the 3-d Heisenberg antiferromagnet, as displayed in fig. 53 below. Also in three dimensions the $\chi_\perp$ firstly decreases with temperature, passing through a minimum as T_c is approached. The increase in $\chi_\perp$ near T_c has been ascribed by Kanamori and Itoh (1968) to contributions of excitations of more than one magnons of non-zero wave-number, which are not taken into account in less sophisticated treatments. This is of importance since for the fully isotropic 2-d case simple spin-wave theory predicts a zero $\chi_\perp$ for any $T>0$, in contrast with the 3-d analogue where for $\alpha=0$ there still is a finite $\chi_\perp$, decreasing to first order quadratically with temperature (Keffer 1966). The anomalous behaviour in two dimensions arises from the unlimited decrease in the sublattice magnetization through the excitation of magnons discussed above, since the term given in eqn. (3.5) enters in the expression for $\chi_\perp(T)$. However, in the case of the $\chi_\perp$ the anomalous decrease might be a result of the approximations made in the simple theory, and could possibly be compensated by mechanisms taken into account in more sophisticated calculations, such as mentioned above. In this respect it is of importance to note that for the 1-d Heisenberg antiferromagnet spin-wave theory is completely in error, predicting a $\chi_\perp$ at $T=0$ that vanishes logarithmically with α in the limit $\alpha \to 0$, in contradiction to the numerical calculations and the experimental findings discussed in § 3.1. Experimentally, the behaviour of the manganese compounds in fig. 28 indicates that at small α (10^{-3}–10^{-4}) a change of a factor 10 in α hardly has an effect upon the temperature dependence of $\chi_\perp$, merely shifting the value of $\chi_\perp(0)$ by a small amount. In view of these considerations one might postulate a susceptibility behaviour of the ideal model quite similar to the experimental curves of fig. 28 (De Jongh 1972 c), bearing in mind that in analogy with the 1-d antiferromagnet the derivative of the ideal χ versus T curve is expected to remain finite everywhere in view of the energy–susceptibility relation discussed in § 1.2. Since a singularity in the specific heat will be reflected in $\partial\chi/\partial T$, it may be argued that the temperature dependence of the latter must be of a similar smooth, non-singular, form as the specific heat in case there is no transition to long-range order, as evidenced by the antiferromagnetic chains. This is clearly a matter that has yet to be solved theoretically. Although the effect of anisotropy does seem to be small, the experiments cannot exclude, for example, a logarithmic decrease of $\chi_\perp$ with α, as is the prediction of the simple spin-wave theory.

There is an interesting aspect of the influence of the anisotropy on the perpendicular susceptibility that we would like to discuss briefly. As seen from eqn. (3.6) $\chi_\perp(0)$ is lowered with respect to $\chi_\perp{}^\circ$ by increasing the anisotropy. At first sight this is in contradiction to the result obtained by Fisher (1960 a) for the $\chi_\perp(0)$ of the 2-d Ising model, which was found to be twice as large as $\chi_\perp{}^0$. But a little reflection shows that this is one of the examples of the fundamental differences that may arise from the particular way in which the anisotropy is introduced in the Hamiltonian, as has been pointed out in § 2.2. In the above spin-wave theory one starts with an isotropic exchange interaction, allowing for the anisotropy by means of an effective anisotropy field H_A, representing, for example, the single-ion or dipolar anisotropy. We may conveniently call this model the anisotropic Heisenberg model. On the other hand, in the Ising model an anisotropic exchange interaction is considered, leading to a lack of antiferromagnetic correlations in the perpendicular direction so that the perpendicular susceptibility above T_c is only lowered little with respect to the paramagnetic behaviour, as shown in fig. 30. This explains the relatively high value of $\chi_\perp(0)$ in the Ising model.

We shall have occasion to return to this point, since it turns out that especially the perpendicular susceptibility of the anisotropic examples of

Fig. 30

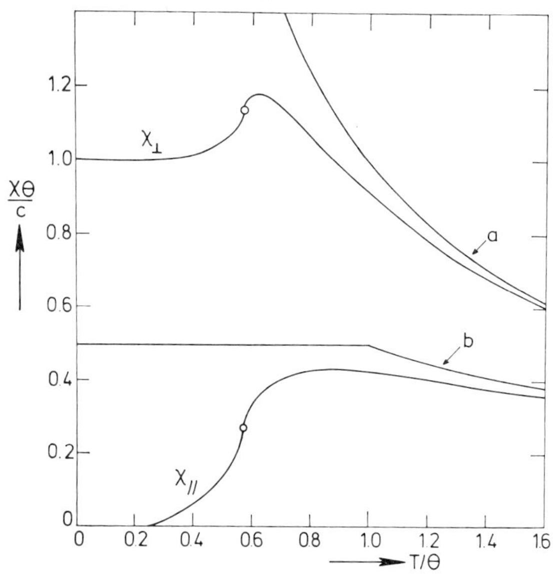

A comparison of the perpendicular ($\chi_\perp$) and the parallel ($\chi_\parallel$) susceptibility of the quadratic Ising lattice (Fisher 1963, Sykes and Fisher 1962) with $S=\frac{1}{2}$. Curve a is the susceptibility of a paramagnetic substance. Curbe b is the molecular field prediction for the antiferromagnetic susceptibility in the paramagnetic region and for the perpendicular part below the transition temperature. Compare also figs. 16 and 17 (a).

2-d antiferromagnetism can mostly be better analysed in terms of the anisotropic Heisenberg model than that of the Ising model. This may be understood by considering that in the experimental examples the anisotropy originates mainly from crystal-field effects, the superexchange mechanism being in most cases quite isotropic.

Quantitative values for the anisotropy can be obtained by various experimental techniques (magnetic torque, magnetic resonance, spin-flop experiments), and in fact for most of the compounds compiled below an estimate of the anisotropy could be given. In case the anisotropy is of orthorhombic symmetry one should distinguish between the different crystallographic directions. In a simple model, we shall denote by x, y and z the preferred, the next preferred and the hardest direction, respectively, and introduce the orthorhombic anisotropy in the Hamiltonian in the form of terms KS_y^2 and LS_z^2, $L>K>0$. For $L \gg K$ the anisotropy becomes of the planar Heisenberg type, there being an easy plane in which the moments are nearly free to rotate, whereas for $K \simeq L$, an Ising-like character is approached by increasing K and L. With the anisotropy constants K and L one may associate the anisotropy fields $H_A{}^{I}=2KS/g\mu_B$ and $H_A{}^{II}=2LS/g\mu_B$. If $H_A{}^{II} \gg H_A{}^{I}$ one may simply take the former for the anisotropy H_A. On the other hand, if $H_A{}^{I}$ and $H_A{}^{II}$ are of the same order and if one does not want to differentiate between them it is not so obvious which quantity should be taken for 'the' anisotropy. In what follows we have occasionally used the quantity $(H_A{}^{I}H_A{}^{II})^{1/2}=H_A$.

Estimates of the interlayer coupling J', which is the interaction in the third dimension, could not be obtained experimentally for most of the examples treated below. The exception is the case of ferromagnetic layers, coupled by an antiferromagnetic J', where it can in principle be calculated from the antiferromagnetic $\chi_\perp$ in the ordered region, $\chi_\perp$ being inversely proportional to J' (see below). For the other structures one must therefore take recourse to estimates of J' by considering the interlayer exchange paths and calculations of the interlayer dipolar coupling. Referring to the discussion in § 2.4, this leads, for example, to values $|J'/J| \simeq 10^{-6}$ in the compounds of the K_2NiF_4 structure and to $|J'/J| \approx 10^{-3}$–10^{-6} in the series $(C_nH_{2n+1}NH_3)_2CuCl_4$ when n is varied from 1 to 10. It should be noted that these values apply to the ideal crystallographic structures and that for instance in the K_2NiF_4 structure the value of $|J'/J|$ will be up to two orders of magnitude larger in the case that lattice defects or distortions of, for example, magnetostrictive origin invalidate the symmetry argument that leads to the decoupling of nearest-neighbouring layers. On the whole, one is apt to think that lattice imperfections and other defects will set a lower bound to the value of $|J'/J|$ that can possibly be achieved. Whenever one considers a $|J'/J|$ as small as 10^{-10}, for example, one should certainly take into account the effects of these imperfections, as well as the fact that the lattice is not rigid, due to the presence of phonons.

In conclusion we want to make a few remarks about the critical behaviour to be anticipated in the experimental examples of 2-d magnetism. Here we must again differentiate between the compounds of the Ising and of the Heisenberg type. For a 3-d array of nearly isolated layers with a highly anisotropic intralayer exchange (J) one expects to find a 2-d Ising-type behaviour, except in a narrow temperature range around T_c where the small interlayer coupling (J') will come into play. The extent of this temperature region will depend on the relative strength of J'. Taking the temperature dependence of the spontaneous magnetization as an example one expects to find a 2-d character with a critical exponent β near to the 2-d value $\frac{1}{8}$, except very close to T_c where β should change to the 3-d value of about $\frac{1}{3}$, according to the universality hypothesis put forward by Griffiths (1970 b) and Kadanoff (1970). This states that close enough to T_c the value of β should be independent of the ratio $|J'/J|$, except at the point $J'=0$, where it should change discontinuously from the 3-d to the 2-d value (Paul and Stanley 1971 b, Citteur 1973). The reason is that for any finite value of J' the system is essentially 3-d.

In the case of the nearly 2-d Heisenberg magnets the situation is different because ideally these systems cannot sustain long-range order for $T>0$. The spontaneous magnetization that is nevertheless observed must therefore be attributed solely to the combined effects of the anisotropy and the inter-layer coupling. In § 3.1.1 we have already pointed out that the difference between the magnetic chains and the 3-d Heisenberg magnets is that in the former only the inter-chain coupling J' can be held responsible for the presence of long-range order, whereas in the latter the anisotropy H_A provides an additional mechanism. Depending on the relative strength of H_A and J' the behaviour will be predominantly of the 2-d Ising type or 3-d in character, again in accordance with the universality hypothesis. Considering again the spontaneous magnetization, one expects that if $g\mu_B H_A \gg J'$ (as is quite often the case) the critical behaviour will be 2-d Ising-like ($\beta \simeq \frac{1}{8}$) over a wide temperature range, changing over to 3-d ($\beta \simeq \frac{1}{3}$) when T_c is approached closely enough. As we will see below indications of such a cross-over of a critical exponent have indeed been found experimentally.

We will now proceed to discuss the examples of the 2-d Ising and planar Heisenberg models. In table 5 we have gathered the antiferromagnetic and ferromagnetic layer-type compounds that receive consideration. It can be seen that these are mostly Co compounds. Subsequently, the approximations of the 2-d Heisenberg magnets will be reviewed.

3.2.2. *Survey of experimental results*

$CoCs_3Br_5$

In § 2.2 we have already discussed the anisotropy in the interaction between the Co^{2+} ions in this compound. As to the origin of the 2-d character it was mentioned in § 2.4 that the cancellation of the interaction

Table 5. Examples of two-dimensional anisotropic magnets. The compounds are grouped according to type and sign of the interaction (Ising or planar Heisenberg (PH) or intermediate; ferro or antiferromagnetic). Listed are the spin value S; the transition temperature T_c; the intralayer exchange J/k; the anisotropy parameter α which is either H_A/H_E or $1-b/a$ with a and b defined by eqn. (1.1) (for the planar Heisenberg type compounds, two anisotropy values have been given, which apply to the anisotropy within the easy plane and the out-of-plane anisotropy, respectively); the ratio $|J'/J|$ of inter to intralayer interaction; and the quantities $kT_c/|J|$ and T_c/θ, where the Curie–Weiss temperatures θ have been calculated from J/k according to the formula: $k\theta=\frac{2}{3}zS(S+1)|J|$. The references to the data are mentioned in the text. We note that for $FeCl_2$ and $CoCl_2$ the 2-d network is the triangular lattice ($z=6$), whereas for the other compounds it is quadratic or quasi quadratic ($z=4$).

Compound	Type of interaction	Sign of interaction	S	T_c	J/k	α	$\|J'/J\|$	$kT_c/\|J\|$	T_c/θ
$CoCs_3Br_5$	I	−	$\frac{1}{2}$	0·282	−0·22	$\simeq 1$	?	1·28	0·64
$Co(HCOO)_2 . 2H_2O$	I	−	$\frac{1}{2}$	5·12	−4·3	0·9	8×10^{-3}	1·19	0·59
Rb_2CoF_4	I	−	$\frac{1}{2}$	101	−91	0·8	$\simeq 10^{-6}$	1·11	0·56
K_2CoF_4	I	−	$\frac{1}{2}$	107	−97	0·7	$\simeq 10^{-6}$	1·10	0·55
$BaFeF_4$	I–PH	−	2	54·2	−6·4	$\simeq 0·2$	$\simeq 10^{-6}$	8·5	0·53
Rb_2FeF_4	I–PH	−	2	56·3	−6·5	$\simeq 0·1$	?	8·7	0·54
$FeCl_2$	I	+	1	23·55	3·4	$\simeq 1$	$7·5\times10^{-2}$	6·9	0·87
$CoCl_2$	PH	+	$\frac{1}{2}$	24·71	$\simeq 10$	$\{\simeq 10^{-3}$, $\simeq 1\}$	$\simeq 0·1$	$\simeq 2·5$	$\simeq 0·8$
$CoCl_2 . 6H_2O$	PH	−	$\frac{1}{2}$	2·29	−1·9	$\{4\times10^{-2}$, $\simeq 0·7\}$	?	1·21	0·60
$CoBr_2 . 6H_2O$	PH	−	$\frac{1}{2}$	3·14	−2·3	$\{4\times10^{-2}$, $\simeq 0·7\}$	?	1·37	0·68
$BaCoF_4$	PH–I	−	$\frac{1}{2}$	69·6	$\simeq -100$	$\{\simeq 0·4$, $\simeq 0·8\}$	$\simeq 10^{-6}$	$\simeq 0·70$	$\simeq 0·35$

in the third dimension is in this case thought to be accidental, since no crystallographic or other arguments can be brought up.

Fig. 31

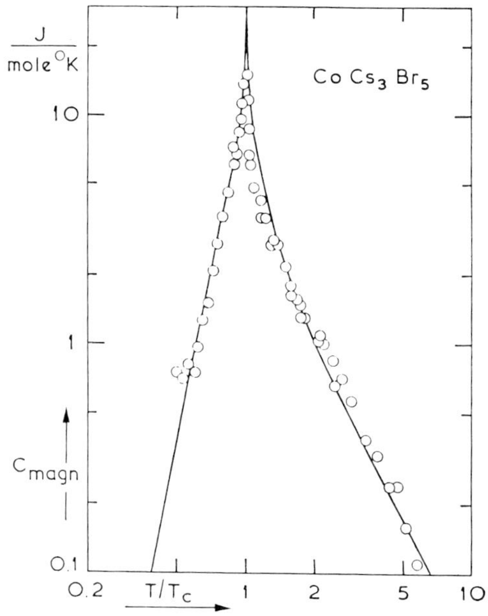

Magnetic specific heat of $CoCs_3Br_5$ plotted versus the temperature relative to T_c. The full curve represents Onsager's exact solution (1944) of the heat capacity of the quadratic, $S=\frac{1}{2}$, Ising lattice. (After Wielinga *et al.* 1967.)

The magnetic lattice of the Co^{2+} ions is quasi simple cubic, the *c* axis of the tetragonal cell being about 10% longer (Figgis *et al.* 1964). The thermal and magnetic properties have been investigated by Wielinga *et al.* (1967) and Mess *et al.* (1967), respectively. In fig. 31 the specific heat is compared with the theoretical curve of Onsager (1944) for the simple quadratic Ising magnet. Although there is a slight disagreement above T_c there is on the whole a striking resemblance. A similarly good agreement with theory is found in the case of the field dependence of the antiferromagnetic transition point, as plotted in fig. 32. The data were obtained by Mess *et al.* (1967) from magneto-thermal experiments, while the theoretical curve has been computed by Bienenstock (1966). This field-dependent behaviour will be treated in more detail in § 4.5. It should be noted that the agreement would be still better if the experimental data could have been corrected for demagnetizing effects, since this would lower the points at the lower temperatures by a few per cent.

Comparing $CoCs_3Br_5$ with the other compounds it may be concluded that its specific heat is up until now the best example of the Onsager curve. Except for $Co(HCOO)_2 . 2H_2O$ the other compounds have transition temperatures that are so high that the subtraction of the lattice heat capacity becomes a very difficult matter, and for the formate the agreement with the theoretical curve is less satisfying (Takeda *et al.* 1971 a). Notwithstanding this, the low value of T_c in the case of $CoCs_3Br_5$ will have the consequence that the effect of dipolar interaction cannot be neglected. As suggested by Wielinga *et al.* (1967) this could account for the difference of the obtained value $T_c/\theta = 0{\cdot}64$ with the theoretical number 0·567, and the other observed discrepancies. Unfortunately, a value for J' cannot be deduced from the existing data.

Fig. 32

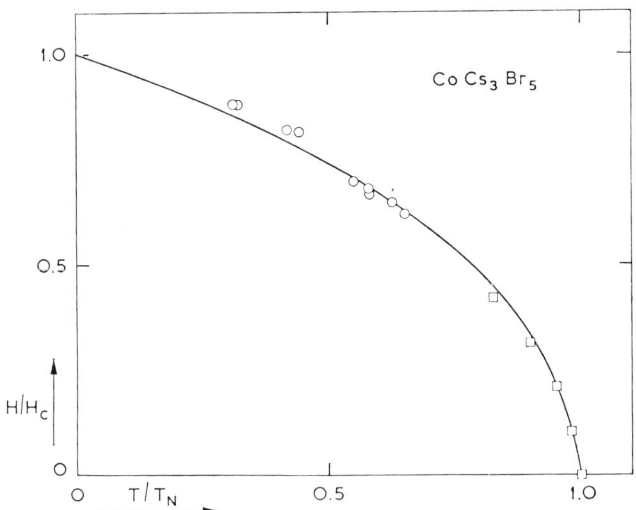

Antiferromagnetic phase diagram of $CoCs_3Br_5$ as determined from magnetocaloric experiments. The full curve has been calculated by Bienenstock (1966) for the quadratic $S=\frac{1}{2}$ Ising antiferromagnet. (After Mess *et al.* 1967.)

$Co(HCOO)_2 . 2H_2O$ *and* $Fe(HCOO)_2 . 2H_2O$

Two-dimensional magnetism in the transition metal formate dihydrates was first discovered in the manganese member of this series of compounds, which will be treated below amongst the Heisenberg compounds. At a later date also the Fe^{2+}, Ni^{2+} and Co^{2+} salts were studied (Hoy *et al.* 1965, Matsuura *et al.* 1970 a, Pierce and Friedberg 1971, Takeda and Matsukawa 1971). Of these the cobaltous and ferrous compounds possess strongly anisotropic properties as required for the Ising model.

In magnetic respect the structure of these formates consists of alternating A and B sheets of metal ions, those in the B sheets remaining paramagnetic down to temperatures much lower than that at which the spins in the A sheets become antiferromagnetically ordered. The 2-d character emanates from the fact that the A sheets are as it were separated from each other by the B sheets. At low enough temperature the spins in the B sheets are gradually ordered under the influence of the field exerted on them by the spins of the A sheets. Evidence for this subdivision in two differently behaving spin systems is provided by various pieces of experimental evidence and may be understood by considering the distinct exchange paths that arise from the difference in environment of the metal ions in the A and B sheets. The existing values deduced for the ratio of inter to intralayer exchange are 8×10^{-3} for the cobaltous and 3×10^{-3} for the manganese compound, these exchange constants having been derived from the heat capacity data.

Fig. 33

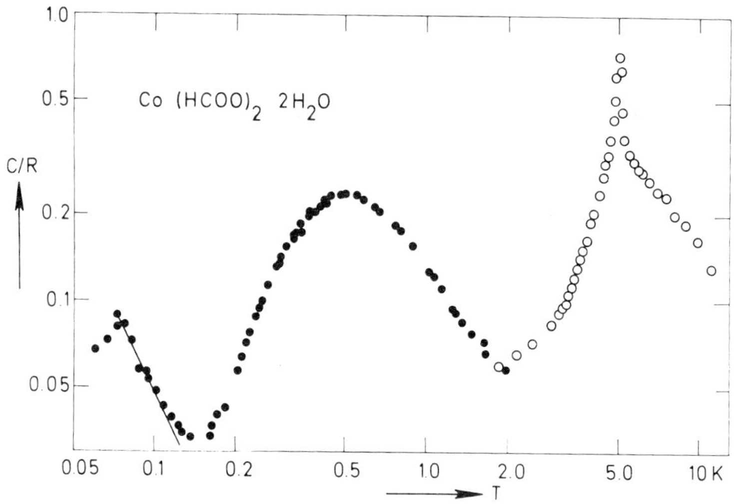

The specific heat of $Co(HCOO)_2 \cdot 2H_2O$ as measured by Matsuura *et al.* (1970 a) and Takeda *et al.* (1971 a).

For the Co formate three maxima in the specific heat are found (fig. 33) (Matsuura *et al.* 1970 a, Takeda *et al.* 1971 a). At 5·12 K a sharp peak is observed due to the 2-d ordering in the A sheets. At lower temperatures the contribution of the ordering within the paramagnetic B sheets is reflected in the heat capacity in the form of a broad, Schottky-type, anomaly with a maximum at 0·50 K. Below 0·1 K the hyperfine coupling between the cobalt spin and its nucleus provides for yet another maximum. This figure illustrates how much information can be obtained from the analysis of a specific heat curve.

The susceptibility (Takeda and Kawasaki 1971) exhibits a large peak at T_c and rises again below T_c as a consequence of the paramagnetic contribution of the B sheets. The features of the peak indicate the existence of a weak ferromagnetic moment due to a canting of the spins in the A sheets, as was also found in the case of the isomorphous manganese salt. A similar anomaly in the susceptibility has been observed for the Fe formate (Hoy *et al.* 1965, Takeda and Kawasaki 1971).

In the specific heat curve of the Fe formate no sign was found of the rounded maximum produced by the B sheets, at least not for $T > 1$ K. The heat capacity decreases exponentially below $T_c = 3{\cdot}75$ K, as expected for an Ising compound (Pierce and Friedberg 1971, Takeda and Kawasaki 1971). Above T_c there is a very large tail in the specific heat that cannot be explained by 2-d short-range-order effects alone. There will probably be a contribution of low-lying excited levels and it has also been suggested by Pierce and Friedberg that the apparent absence of the Schottky anomaly (although this could still appear at temperatures much below 1 K) might be explained by assuming that the ions in the B sheet already lose their entropy above T_c. This would arise in the following way. From entropy considerations and the earlier Mössbauer results (Hoy and Barros 1965), Pierce and Friedberg concluded that most probably the A ions have an effective $S = \frac{1}{2}$, whereas the B ions have $S = 2$. The Mössbauer experiments of Shinohara *et al.* (1972) confirm these assignments of spins and indicate that the zero-field splitting of the B ion leaves a singlet ground state, sufficiently separated from the remaining components of the $S = 2$ manifold, so that the B spins do not order spontaneously at any temperature. This implies that J_{AB} is rather small in this compound, as in the other isomorphs. The anomalously large specific heat above T_c is therefore the result of short-range-order effects within the A sheets, combined with a Schottky-type heat capacity due to the B sheets.

In view of the uncertainties that still remain we have refrained from entering this compound in the table.

Rb_2CoF_4 *and* K_2CoF_4

The anisotropy in these salts, which have the K_2NiF_4 structure, has already been discussed in § 2.2. The values for H_A/H_E listed in table 5 stand in this case for the quantity $1 - b/a$, where b and a have been defined in eqn. (1.1).

After a preliminary experiment by Srivastava (1963) on K_2CoF_4 the susceptibilities of these compounds have been investigated *in extenso* by Breed *et al.* (1969), and have been reproduced in fig. 34 (*a*). Apparently there is a large Van Vleck contribution, due to the presence of the higher energy levels, as may be inferred from the fact that the parallel susceptibilities reach finite values at $T = 0$. The large uniaxial anisotropy favouring the c axis is clearly seen in the behaviour of the susceptibility measured perpendicular to this direction. The result for K_2CoF_4, after subtraction of the Van Vleck contribution, is compared with the calculations of Sykes and Fisher (1962) in fig. 34 (*b*). For this purpose we have assumed the Van Vleck term (determined from $\chi_{\parallel}(0)$) to be temperature independent, which is certainly not justified since the distance from the lowest-lying doublet to the next excited level has been calculated by Folen *et al.* (1968) to be about 400 K. As an additional complication T_c (=107 K) and J/k (= −97 K) are of this order of magnitude, which makes a more sophisticated correction for the Van Vleck contribution unfeasible. For these reasons it is easily understood why the data in the high-temperature region deviate from the theoretical curve. Notwithstanding this it may be seen that a reasonable agreement is obtained at lower temperatures, using the J/k and g values derived by Breed *et al.* (1969). It is worth mentioning here that the theoretical prediction that T_c should coincide with the temperature at which the derivative of the $\chi_{\parallel}$ versus T curve reaches infinity (under experimental conditions: its maximum value), could also be verified by these authors, since they were able to locate T_c independently from fluorine N.M.R. experiments similar to those described by Maarschall *et al.* (1969) for K_2NiF_4. It can be seen from fig. 34 (*a*) that at T_c the perpendicular susceptibilities have a maximum temperature derivative too. In fact the $\chi_{\perp}$ can also be fitted to the theoretical curve up to $T \simeq 1{\cdot}1\ T_c$ by using the appropriate g value, but in order to accomplish this the Van Vleck contribution in the perpendicular direction has to be taken to be about two-thirds of its value parallel to the c axis. In view of this uncertainty we have refrained from

Fig. 34

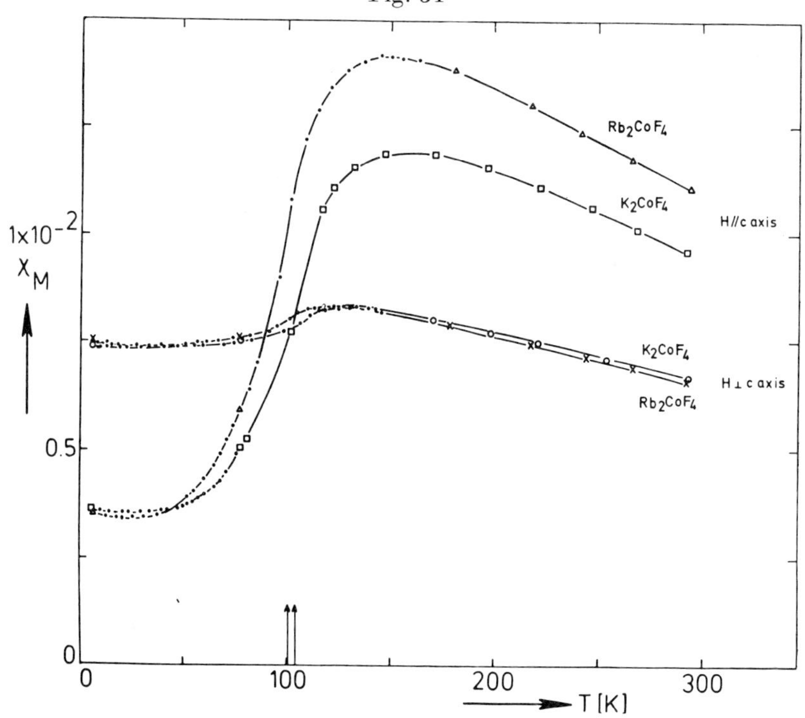

(*a*)

Fig. 34 (*continued*)

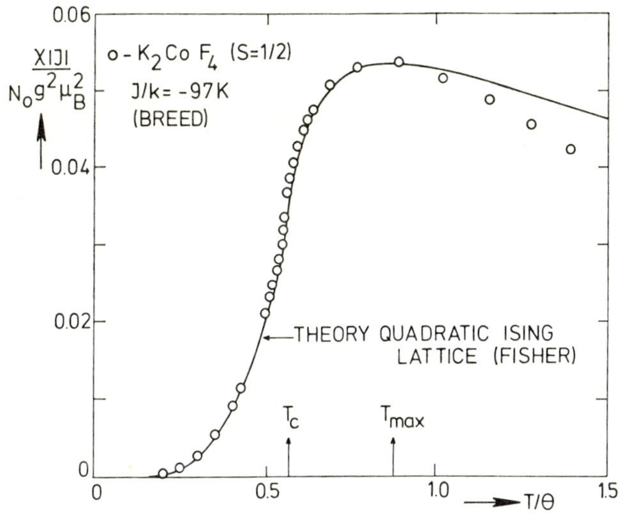

(*b*)

(*a*) Magnetic susceptibilities of K_2CoF_4 and Rb_2CoF_4 (Breed *et al.*, 1969), which are approximants of the quadratic $S=\frac{1}{2}$, Ising antiferromagnet. The positions of T_c have been indicated by the arrows. The large uniaxial anisotropy favouring the c axis can be inferred from the behaviour of the susceptibility measured perpendicular to this direction. The large Van Vleck contribution to the susceptibility is seen for instance from the fact that the parallel susceptibilities do not decay to zero as $T\to 0$. (After Breed *et al.*, 1969.) (*b*) The susceptibility of K_2CoF_4 compared with the theoretical prediction of Sykes and Fisher (1962) for the quadratic, $S=\frac{1}{2}$, Ising antiferromagnet. In order to correct for the Van Vleck contribution the value of the parallel susceptibility extrapolated to $T=0$ has been subtracted from the data.

producing the result for $\chi_\perp$ here, all the more since one may seriously doubt the applicability of the Ising model for the perpendicular susceptibility of experimental examples, as outlined in § 3.2.1.

These compounds are also currently being studied by Samuelsen with neutron diffraction. As preliminary values for the spontaneous magnetization parameters we quote for e.g. Rb_2CoF_4: $\beta=0{\cdot}119\pm 0{\cdot}008$ and $B\simeq 1{\cdot}16\pm 0{\cdot}03$, in the range $5\times 10^{-4}<1-T/T_c<1\times 10^{-1}$ (E. J. Samuelsen 1973 and private communication). These numbers are within the experimental uncertainties equal to the theoretical predictions for the $S=\frac{1}{2}$ quadratic Ising lattice, $\beta=0{\cdot}125$ and $B=1{\cdot}24$ (for $1-T/T_c<2\times 10^{-2}$).

$BaFeF_4$

This compound belongs to the $BaMF_4$ group of fluorides (M = divalent transition ion). Other members of this series (M = Co, Mn and Ni) will be encountered below. The orthorhombic crystal structure consists of puckered sheets of MF_6 octahedra, separated by a layer of nonmagnetic Ba^{2+} ions (Keve *et al.* 1969). The 2-d lattice is thus a puckered, quasi quadratic network. One does not expect the rumpling of the lattice to have substantial influence on the 2-d properties. We note that also in this structure the interaction between neighbouring layers is cancelled because of symmetry. Since the structure is rather complicated, it is difficult to estimate the interlayer coupling. However, one may expect it to be of the same order of magnitude as in the K_2NiF_4-type compounds, so we have likewise listed the value $|J'/J| \simeq 10^{-6}$ in table 5.

Magnetic measurements have been performed by Eibschütz *et al.* (1970). As in the case of Rb_2FeF_4, the data are consistent with an effective spin $S=2$. This may be brought about by a crystal field of low symmetry, that quenches the orbital angular momentum, leaving a ground state that possesses only the five-fold spin degeneracy. These levels will be split up even in the absence of a field by the spin–orbit coupling. If the exchange energy is sufficiently large as compared with the splitting of this manifold, one may expect a spin Hamiltonian to be applicable with effective $S=2$ and considerable single-ion anisotropy. The difference with Rb_2FeF_4 is that in $BaFeF_4$ the moments are parallel to the c axis, while in the former they are oriented within the magnetic layer.

From the temperature $T(\chi_{\parallel\,\max})$ at which the maximum in $\chi_\parallel$ occurs we obtain $J/k = -6{\cdot}4$ K, using the results for the Heisenberg model of table 4, there being no theoretical prediction available for the Ising model with $S=2$. On the other hand, if we assume the percentage decrease in $T(\chi_{\max})$ in going from $S=\frac{1}{2}$ to $S=2$ for the Ising model to be the same as for the Heisenberg case, this would result in $kT(\chi_{\max})/|J|S(S+1) \approx 1{\cdot}90$ for the Ising model with $S=2$. Using this value, the experimentally observed maximum yields $J/k = -7{\cdot}0$ K. With the molecular field result $\chi_\perp(0) = Ng^2\mu_B{}^2/4z|J|$ for the perpendicular susceptibility of the Heisenberg model at $T=0$, the value $J/k = -5{\cdot}7$ K is obtained with $g=2{\cdot}1$, as derived from the value of the magnetic moment. Correcting this for the anisotropy, estimated below, gives $J/k = -6{\cdot}3$ K. If the Ising prediction for $\chi_\perp(0)$ is used, $J/k = -11{\cdot}4$ K results. We note that taking $|J/k| > 7$ K gives $T_c/\theta < 0{\cdot}48$, which seems to be unreasonable since the quantity T_c/θ will be increasing with spin value (the differences with molecular field theory become less if the spin value is increased) and because for $S=\frac{1}{2}$ and $S=1$ the theoretical Ising values are 0·57 and 0·63, respectively (Fisher 1967, Guttmann *et al.* 1970). Therefore the exchange constant $J/k \simeq -6{\cdot}4$ K, obtained by applying the anisotropic Heisenberg model, seems to be a better estimate. Clearly we have here an example of an anisotropic compound that approximates the anisotropic Heisenberg model rather than the Ising model. Although in many of its properties the difference will not be obvious, e.g. $\chi_\parallel$, especially in the perpendicular susceptibility the inapplicability of the latter model is exposed.

The anisotropy estimate has been obtained by comparing the different values of the two perpendicular susceptibilities and by comparing the $\chi_{\parallel}$ with spin-wave theory. We estimate $H_A/H_E \simeq 0{\cdot}2$, where $H_A = (H_A{}^I H_A{}^{II})^{1/2}$. Since the anisotropies $H_A{}^I$ and $H_A{}^{II}$ differ by a factor of about 5 we have listed the compound as being intermediate between the Ising and the planar Heisenberg model. Note that the maxima in the two different perpendicular susceptibilities occur at different temperatures. Lastly, we mention that the temperature dependence of the sublattice magnetization has been derived from a Mössbauer study by Eibschütz *et al.* (1970). The best fit of the data to the power law

$$M_s(T)/M_s(0) = B(1 - T/T_c)^\beta \tag{3.7}$$

was obtained with $B = 1{\cdot}18$ and $\beta = 0{\cdot}17$ in the relative temperature range $0{\cdot}80 \leqslant T/T_c \leqslant 0{\cdot}985$. These values compare rather favourably with $B = 1{\cdot}22$ and $\beta = \frac{1}{8} = 0{\cdot}125$ as predicted for the quadratic $S = \frac{1}{2}$ Ising lattice (Fisher 1967) and are certainly quite different from the 3-d $S = \frac{1}{2}$ Ising values, which are $B \simeq 1{\cdot}52$ and $\beta = 0{\cdot}312$ (Fisher 1967). One does not expect the critical exponent to be different for $S = 2$. In any case Guttmann *et al.* (1970) found identical β values for $S = 1$ as for $S = \frac{1}{2}$. The value of B, however, decreases slightly for increasing S. The data obtained have been reproduced in fig. 48, and will be discussed at the end of this section, together with those obtained on other compounds.

Rb_2FeF_4

The sublattice magnetization of this member of the K_2NiF_4 family has been measured by Birgeneau *et al.* (1970 b), using neutron diffraction techniques. In the range $0{\cdot}7 < T/T_c < 0{\cdot}98$, about the same result was obtained as in $BaFeF_4$ ($\beta \simeq 0{\cdot}2$). Closer to the transition point, however, the exponent β was found to increase suddenly which, according to the arguments presented in § 3.2.1, can be explained by assuming that for $T/T_c > 0{\cdot}98$ the phase transition has become 3-d in character. Although the symmetry in Rb_2FeF_4 is indeed slightly distorted below T_c by magnetostriction (Wertheim *et al.* 1968), it is not clear how this may destroy the symmetry argument for the cancellation of the interactions between nearest neighbouring layers. For instance, the argument would not be invalidated if the symmetry would be lowered from tetragonal to orthorhombic, as is very likely. The apparent changeover in β might also be explained by the observed spread in the transition temperature of about 2 K (Wertheim *et al.* 1968, Birgeneau *et al.* 1970 b).

As mentioned above, the spins lie in the basal plane. This was already concluded from the susceptibility measurements (Wertheim *et al.* 1968), which could be explained by assuming a domain structure in which the spins belonging to different domains are in two mutually perpendicular directions in this plane. Concerning the anisotropy value we can only make a guess, since there is as yet no experimental information available to this end. Comparing the susceptibility curves with those of $BaFeF_4$ one may infer that the anisotropy is somewhat smaller, so that the value $H_A/H_E \simeq 0{\cdot}1$ seems to be as reasonable an estimate as can be made. One may, however, draw the conclusion from the χ behaviour that the anisotropy within the layer is much smaller than the out-of-plane anisotropy, which makes this compound also intermediate between the Ising and the planar Heisenberg model, the latter being probably the most appropriate.

$CoCl_2$ *and* $FeCl_2$

For more than 50 years the peculiar properties of $FeCl_2$ and $CoCl_2$ have drawn the attention of experimental and theoretical physicists (for a review of the earlier data see Wilkinson *et al.* 1959). It gradually became clear (Landau 1933) that they could be explained by assuming the magnetic structure to consist of ferromagnetic layers, with an intralayer exchange J, coupled by a much weaker antiferromagnetic interaction J'. Since $J > |J'|$, these systems will have positive Curie–Weiss temperatures, whereas at low temperatures and in low fields they will behave as antiferromagnets. Because of the small value of the antiferromagnetic coupling the magnetization already reaches near-saturation values in moderate fields ($g\mu_B H \simeq 2z'|J'|S$), the crystals becoming essentially ferromagnetic. More about the field-dependent behaviour will be said in § 4.5.

The crystal structure of these compounds (and also of $NiCl_2$) is of the $CdCl_2$ type, with hexagonal layers of metal ions, separated by two hexagonal layers of chlorine anions. They are different in that in $FeCl_2$ the (strong) anisotropy is uniaxial, favouring the hexagonal c axis and thus making the compound of the Ising type, whereas in $CoCl_2$ there is a strong anisotropy constraining the moments within the layer, the in-plane anisotropy being very small, so that it qualifies for the planar Heisenberg model.

The fact that an effective $S = 1$ may be assigned to the Fe^{2+} ion in $FeCl_2$ has been discussed by Ono *et al.* (1964), and Birgeneau *et al.* (1972 b). With reference to the discussion of $BaFeF_4$ we note that this can be expected if the departure from cubic symmetry of the crystal field is small, leaving an orbital triplet lying lowest. Spin–orbit coupling will split up this manifold in three levels, having $\mathcal{J} = 1$, 2 and 3. Provided the effects of non-cubic components of the crystal field are small, the lowest $\mathcal{J} = 1$ level may be treated with an effective spin $S = 1$ in the Hamiltonian. The second proviso for this, namely, that the spacing of the $\mathcal{J} = 1$ and the next level ($\mathcal{J} = 2$) is large as compared to the transition temperature is reasonably well met, since the distance is about $5T_c$ in $FeCl_2$. An evaluation of the exchange constants has been accomplished by Birgeneau *et al.* (1972 b). From a fit to spin-wave theory of the dispersion curve of the ferromagnetic (planar) spin waves, as measured with neutron diffraction, they obtained the exchange between nearest and next-nearest neighbours within the plane as well as the anisotropy, finding $+3{\cdot}9$ K, $-0{\cdot}52$ K and $D=-17$ K, respectively, within the effective $S=1$ formalism. The antiferromagnetic coupling between the layers may be determined from the metamagnetic transition field H_c as measured by Jacobs and Lawrence (1967). Using $2z|J'|S=g\mu_B H_c$ and $H_c = 1{\cdot}1\times 10^4$ Oe, $g=4{\cdot}1$, we obtain $J'/k=-0{\cdot}25$ K.

Another value may be obtained from the susceptibility measurements of Bizette *et al.* (1965 a, b). The parallel susceptibility shows the sharp peak characteristic for this type of antiferromagnetic arrays, reaching $\chi_m = 0{\cdot}89$ cm^3/mole at T_c. Taking, as in the molecular field approximation, this value to be equal to the (isotropic) perpendicular susceptibility at $T=0$, we calculate, using $\chi_\perp(0)=Ng^2\mu_B{}^2/4z|J'|$, $J'/k=-0{\cdot}29$ K. It is not clear whether the authors corrected the susceptibility for demagnetizing effects. If not, the corrected $\chi_\perp(0)$ would be a few per cent higher, improving the agreement between the two J'/k values. We have therefore taken $J'/k=-0{\cdot}25$ K. For the effective intraplanar exchange we have taken the sum of nearest and next-nearest neighbour interactions per nearest neighbour, assuming these to be additive as in molecular field calculations.

The critical behaviour was found to be 3-d in character (Yelon and Birgeneau 1972), which may be understood from the relatively large value of the inter-layer coupling. In fact the sublattice magnetization and the susceptibility, as determined by neutron diffraction, show properties appropriate to the 3-d Ising antiferromagnet with $\beta=0{\cdot}29$ and $B=1{\cdot}47$ (eqn. (3.6)). We will return to this matter at the end of this section and now proceed to discuss $CoCl_2$.

From the value of the spin-flop field $H_{SF}\simeq 2$ kOe (Wilkinson *et al.* 1959), and the field needed to saturate the sample, $H_c\simeq 33$ kOe (Jacobs *et al.* 1968), one obtains (with $H_{SF}{}^2=2H_E H_A{}^I$ and $H_c=2H_E-H_A{}^I$) for the antiferromagnetic interlayer coupling J' and the in-plane anisotropy $H_A{}^I$ the values $J'/k=-1{\cdot}08$ K and $H_A{}^I\simeq 10^2$ Oe. From the parallel susceptibility at T_c (Bizette *et al.* 1956), $\chi\simeq 0{\cdot}4$ cm^3/mole, it follows that $J'/k\simeq -1{\cdot}4$ K. Here we have used $g=6{\cdot}0$, as obtained by Jacobs *et al.* (1968). The intralayer exchange cannot be derived from the existing data. From the work of Lines (1964), who obtained an approximate relationship between $|J'/J|$ and T_c, one may estimate $|J'/J|\simeq 0{\cdot}1$, giving $J/k\simeq 10$ K and $T_c/\theta\simeq 0{\cdot}8$, which does seem reasonable. The out-of-plane anisotropy was estimated to be of the same order as J.

The transition temperature $T_c=24{\cdot}71$ K was derived from the heat capacity measurements of Chisholm and Stout (1962), who devised an ingenious way to separate the lattice contribution. For the ideal 2-d planar Heisenberg model one expects a broad maximum in C_m as in the case of the 2-d Heisenberg model, since both cannot sustain long-range order. No sign of such a maximum is found in the specific heat, which only shows the sharp anomaly due to the onset of long-range order, although there is an appreciably large high-temperature tail.

Apart from the fairly large interlayer interaction J'/k, we may also consider the small anisotropy $H_A{}^I$ within the layer (about 10^{-3} of J/k) as a possible mechanism for establishing the long-range order. The out-of-plane anisotropy $H_A{}^{II}$ most probably does not come into play, since this brings about the planar Heisenberg character and therefore does not help in bringing about a transition. According to whether $|J'| \gg g\mu_B H_A{}^I$, or conversely, one expects to find a 3-d or a 2-d Ising character. Since in the present case $|J'| \simeq 0{\cdot}1J$, we are apt to interpret the heat capacity as being of a 3-d nature, albeit with enhanced short-range-order contributions. This view is supported by the fact that in examples of the 2-d Heisenberg model to be treated below, which have anisotropies also of the order of $10^{-3} J/k$ but J'/k values of about 10^{-5} of J/k, the transition to long-range order is indeed only reflected in the heat capacity as a small spike, sitting on the flank of the broad maximum. Notwithstanding this, we shall see below that minute amounts of anisotropy can put a considerable amount of Ising character into the system.

$CoCl_2 . 6H_2O$ *and* $CoBr_2 . 6H_2O$

The crystal structure (Mizuno 1960, 1961) of these compounds (and also of $NiCl_2 . 6H_2O$) is monoclinic (space group C2/m). In the case of the Ni salt the moments are perpendicular to the *c* axis, whereas for the Co compounds they are parallel to it. The behaviour of the Ni salt is definitely 3-d, as can be concluded from the susceptibility behaviour (Haseda *et al.* 1959). Conversely, the susceptibility of $CoCl_2 . 6H_2O$ shown in fig. 35 (*a*) (Haseda 1960) exhibits 2-d characteristics, the broad maximum occurring at 3·3 K, while the transition temperature, as determined from heat capacity experiments, equals $T_c = 2{\cdot}29$ K (Robinson and Friedberg 1960, Skalyo and Friedberg 1964). Hence, this compound is one of the first 2-d antiferromagnets on which experiments were performed. Robinson and Friedberg (1960) were certainly very near to the truth when they remarked that the amount of entropy gained above T_c obtained by them (52% of the total ΔS) was comparable with the theoretical result for the quadratic Ising model (56%).

The maximum in the temperature derivative of the $\chi_\parallel$ coincides with T_c as can be seen in fig. 35 (*b*) taken from Skalyo *et al.* (1967). In this paper an extensive experimental verification of Fisher's relation between the energy and the parallel susceptibility, as outlined in § 1.2, may be found. Since the interaction in $CoCl_2 . 6H_2O$ is of the planar Heisenberg type, this illustrates the general validity of this relation.

The difference of the planar Heisenberg antiferromagnet from the Ising antiferromagnet (cf. fig. 34 (*a*)) is apparent, since in the latter the susceptibilities in the perpendicular directions are very nearly equal, whereas for the present compound (fig. 35 (*a*)) the $\chi_\perp$ measured within

Fig. 35

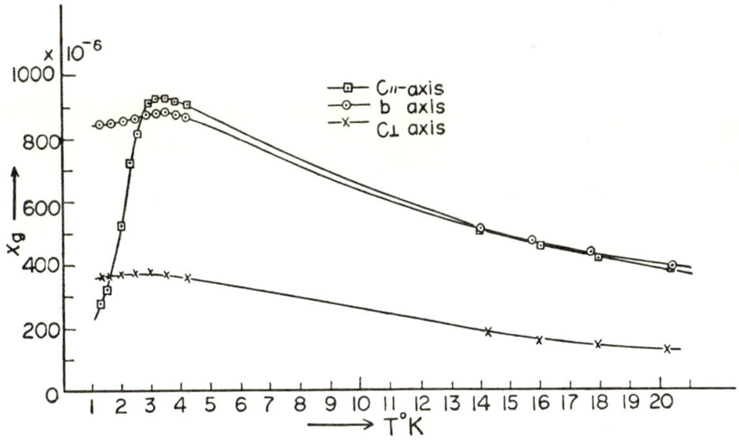

(*a*)

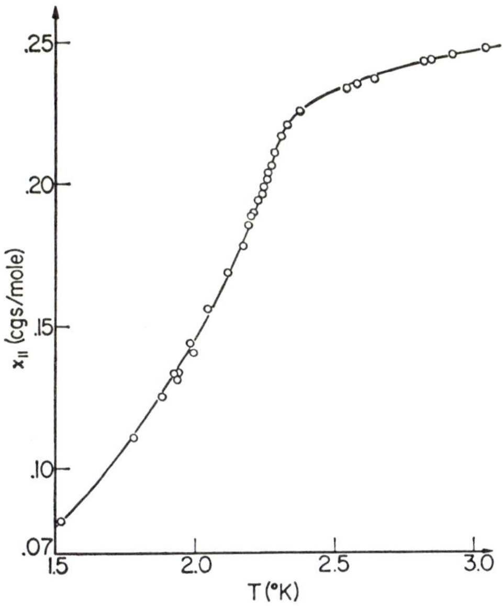

(*b*)

(*a*) Susceptibility of $CoCl_2 . 6H_2O$, which may serve as an example of a 2-d planar Heisenberg antiferromagnet. The planar anisotropy clearly follows from the fact that above the maximum, the perpendicular susceptibility as measured within the easy plane nearly coincides with the parallel susceptibility, whereas the perpendicular susceptibility in the direction out of the easy plane is much lower. (After Haseda 1960.) (*b*) The parallel susceptibility of $CoCl_2 . 6H_2O$ in the region around $T_c = 2{\cdot}29$ K, showing that the maximum in $d\chi/dT$ indeed very nearly coincides with the transition temperature which was determined from the heat capacity. The temperature at which $\chi_{\parallel}$ reaches its maximum is $T = 3{\cdot}3$ K. (After Skalyo *et al.* 1967.)

the easy plane is for $T > T_{max}$, not much different from $\chi_{\parallel}$, in contrast with the susceptibility perpendicular to this plane, which is lowered substantially as a consequence of the planar anisotropy.

As in $CoCl_2$, the form of the specific heat curve does not resemble a system that ideally would have no transition to long-range order. In fact $CoCl_2 . 6H_2O$ has been mentioned as an example of the 2-d Ising model. For $|T-T_c|/T_c < 0{\cdot}05$, Skalyo and Friedberg (1964) found that the specific heat behaved as $C/R = -0{\cdot}271 \ln |T-T_c| + \Delta$, with $\Delta = -0{\cdot}015$ for $T > T_c$ and $\Delta = 0{\cdot}559$ for $T < T_c$. Wielinga *et al.* (1967) have observed that for $|T-T_c|/T_c < 0{\cdot}1$ the exact Onsager solution for the quadratic Ising lattice agrees within a few per cent with

$$C/R = -0{\cdot}49 \ln |T-T_c| - 0{\cdot}29$$

for $T < T_c$ as well as $T > T_c$. One may conclude therefore that although the qualitative behaviour is similar to that of the 2-d Ising model, in quantitative respect there is a rather poor agreement. It should also be noted that the measured heat capacity shows considerable rounding for $|T-T_c|/T_c < 10^{-3}$–10^{-2}.

Numerical results for the intralayer interaction can be obtained from the specific heat and the susceptibility data. In both cases $z = 4$ has been assumed. From the total energy involved in the magnetic ordering Robinson and Friedberg (1960) derived $J/k = -2{\cdot}45$ K.

In principle one could also determine J/k from the χ maximum by fitting to the high-temperature susceptibility. Unfortunately, although Betts *et al.* (1971) have calculated the susceptibility for the ferromagnetic 2-d XY lattice, there is no comparable series for the antiferromagnetic quadratic XY lattice as yet.

On the other hand, the magnetic phase diagram has been mapped completely by Metselaar and De Klerk (1973 a), extending the earlier work of Van der Lugt and Poulis (1960), Schmidt and Friedberg (1967) and McElearny *et al.* (1969). The results for the spin-flop field are in fair agreement, yielding an extrapolated value at $T = 0$ of $H_{SF}(0) \simeq 6{\cdot}5 \times 10^3$ Oe. The saturation field at $T = 0$ has been determined as $H_c(0) = 4{\cdot}6 \times 10^4$ Oe (Metselaar and De Klerk 1973 a). Using the formulae

$$2H_E H_A{}^I - (H_A{}^I)^2 = H_{SF}{}^2$$

and $2H_E - H_A{}^I = H_c$, one obtains $H_E = 2{\cdot}25 \times 10^4$ Oe and $H_A{}^I = 920$ Oe (in-plane anisotropy). With $z = 4$ this yields $J/k = -1{\cdot}9$ K. A larger value is calculated from the perpendicular susceptibility. Taking $\chi_\perp(0) = 0{\cdot}25$ cm^3/mole (Metselaar and De Klerk 1973 a, Flippen and Friedberg 1960, Haseda 1960) one obtains with $\chi_\perp(0) = N_0 g^2 \mu_B{}^2/4z|J|$ the value $J/k = -2{\cdot}3$ K, after correcting for an estimated Van Vleck contribution of $0{\cdot}01$ cm^3/mole and adopting $g = 4{\cdot}9$ (Ûryu and Friedberg 1965). In our opinion the seeming agreement with the result from the heat capacity is fortuitous. Instead we take the value $J/k = -1{\cdot}9$ K, obtained from H_c, to be most reliable in view of the uncertainties involved in the J/k determination from the specific heat. The larger value following from $\chi_\perp(0)$ is ascribed to the effect of zero-point spin deviations, that will also exist in planar antiferromagnets (see, e.g., Semura and Huber 1971) and will lower the $\chi_\perp(0)$. In the present case the experiment suggests a reduction of about 20% for the $S=\frac{1}{2}$ planar Heisenberg quadratic antiferromagnet. This may be compared to the $\chi_\perp(0)$ reduction of about 55% predicted for the $S=\frac{1}{2}$ Heisenberg quadratic antiferromagnet.

Since the anisotropy is for most part due to the g tensor, one may obtain a rough estimate of the out-of-plane anisotropy by putting

$$\alpha = 1 - J_\perp/J_\parallel \simeq 1 - (g_\perp/g_\parallel)^2 \simeq 0{\cdot}7$$

(note that the measured in-plane anisotropy of about 0·04 correlates well with $1-(g_b/g_c)^2 \simeq 0{\cdot}04$). A value for the interlayer coupling cannot be obtained from the existing experimental data.

Of the many investigations performed on this compound we mention further the resonance work of Date (1961) as well as the neutron investigation of the magnetic structure by Kleinberg (1970). As concerns the sublattice magnetization, we have analysed the results obtained by Van der Lugt and Poulis (1960) with an N.M.R. technique. For $(T_c - T)/T_c < 0{\cdot}4$ we find a critical index $\beta \approx 0{\cdot}18$, implying a fairly low value for $|J'/J|$ indeed (see the discussion in § 3.2.3). Unfortunately their results did not extend nearer to T_c than $(T_c - T)/T_c = 0{\cdot}04$.

As in the case of $CoCl_2$, we may ask the question whether it is the in-plane anisotropy $H_A{}^I$ of the interlayer interaction J'/k that is most important for establishing the long-range order in $CoCl_2 \,.\, 6H_2O$. From the low value of β, which is nearer to the 2-d Ising value $\frac{1}{8}$ than to the 3-d result $\frac{1}{3}$, one would conclude that $|J'| \ll g\mu_B H_A{}^I$ in this salt, which is not unreasonable in view of the rather large $H_A{}^I$ value (one order of magnitude larger than in $CoCl_2$). A more elaborate study of the sublattice magnetization would be very welcome to substantiate this, especially closer to T_c, where one would expect to observe a changeover from 2-d Ising to 3-d behaviour. In this picture the critical behaviour of the specific heat for $|T - T_c|/T_c < 0{\cdot}05$, mentioned above, can be understood as being due to a mixture of the effects of the in-plane anisotropy and the interlayer coupling.

A specific heat measurement on the isomorphous bromine compound has been performed by Forstat *et al.* (1959), who found $T_c \simeq 3{\cdot}07$ K and obtained 38% for the amount of entropy gained above T_c, considerably less than in the chlorine compound. The magnetic structures of the two salts were found to be most probably identical by Spence *et al.* (1964).

We may use the g values obtained by Murray and Wessel (1968) to estimate the intralayer exchange constant from the susceptibility measurements of Garber (1960), which also indicate $T_c \simeq 3{\cdot}1$ K (see also

Metselaar and De Klerk 1973 b). With $\chi_\perp(0) \simeq 0{\cdot}19$ cm^3/mole and $g \simeq 5{\cdot}1$, we calculate in the same way as for the bromide $J/k \simeq -3{\cdot}4$ K. The phase diagram has been studied by McElearney *et al.* (1969) and by Metselaar and De Klerk (1973 b), yielding an extrapolated value of $7{\cdot}5 \times 10^3$ Oe for $H_{SF}(0)$, and $H_c(0) = 5{\cdot}4 \times 10^4$ Oe. Accordingly one calculates $H_E \simeq 2{\cdot}7 \times 10^4$ Oe, $J/k = 2{\cdot}3$ K and $H_A{}^I = 1{\cdot}0 \times 10^3$ Oe. Comparing the J/k values, one again observes an apparent reduction of about 30% of the $\chi_\perp(0)$, that is most probably due to zero-point motions.

From the specific heat and the susceptibility results one may conclude that the 2-d character is less pronounced in the bromine as in the chlorine compound (compare also the T_c/θ values). It is further remarked that in these structures the cancellation of the interaction in one direction must be accidental, in view of the 3-d behaviour of the isomorphous nickel compound. Although in crystallographic respect the crystals do have a layered structure, with perfect cleavage along the (001) plane (Mizuno 1960, 1961), using this as an argument would be fallacious since it would also apply to the Ni^{2+} compound.

$BaCoF_4$

This orthorhombic compound has the same structure as $BaFeF_4$ (see above) and has been investigated by Eibschütz *et al.* (1972 b), who performed susceptibility and neutron diffraction measurements. Although the anisotropy is of orthorhombic symmetry, it is not so good an example of the planar model, since the anisotropy within the easy plane is relatively larger than in $CoCl_2 . 6H_2O$, as can be seen from the behaviour of the susceptibility. It is therefore intermediate between the planar and the Ising model.

In order to obtain the exchange we only have the susceptibility available and again we must take recourse to the relation $\chi_\perp(0) = N_0 g^2 \mu_B{}^2/4z|J|$. However, in the present case, also the $\chi_\perp(0)$ within the easy plane is considerably lowered as a consequence of the anisotropy. An estimate of J/k can therefore only be obtained in the following way. From inspection of fig. 35 (*a*) it can be inferred that most probably the $\chi_\perp(0)$ of the planar model is for $S = \frac{1}{2}$ about the same as the value attained at the rounded maximum. Substituting the latter, after subtraction of the Van Vleck contribution, in the MF relation for $\chi_\perp(0)$ and allowing for a probable reduction of about 20% due to the effects of zero-point spin deviation, as found in $CoCl_2 . 6H_2O$, we obtain for the exchange $J/k \simeq 100$ K. Here the g value was taken to be about 6·8, as following from the neutron diffraction work. Evidently, this result is only a crude estimate. Note, however, that the J/k value so derived is of the same order as those obtained for the other two cobalt fluorine compounds in table 5.

The in-plane anisotropy following from the apparent reduction of the $\chi_\perp(0)$ in the ‘ easy ’ plane gives $\alpha \simeq 0{\cdot}4$. It is thus of the same order as the out-of-plane anisotropy, which amounts to $\alpha \simeq 0{\cdot}8$, as may be inferred from the susceptibility and from the g anisotropy, which in view of the value of $g_{\parallel}$ may be anticipated to be about the same as in the other cobalt–fluorine complexes.

Having surveyed the strongly anisotropic 2-d magnets, we now proceed to the examples of the 2-d Heisenberg antiferromagnet, which have been compiled in table 6. It can be seen that numerous approximations have already been discovered, many of which have the K_2NiF_4 structure described in § 2.4. We have mentioned above that in the case of 2-d antiferromagnets it is all but impossible to obtain quantitative information about the interlayer coupling J'. This mainly arises from the fact that J' is usually so small as compared to J that its effect on the thermodynamic properties cannot be measured quantitatively. Only in $Mn(HCOO)_2 \,.\, 2H_2O$ could an experimental value for J' be deduced (from heat capacity measurements). For the compounds with the K_2NiF_4 and the $BaNiF_4$ structure we have put $|J'/J| = 10^{-6}$, which is the earlier mentioned estimate, obtained by considering that the next-nearest neighbouring layers are decoupled because of the crystal symmetry. However, deviations from the ideal crystal structure (distortions, lattice defects, phonon effects) may (partly) invalidate the argument leading to the cancellation of the coupling between nearest neighbouring planes. In that case the ratio $|J'/J|$ can be up to two orders of magnitude larger, as follows by considering, for example, the dipolar interlayer coupling.

DPAN

This four-letter word stands for the aromatic free radical $(H_3COC_6H_4)_2$ $\dot{N}O$ (di-*p*-anisylnitrosyl). In this compound, studied by Duffy *et al.* (1969), the magnetic moments are situated at the sites of the nitrogen atoms, which have an unpaired electron. The crystal symmetry is orthorhombic (space group Aba2) and the nitrogen radicals form layers perpendicular to the b axis, the nearest-interplane-neighbour distance being about three times larger than the nearest-neighbour distance within the plane. As mentioned in § 2.2, free radical solids can be considered to approximate the Heisenberg $S = \frac{1}{2}$ model.

The value of the exchange constant was deduced by Duffy *et al.* from various experimental data. The compounds DPAN and $CuF_2 \,.\, 2H_2O$ constitute the few examples of the 2-d Heisenberg antiferromagnet that have transition temperatures low enough to enable a more or less reliable separation of the magnetic specific heat from the lattice contribution. The specific heat reported by Duffy *et al.* is indeed a smooth, non-anomalous curve, as expected for the isotropic 2-d lattice which ideally cannot sustain long-range order. At the transition temperature, $T_c = 2{\cdot}7$ K, as determined from the discontinuity in $\partial\chi/\partial T$, there is only a very slight indication of a heat capacity spike due to interplanar coupling and anisotropy effects. However, a comparison with the results obtained for $CuF_2 \,.\, 2H_2O$

Table 6. Properties of experimental examples of the 2-d Heisenberg antiferromagnets of different spin value. The listed quantities have already been defined in the heading of table 5. With the exception of $MnTiO_3$ the 2-d network in these compounds is quadratic or quasi quadratic ($z=4$). In $MnTiO_3$ it is the honeycomb lattice ($z=3$).

Compound	S	T_c (K)	J/k (K)	$\alpha = H_A/H_E$	$\|J'/J\|$	$kT_c/\|J\|$	T_c/θ
DPAN	$\frac{1}{2}$	2·7	−2·4	?	?	1·13	0·56
$CuF_2 . 2H_2O$	$\frac{1}{2}$	10·92	−13	$3{\cdot}7 \times 10^{-3}$	?	0·84	0·42
$Cu(HCOO)_2 . 4H_2O$	$\frac{1}{2}$	16·57	−30	$\simeq 1 \times 10^{-3}$	$\simeq 1 \times 10^{-3}$	0·55	0·28
Tl_2NiF_4	1	100·8	−45	$?(>1{\cdot}4 \times 10^{-2})$	$\simeq 10^{-6}$	2·24	0·42
$BaNiF_4$	1	70	−32	2×10^{-2}	$\simeq 10^{-6}$	2·19	0·41
Rb_2NiF_4	1	91	−42	1×10^{-2}	$\simeq 10^{-6}$	2·17	0·41
K_2NiF_4	1	97·23	−50	$2{\cdot}0 \times 10^{-3}$	$\simeq 10^{-6}$	1·94	0·36
$Ni(HCOO)_2 . 2H_2O$	1	15·6	?	?	?	?	?
Ca_2MnO_4	$\frac{3}{2}$	114	−28·6	?	$\simeq 10^{-6}$	3·99	0·40
$CsFeF_4$	$\frac{5}{2}$	160	−14·2	7×10^{-3}	$?(\simeq 10^{-3})$	11·3	0·48
$RbFeF_4$	$\frac{5}{2}$	133	−12·0	$6{\cdot}5 \times 10^{-3}$	$?(\simeq 10^{-3})$	11·1	0·48
$KFeF_4$	$\frac{5}{2}$	137	−12·3	$6{\cdot}5 \times 10^{-3}$	$?(\simeq 10^{-4})$	11·1	0·48
Rb_2MnF_4	$\frac{5}{2}$	38·4	−3·73	$4{\cdot}7 \times 10^{-3}$	$\simeq 10^{-6}$	10·3	0·44
K_2MnF_4	$\frac{5}{2}$	42·3	−4·20	$3{\cdot}9 \times 10^{-3}$	$\simeq 10^{-6}$	10·1	0·43
Rb_2MnCl_4	$\frac{5}{2}$	57	−6·2	$1{\cdot}5 \times 10^{-3}$	$\simeq 10^{-6}$	9·2	0·39
$Mn(HCOO)_2 . 2H_2O$	$\frac{5}{2}$	3·686	−0·37	$?(\simeq 10^{-3}–10^{-2})$	3×10^{-3}	10·0	0·43
$(CH_3NH_3)_2MnCl_4$	$\frac{5}{2}$	47	−5·0	$1{\cdot}1 \times 10^{-3}$	$\simeq 10^{-8}$	9·4	0·40
$BaMnF_4$	$\frac{5}{2}$	24–25	−2·75	$3{\cdot}1 \times 10^{-4}$	$\simeq 10^{-6}$	8·9	0·38
$MnTiO_3$	$\frac{5}{2}$	62·5	$\simeq -9{\cdot}0$	$1{\cdot}2 \times 10^{-3}$	?	$\simeq 6{\cdot}9$	0·40

and for the ferromagnetic 2-d Cu salt, shows that the near non-existence of the spike in DPAN does not point to a very small J' and anisotropy, but is probably due to a limited experimental resolution or to sample imperfections. In these other examples the contribution of the deviations from the ideal model to the heat capacity is clearly resolved from the rounded maximum, whereas from inspection of the T_c/θ and $T_c/T(C_{max})$ values ($T(C_{max})$ denotes the temperature of the broad maximum) it follows that these deviations from ideality must be considerably larger in DPAN.

Comparing the specific heat at low temperatures with Kubo's (1952) spin-wave prediction ($C_m \sim T^2$), we found no correlation with theory. The experimental specific heat lies above Kubo's curve instead of below, as would be expected (see $CuF_2 . 2H_2O$). The apparent fit reported by Duffy *et al.* is erroneous and is due to a misinterpretation of the theoretical coefficient of the quadratic spin-wave term in the specific heat, as calculated by Kubo. In conclusion one may state that DPAN unfortunately turns out to be a rather poor example of 2-d antiferromagnetism.

$CuF_2 . 2H_2O$

It is a pity that this interesting compound has nearly escaped the attention of experimenters, since, with its low spin value, it constitutes an interesting object of inquiry.

As far back as 1961, Shulman and Wyluda observed that the T_c deduced from their fluorine resonance experiments (10·92 K) did not at all correlate with the maximum in the (powder) susceptibility that was found by Bozorth and Nielsen (1958) to be 26 K. The susceptibility measurements on a single crystal by Tazawa *et al.* (1965) confirmed these findings. It seems therefore that $CuF_2 . 2H_2O$ is the earliest investigated quasi 2-d isotropic antiferromagnet. We have used both susceptibility measurements in constructing the curve shown in fig. 28.

Although similar in chemical formula to $CuCl_2 . 2H_2O$, the fluorine compound has a monoclinic crystal structure, whereas $CuCl_2 . 2H_2O$ is orthorhombic. Accordingly, the two compounds are quite different in magnetic respect, which is immediately obvious from a comparison of the specific heats as measured by Clay and Staveley (1966). For a discussion of the magnetic structures see, e.g., Nagai (1963). $CuCl_2 . 2H_2O$ consists of ferromagnetic layers coupled by an antiferromagnetic interaction along the *c* axis (Shirane *et al.* 1965, Poulis and Hardeman 1952), the latter being probably a bit larger than the ferromagnetic exchange within the layer, thereby forming antiferromagnetic chains along the *c* axis. This results in a slight enhancement of the short-range-order effects in this otherwise 3-d crystal. On the other hand, the structure of $CuF_2 . 2H_2O$ (Geller and Bond 1958, Abrahams and Prince 1962) may be regarded as consisting of antiferromagnetic layers of CuF_2O_2 groups, parallel to the (101) plane, the (superexchange) connection between the planes being made by long Cu–F bonds. There might have been a direct exchange along the c axis between Cu ions of neighbouring planes, but the large Cu–Cu distance of 3·244 Å (in Cu metal 2·55 Å) and the fact that along this axis the spins are parallel ordered implies that this interaction must be very small, since a direct exchange would favour an antiferromagnetic orientation.

From the fit of the high-temperature susceptibility (fig. 28) we obtain $J/k = -13$ K, after correcting for the diamagnetic contribution, estimated as $0{\cdot}6 \times 10^{-4}$ from the known susceptibilities of H_2O, CaF_2 and ZnF_2. Using this J/k value we may derive the anisotropy from the AFMR frequency at $T=0$ in zero field, measured by Peter and Moriya (1962). With the aid of the well-known formula $h\nu = g\mu_B(2H_EH_A + H_A{}^2)^{1/2}$ $(H=0)$, it follows from $\nu = 9{\cdot}6 \times 10^{10}$ Hz that $H_A \simeq 1300$ Oe $(\alpha = 3{\cdot}7 \times 10^{-3})$. These authors found the anisotropy to be uniaxial and determined the direction of the magnetic moments to be about 3·5° away from the c axis.

The specific heat between 1 and 80 K has been measured by Clay and Staveley (1966). We have attempted to separate the magnetic part from the lattice contribution in the following way. Using $J/k = -13$ K, we have calculated the magnetic specific heat from the high-temperature series expansion for the quadratic $S=\frac{1}{2}$ lattice (Baker *et al.* 1967 a, b), which in this case is reliable down to $T \simeq 30$ K. Subtracting this magnetic part from the total measured heat capacity we obtain the lattice contribution for $T > 30$ K. This has been fitted to a Debye function $(\theta_D \simeq 135$ K)

Fig. 36

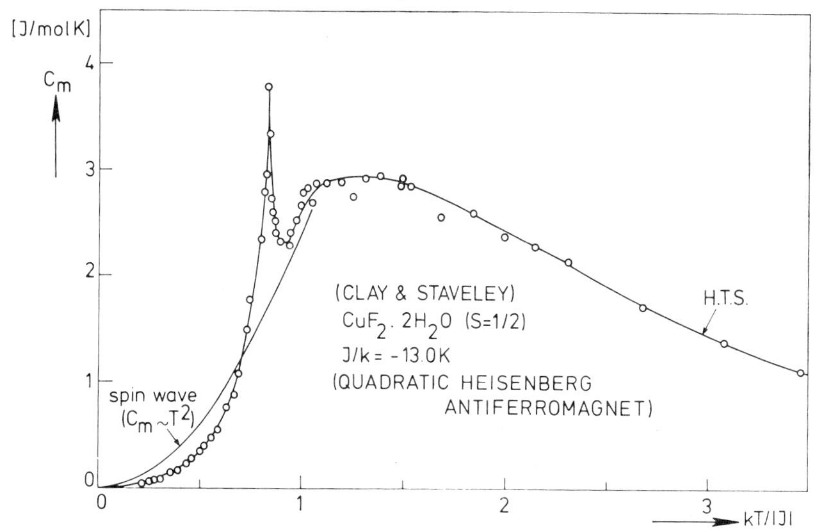

Magnetic specific heat of $CuF_2 . 2H_2O$, as derived from data of Clay and Staveley (1966). Included are the high-temperature series expansion prediction and the spin-wave result for a 2-d antiferromagnet with $S=\frac{1}{2}$, both calculated with $J/k = -13$ K, which is the value obtained from the fit of the susceptibility to the high-temperature series expansion, as shown in fig. 28.

and thereby a rough estimate of the lattice specific heat over the lower temperature range is obtained. The resulting magnetic part has been plotted in fig. 36, together with the series expansion prediction and the theoretical spin-wave heat capacity, as calculated by Kubo (1952). The rounded maximum is clearly resolved, although its exact height and position are still uncertain, due to the large error possibly involved in the subtraction of the lattice contribution. From the fact that T_c occurs not very far below $T(C_{max})$ it can be inferred that the interlayer interaction is relatively large (from a comparison with the ferromagnetic Cu salts discussed below, which have comparable anisotropies, the value $|J'/J| \simeq 1{\cdot}5 \times 10^{-2}$ may be deduced). As expected, the experimental specific heat at low temperatures lies below the spin-wave prediction, since in this region the energy involved in the 3-d ordering has to be 'paid back'. The relatively large value of $|J'/J|$ in this compound may explain why the experimental value of $\chi_{\perp}(0) = 5{\cdot}26 \times 10^{-3}$ cm³/mole is considerably larger (26%) than the spin-wave value, since the presence of J' will tend to reduce the effects of zero-point spin deviation (which lower the $\chi_{\perp}(0)$ with respect to the MF value). However, one cannot exclude the possibility that, especially for $S = \frac{1}{2}$, the spin-wave prediction may be quantitatively in error in two dimensions (De Jongh 1972 c).

Another interesting experiment performed on this compound is the study of the temperature dependence of the paramagnetic resonance line-width by Nagata and Date (1964). A comparison of the behaviour in MnF_2 shows the enhancement of short-range order in the lower dimensional compound.

$Cu(HCOO)_2 . 4H_2O$

Two-dimensional magnetism in cupric formate tetrahydrate—not to be confused with the dihydrate—was first suggested by Martin and Waterman (1959) from crystallographic consideration. In the monoclinic structure the Cu^{2+} ions, linked together by formate groups, form layers that are separated from one another by layers of water molecules. Their preliminary experiments were soon followed by the susceptibility measurements of Kobayashi and Haseda (1963) and of Flippen and Friedberg (1963). The susceptibility is characterized by the occurrence of a broad maximum near 60 K, and a transition to the ordered state at about 17 K, accompanied by the appearance of a weak ferromagnetic moment. The latter authors also reported that no pronounced anomaly could be detected in the specific heat at neither of these temperatures.

Taking the maximum in the susceptibility to be at 65 K in view of the findings of Seehra (1969), we calculate (table 4) $J/k = -34$ K. From $\chi_{max} = 3{\cdot}0 \times 10^{-3}$ cm³/mole (Kobayashi and Haseda 1963) one obtains with the series result: $J/k = -30$ K. We have adopted the value $J/k = -30$ K.

The magnetic structure has been investigated by N.M.R. techniques by Van der Leeden *et al.* (1967) and by Dupas and Renard (1970 a). The latter authors also studied the temperature dependence of the sublattice magnetization by N.M.R. (Dupas and Renard 1970 b) and performed an accurate determination of T_c ($=16{\cdot}57$ K). The value found for the critical index for the magnetization was $\beta=0{\cdot}32$. Of great interest is the large spin reduction of 47%, derived by them by comparing observed and calculated values of the proton frequencies, a reduction that is of the same order as the theoretical value of 36%. The difference may be attributed to the crudeness of the dipole model used in calculating the frequencies.

The same authors also derived a value for the anisotropy by fitting the magnetization at low temperature to the calculations of Lines (1970). Combining their result with those of Seehra and Castner (1970) we obtain the estimate $\alpha=1\times10^{-3}$.

The value for $|J'/J|$ in table 6 has been estimated by Kobayashi and Haseda (1963) in view of the different exchange paths. A similar result has recently been obtained by Ajiro and Terata (1970).

Apart from susceptibility measurements (Flippen and Friedberg 1963), the dihydrate has until now not been studied. It will most likely have the $Mn(COOH)_2 . 2H_2O$ structure and thus will exhibit 2-d properties too. From the low θ value (2 K) one expects the maxima in χ and C_m to occur at a much lower temperature ($T<4{\cdot}2$ K) than in the tetrahydrate.

$BaNiF_4$

This compound has essentially the $BaMnF_4$ structure already described under $BaFeF_4$. Neutron diffraction and susceptibility experiments have been performed by Cox *et al.* (1970). The magnetic moment was found to be $2{\cdot}0\ \mu_B$ per Ni ion. Although this is fairly low considering that the g values in Ni salts are usually about 2·25, the experimental error of $0{\cdot}2\ \mu_B$ prevents a definite conclusion regarding the spin reduction (which would be $0{\cdot}14\ \mu_B$ theoretically) to be drawn. From the fit of the high-temperature susceptibility, shown in fig. 28, the value $J/k=-32$ K is derived. Here we have corrected for a diamagnetic contribution $\chi_{dia}=-0{\cdot}1\times10^{-3}$ cm^3/mole, which, following Breed (1967), has been estimated from the known susceptibilities of BaF_2, CaF_2 and ZnF_2. Inspection of the $\chi_{\parallel}$ curve shows that $T_c=70\pm5$ K. An estimate for α has been obtained by comparing the $\chi_{\parallel}$ curve with spin-wave theory, using $J/k=-32$ K. As shown in fig. 28 (see also § 4.2), the value of $\chi_{\perp}(0)$ attained at $T=0$ is in good agreement with the spin-wave prediction.

Rb_2NiF_4 *and* Tl_2NiF_4

These compounds are both isomorphous with K_2NiF_4. We have re-interpreted the susceptibility and spin-flop measurements of Matsuura *et al.* (1970 b) on Rb_2NiF_4. From $T(\chi_{\parallel\,max})=210$ K we calculate $J/k=-47$ K. After subtraction of the Van Vleck contribution, $\chi_{\perp}(0)$ is obtained as $2{\cdot}1\times10^{-3}$ cm^3/mole. In the same way as above we have estimated the diamagnetic susceptibility $\chi_{dia}=-0{\cdot}1\times10^{-3}$ cm^3/mole from the known susceptibilities of RbF, CaF_2 and ZnF_2. With $\alpha = 9{\cdot}5 \times 10^{-3}$, $\Delta S(\alpha) = 0{\cdot}16$ and $e(\alpha) = 0{\cdot}632$ (cf. eqn. (3.6)) it follows that $J/k = -42$ K. From the value $\chi_{\max} = -2{\cdot}5 \times 10^{-3}$ cm³/mole at $T(\chi_{\parallel \max})$ we also obtain $J/k = -42$ K (with $g = 2{\cdot}27$). Since the first result is the most uncertain we adhere to the value of the latter two, $J/k = -42$ K, and from the spin-flop field at $T = 4{\cdot}2$ K of 350 kOe we subsequently calculate $H_A/H_E \simeq 1 \times 10^{-2}$.

The transition temperature has been located by Maarschall *et al.* (1969) at 90·4 K. This has been checked by additional susceptibility measurements in the region $80 < T < 120$ K (De Jongh, unpublished) in which the maximum in $\partial\chi/\partial T$ was found to occur at $T \simeq 92$ K, so that we have listed $T_c = 91$ K in table 6.

Maarschall *et al.* (1969) also obtained $T_c = 100{\cdot}8$ K for Tl_2NiF_4. Susceptibility measurements on their (powdered) sample (De Jongh, unpublished) confirm this value. The maximum in $\chi_{\parallel}$ was found at 220 K, giving $J/k = -50$ K. From $\chi_{\max} = 2{\cdot}31 \times 10^{-3}$ cm³/mole, likewise corrected for $\chi_{\text{dia}} = -0{\cdot}12 \times 10^{-3}$cm³/mole, we obtain $J/k = -41$ K. The difference in both values is explained by the fact that there was an impurity present in the sample. Although its contribution to the susceptibility was only substantial at temperatures considerably lower than $T(\chi_{\parallel \max})$, it will nevertheless heighten the value of $\chi_{\max}$ by a small amount, thereby lowering the J/k value derived from it. With an eye on the corresponding results for Rb_2NiF_4 we have listed the value $J/k = -45$ K in table 6. For $H < 400$ kOe no spin-flopping could be observed, implying that $H_A/H_E > 1{\cdot}4 \times 10^{-2}$. This is in accordance with the $kT_c/|J|$ value, which is considerably higher than that of Rb_2NiF_4. A value of α from the $\chi_{\parallel}$ curve could not be derived, there being only a powder specimen available.

K_2NiF_4

The compound K_2NiF_4 is certainly the most extensively investigated 2-d antiferromagnet. A short historical survey therefore seems appropriate.

The first clue to the 2-d properties was given in the neutron investigation by Legrand and Plumier (1962 a, b), who pointed out that in the antiferromagnetic state the exchange and the dipolar interaction between neighbouring Ni^{2+} sheets will cancel. They observed antiferromagnetic correlations below a temperature $T \simeq 180$ K. The subsequent susceptibility measurements of Srivastava (1963) showed the familiar 2-d χ curve, with the characteristic maximum at a temperature about twice as high as that at which the χ becomes anisotropic ($\simeq 100$ K).

In 1967 Lines collected and discussed the various pieces of information then available and showed that the presence of a broad maximum in the paramagnetic susceptibility indeed emerges from series expansion calculations. He also explained that the transition temperature T_c could very well occur far below this maximum, thus reconciling the neutron and the susceptibility results.

Maarschall *et al.* (1969) re-measured the susceptibility and were able to locate the transition temperature T_c by studying the temperature dependence of the fluorine resonance linewidth, thereby proving that T_c indeed coincides with the temperature at which the derivative of the $\chi_\parallel$ curve reaches its maximum, instead of with that of the susceptibility maximum.

At about the same time Birgeneau *et al.* (1969) had performed their first neutron experiments, showing the magnetic scattering above T_c to be of 2-d nature. The important work of this group will be discussed further in the next section. Below T_c long-range 3-d ordering was found to set in. They also studied the temperature dependence of the sub-lattice magnetization, obtaining $\beta = 0{\cdot}15$ (compare with the 2-d Ising value $\beta = 0{\cdot}125$!) and $T_c = 97{\cdot}1$ K. Short-range-order effects were found to persist up to $T = 2T_c$. The magnetization curve in the critical region is displayed in fig. 47. In subsequent papers of this group of workers, neutron diffraction techniques were used to their full extent to derive as much information as possible (Birgeneau *et al.* 1970 b). Skalyo *et al.* (1969 a) measured the spin-wave dispersion curve shown in fig. 37. The lack of dispersion found in the direction perpendicular to the layers in reciprocal space (crosses in fig. 37), implies that $|J'/J| < 3{\cdot}7 \times 10^{-3}$, the limit being set by the experimental resolution. Furthermore, the temperature dependence of the dispersion revealed the important fact that renormalization effects do not come into play up to $T \simeq 1{\cdot}1\ T_c$, quite different from the behaviour in 3-d antiferromagnets where they are usually already seen at $T \simeq 0{\cdot}3\ T_c$ (cf. § 4.2). In subsequent work (Birgeneau *et al.* 1970 b) the magnetization measurements were refined, yielding $\beta = 0{\cdot}138 \pm 0{\cdot}004$ and $B = 0{\cdot}973$ in the temperature region $3 \times 10^{-4} < 1 - T/T_c < 0{\cdot}2$, with $T_c = 97{\cdot}23$ K (see fig. 47). The low value of β is certainly very near to the 2-d Ising value $\beta = 0{\cdot}125$ (see discussion in § 3.2.3). Contrastingly, the critical exponents γ and ν of the staggered susceptibility and the correlation length, respectively, appear to have classical rather than 2-d values (Birgeneau *et al.* 1971 b). We will return to this matter in § 4.4.

Values for the intralayer exchange constant are available from various experiments. From the magnon dispersion curve, Skalyo *et al.* (1969 a) obtained $J/k = -56$ K at 4·2 K. From Raman scattering experiments Chinn *et al.* (1971) have deduced $J/k = -55{\cdot}5$ K. Yamaguchi and Sakamoto (1969) found $J/k = -60 \pm 5$ K from measurements of the susceptibility of Ni-doped K_2MgF_4. These authors also determined the next-nearest neighbour interaction within the plane to be $J_2/k \simeq -0{\cdot}5$ K ($J_2/J_1 \simeq 10^{-2}$). From $T(\chi_{max}) = 230$ K (Maarschall *et al.* 1969, Matsuura *et al.* 1970 b) it follows that $J/k = -52$ K with table 4. Values for $\chi_\perp(0)$ have been measured by Srivastava (1963), Maarschall *et al.* (1969) and Matsuura *et al.* (1970 b), who obtained $8{\cdot}35 \times 10^{-6}$, $9{\cdot}3 \times 10^{-6}$ and $9{\cdot}8 \times 10^{-6}$ cm³/g, respectively. Also in case of the value χ_{max} attained at T_{max} there is a similar disagreement since Srivastava reported a value of

Fig. 37

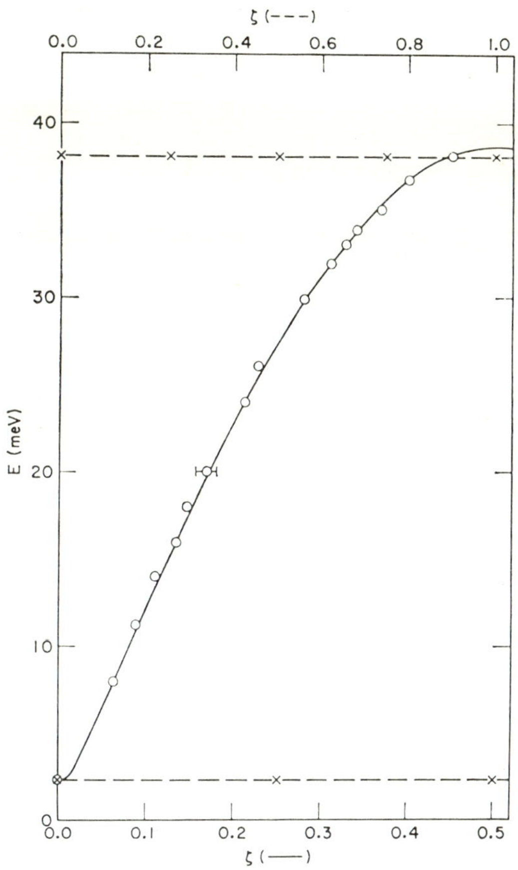

Spin-wave dispersion curve of K_2NiF_4, measured with neutron diffraction by Skalyo *et al.* (1969 a). The 2-d character follows from the apparent lack of dispersion in the direction in reciprocal space perpendicular to the magnetic layers (dashed curves). The energy gap found for zero wave-vector ζ reflects the small anisotropy present in this compound.

$10{\cdot}1 \times 10^{-6}$, whereas Maarschall *et al.* give 10·6 and Matsuura *et al.* $10{\cdot}8 \times 10^{-6}$. Legrand and Van den Bosch (1969) have published $10{\cdot}1 \times 10^{-6}$, in agreement with Srivastava. Since these differences may be due to various sources such as impurity contributions, misorientation, or experimental uncertainties arising from the fact that the susceptibility is very small, we took mean values, corrected for an estimated diamagnetic contribution of $0{\cdot}4 \times 10^{-6}$ cm³/g and a Van Vleck contribution of $1{\cdot}0 \times 10^{-6}$ cm³/g. The latter has been deduced from $\chi_{\parallel}(0)$ and assumed to be the same for $\chi_{\perp}(0)$ and for $\chi_{\parallel}(\max)$. In this way we obtain $\chi_{\parallel}(\max) = 9{\cdot}8 \times 10^{-6}$ and $\chi_{\perp}(0) = 8{\cdot}6 \times 10^{-6}$ cm³/g, yielding (with $g = 2{\cdot}27$) $J/k = -48$ K and -49 K, respectively, in the latter case assuming a spin reduction $\Delta S(\alpha)=0{\cdot}18$ and taking $e(\alpha)=e(0)=0{\cdot}632$. The best fit of the high-temperature susceptibility data shown in fig. 28 was obtained with $J/k=-49$ K and $g=2{\cdot}27$. We note that in an earlier calculation (De Jongh *et al.* 1972 a) $J/k=-57$ K was deduced from the $\chi_{\perp}(0)$ obtained by Matsuura *et al.* This originates from the fact that no correction for χ_{dia} was made and the $e(\alpha)$ term was not accounted for. In summing up we may say that the susceptibility data yield $J/k=-49$ K. The apparent large difference with the other determinations, e.g. from the spin-wave dispersion curve, can be removed since Skalyo *et al.* did not take into account renormalization effects in their calculation. Using a renormalized spin-wave theory, De Wijn *et al.* (1973 b) obtained $J/k=-51\pm 0{\cdot}4$ K. We conclude therefore that the value of -50 K will be trustworthy within about 3% over the temperature range 1–300 K.

The anisotropy has been obtained by Birgeneau *et al.* (1970 a) from AFMR experiments and by Matsuura *et al.* (1970 b) and Yamazaki *et al.* (1972) from the value of the spin-flop field. All results agree to $\alpha=2{\cdot}0\times 10^{-3}$. A comparison with the dipolar anisotropy, which yields $\alpha=4{\cdot}8\times 10^{-4}$ only (Colpa, private communication), shows that the observed anisotropy must for the most part be attributed to the single-ion mechanism.

Values for the spin reduction can be deduced from $\chi_{\perp}(0)$ as well as from measurements of the sublattice magnetization by neutron (Birgeneau *et al.* 1970 a) and magnetic resonance techniques (De Wijn *et al.* 1973 b). Birgeneau *et al.* found a $15\pm 5\%$ reduction of the magnetic moment (including the effects of covalency), De Wijn *et al.* found a reduction of $20\pm 3\%$. Furthermore, an upper limit for $|J'/J|$ was estimated from the resonance study, viz. $|J'/J|<2\times 10^{-4}$. This is in accordance with the already mentioned estimate of about 10^{-6}, following from a comparison of the intra and interlayer superexchange paths. The interlayer dipolar coupling has been calculated by Colpa (private communication) to be a mere $6{\cdot}8\times 10^{-9}$ of the intralayer exchange.

Lastly we mention a specific heat measurement of Salamon and Hatta (1971), who detected a small anomaly at 98·7 K, which result is intermediate between the $T_c=97{\cdot}23$ of Birgeneau *et al.* (1970 b) and $T_c=100{\cdot}5$ K reported by Maarschall *et al.* (1969). These small differences are not so surprising and may, for instance, be due to calibrational errors or small chemical impurities, as can be understood by realizing that replacement of Ni by other elements results in widely different T_c's.

$Ni(HCOO)_2 \cdot 2H_2O$

This Ni salt has the manganese formate structure already discussed above. As in the case of Co and Mn formate, the specific heat shows a Schottky anomaly at low temperature, ascribed to the gradual ordering of the paramagnetic B sheets, and a broad maximum at more elevated temperatures due to the antiferromagnetic ordering within the A sheets.

The heat capacity has been measured by Pierce and Friedberg (1971) and Takeda and Kawasaki (1971). Also in this case the experimental resolution has most probably not been sufficient to resolve the peak due to the 3-d ordering from the broad maximum.

The transition temperature, as determined from the susceptibility experiment of Hoy *et al.* (1965) is $T_c = 15{\cdot}6$ K. Unfortunately these measurements did not extend to high enough temperatures to detect the high-temperature maximum (which should occur at 40–50 K) predicted by the series expansions and as observed in Mn formate. As a consequence we do not have experimental information from which an estimate of J/k can be made. In view of the results for the other Ni salts listed in table 6, we may however, expect $|J|/k$ to be in the range 6·5–7·5 K.

Obviously, a value for H_A/H_E is also not available at present. We have classified the compound as being of the Heisenberg type in view of the small anisotropies found in the other Ni^{2+} salts.

Ca_2MnO_4

This is another compound with the K_2NiF_4 structure. From the susceptibility measurements of Davis (see MacChesney *et al.* 1967), one finds $\chi_{max} = 2{\cdot}76 \times 10^{-3}$ cm³/g and $T(\chi_{max}) \simeq 220$ K, giving (table 4) $J/k = -28{\cdot}6$ and $J/k = -27{\cdot}9$ K, respectively, accounting for an estimated diamagnetic contribution of $0{\cdot}07 \times 10^{-3}$ cm³/g. For reasons already explained we take the former value. Also $T_c = 114$ K was found in this work. Unfortunately there is no experimental result for the anisotropy. Neutron diffraction studies (Cox *et al.* 1969, Ollivier and Buisson 1971) revealed a magnetic unit cell that is doubled in the *c* direction as compared to that of K_2NiF_4. The 2-d character was found to be not as pronounced as in the case of K_2NiF_4 (e.g. $\beta \simeq 0{\cdot}3$). The value of the magnetic moment was $2{\cdot}0 \pm 0{\cdot}3\ \mu_B$, much lower than the expected $3\ \mu_B$, even after correcting this for the spin reduction of $0{\cdot}4\ \mu_B$ from zero-point spin deviations. Covalency effects may play a role. Another explanation suggested is the fact that the model of localized electron spins may be not wholly appropriate for this material.

$RbFeF_4$, $CsFeF_4$ *and* $KFeF_4$

Although also orthorhombic, the crystal structure of these compounds differs considerably from that of $BaNiF_4$. Also in this case the layers are puckered, but it is not the 2-d magnetic lattice built up by the Fe^{3+} ions that is rumpled, the washboard effect being caused by the tipping out of the layers of the F–Fe–F bonds (Heger *et al.* 1971). The sheets of FeF_6 octahedra are separated by layers of Rb^+ ions. Of importance is the fact that the symmetry argument leading to the cancellation of the interactions between nearest neighbouring layers in the K_2NiF_4 structure is also valid for the $KFeF_4$ structure, but not so for $RbFeF_4$ and $CsFeF_4$, as can be derived from the magnetic structure proposed by Eibschütz *et al.* (1972 a) on the basis of their neutron diffraction measurements. Estimates of the superexchange interaction between the layers yield $|J'/J| \simeq 10^{-4}$ and 10^{-2}–10^{-3} for the $KFeF_4$ and for the $RbFeF_4$ structure, respectively.

Magnetic susceptibility, Mössbauer effect and neutron diffraction experiments on $RbFeF_4$ have been performed by Eibschütz *et al.* (1971, 1972 a). With the aid of table 4 we calculate $J/k = -12{\cdot}0$ K from $T(\chi_{max}) = 215 \pm 10$ K. The transition temperature indicated by $\partial\chi/\partial T$ was $T_c = 133 \pm 2$ K. From the value $\chi_{\parallel max} = 6{\cdot}71 \times 10^{-3}$ cm³/mole and $\chi_\perp(0) = 7{\cdot}25 \times 10^{-3}$ cm³/mole, allowing for $\chi_{dia} = -0{\cdot}66 \times 10^{-3}$ cm³/mole, $\Delta S(\alpha) = 0{\cdot}16$, $e(\alpha) = 0{\cdot}632$, we calculate $J/k = -12{\cdot}2$ K and $J/k = -11{\cdot}7$ K. Consequently we take J/k to be about -12 K. An estimate for the anisotropy was obtained by Eibschütz *et al.* (1971) by fitting the magnetization curve at low temperatures to spin-wave theory, thereby deriving an energy gap of 30 ± 5 K. A 16% lower result was obtained by De Rosa (private communication to Eibschütz *et al.*) from AFMR, which yields the α value listed in table 6. Near to T_c $(0{\cdot}40 < T/T_c < 0{\cdot}99)$ the magnetization was found to follow a power law with $\beta = 0{\cdot}245 \pm 0{\cdot}005$.

Similar results were obtained from the Mössbauer study of $KFeF_4$ (Eibschütz *et al.* 1972 a). For $0{\cdot}72 < T/T_c < 0{\cdot}99$ they found $\beta = 0{\cdot}185 \pm 0{\cdot}005$. These β values are discussed in § 3.2.3. For T_c the authors obtained $137{\cdot}2 \pm 0{\cdot}1$ K, in good agreement with the value $T_c = 137 \pm 1$ K derived by Heger *et al.* (1971) from Mössbauer and neutron studies†.

Heger *et al.* also measured the susceptibility of a powdered sample of $KFeF_4$. Unfortunately there was a contribution from impurities at $T < T_c$, so that we only have the high-temperature χ from which to obtain J/k. The temperature $T(\chi_{max}) = 222$ K yields $J/k = -12{\cdot}4$ K. From the value $\chi_{max} = 6{\cdot}68 \times 10^{-3}$ cm³/mole, corrected for $\chi_{dia} = -0{\cdot}06 \times 10^{-3}$ cm³/mole, $J/k = -12{\cdot}3$ K follows. The latter result has been listed in table 6. We note that this J/k value may be a little too small due to the impurity contribution that will increase the susceptibility of the maximum by a small amount. Also for this compound Eibschütz obtained an energy gap of 30 ± 6 K from the fit of the magnetization to spin-wave theory. In view of the just mentioned findings of De Rosa, we conclude that $KFeF_4$ most probably has about the same anisotropy value as $RbFeF_4$.

In the case of $CsFeF_4$ we derive the exchange from the susceptibility measurements of Eibschütz *et al.* (1972 a). The temperature of the susceptibility maximum $T(\chi_{max}) = 235$ K gives $J/k = -13{\cdot}1$ K. From the value $\chi_{max} = 5{\cdot}75 \times 10^{-3}$ cm³/mole at the maximum, and from $\chi_\perp(0) = 6{\cdot}0 \times 10^{-3}$ cm³/mole, both corrected for a diamagnetic contribution $\chi_{dia} = -0{\cdot}06 \times 10^{-3}$ cm³/mole, the values $J/k = -14{\cdot}2$ K and $J/k =$

† Heger and Geller (1972) have subsequently reported the considerably higher transition temperature $T_c = 148$ K.

$-14{\cdot}1$ K are derived, respectively (taking $\Delta S(\alpha)=0{\cdot}16$). We conclude to $J/k=-14{\cdot}2$ in view of the large uncertainties involved in the first determination. An estimate of the anisotropy may be obtained by comparing the $\chi_{\parallel}$ curves of $CsFeF_4$ and $RbFeF_4$ on a relative temperature scale, from which it is concluded that α must be about the same.

Rb_2MnF_4 *and* K_2MnF_4

These fluorine compounds are also amongst the most extensively investigated members of the K_2NiF_4 family. The earliest inquiries were susceptibility and spin-flop experiments by Breed (1966, 1967, 1969), from which the 2-d character was established and a determination of the anisotropy could be made. Neutron diffraction investigations were subsequently carried out by Loopstra *et al.* (1968) and by Birgeneau *et al.* (1970 b). The latter authors found the Rb_2MnF_4 crystals to consist of two phases, one with the K_2NiF_4 and the other with the Ca_2MnO_4 structure, both phases occurring even in the same single crystal. This is an important discovery, for the following reason. The fact that the phases with a ferromagnetic and an antiferromagnetic alignment along the c axis are both observed in the same sample may be interpreted in the sense that there exists a subtle balance between the different kinds of coupling between the (next-nearest) layers, as there are the dipolar and the superexchange interaction. Which of the two phases will occur in a particular part of the crystal would then be determined by lattice imperfections. Since the dipolar interlayer coupling favours a ferromagnetic alignment along the c axis, the superexchange part would tend to establish the antiferromagnetic alignment.

The importance of this intuitive picture is that it implies that the two different kinds of interlayer coupling must be about equal in magnitude. Since the dipolar part can be evaluated numerically, we obtain in this way an indication as to the quantitative value of the interlayer interaction. Calculations of the dipolar interlayer coupling by Colpa (private communication) yield $|J'/J|=5{\cdot}8\times10^{-8}$ and $8{\cdot}2\times10^{-8}$ for Rb_2MnF_4, respectively, indicating that the estimate $|J'/J|\simeq10^{-6}$ (at highest) for the compounds of the K_2NiF_4 structure may indeed be correct.

The temperature dependence of the sublattice magnetization, in particular in the critical region, has been studied with neutron diffraction by Birgeneau *et al.* (1970 b) and by Ikeda and Hirakawa (1972) for Rb_2MnF_4 and K_2MnF_4, respectively. The former authors derived $\beta=0{\cdot}18$ and $B=1{\cdot}02$ for both phases of Rb_2MnF_4 from measurements in the region $1-T/T_c>3\times10^{-3}$. It is a pity that their work does not extend nearer to T_c because just at $1-T/T_c\simeq4\times10^{-3}$, Ikeda and Hirakawa (1972) observed a change-over of the β value of K_2MnF_4 from $\beta=0{\cdot}188$ $(1-T/T_c<4\times10^{-3})$ to a 3-d value (in the range $2{\cdot}5\times10^{-4}<1-T/T_c<4\times10^{-3}$). Both measurements have been reproduced in fig. 47 and will be discussed in § 3.2.3. We remark that the magnetization curve of K_2MnF_4 just mentioned applies to a phase with the magnetic K_2NiF_4 structure. Ikeda and Hirakawa also reported the existence of a second magnetic phase with a different T_c in their crystals, which they identified as a K_2MnF_4 phase having the magnetic Ca_2MnO_4 structure. We have strong reasons to doubt this identification, since in Rb_2MnF_4, Rb_2MnCl_4 and Cs_2MnCl_4 (see below), these two phases were reported to have the same transition temperature within the experimental resolution. This is what one would anticipate, since T_c is to a very high degree only determined by the exchange within the layer (see discussion in § 3.2.3), which is not expected to be different for both phases. On the other hand, Ikeda and Hirakawa report $T_c = 42{\cdot}37$ K and $T_c = 58{\cdot}0$ K for K_2MnF_4 of the K_2NiF_4 and the Ca_2MnO_4 magnetic phase, respectively. An explanation would be the presence of an impurity in their crystals. For instance, the fact that Cs_2MnCl_4 and Rb_2MnCl_4 both have $T_c \simeq 55$ K (see below) is very suggestive†.

In the lower temperature region the sublattice magnetization has been accurately measured by N.M.R. techniques (de Wijn *et al.* 1971, 1973 b). The excellent fit of the data on K_2MnF_4 and K_2NiF_4 to spin-wave theory up to $T \simeq 0{\cdot}5\ T_c$, is shown in fig. 38. Furthermore, various authors have reported on the observation of the effects of zero-point spin deviations in these compounds (Breed 1967, Loopstra *et al.* 1968, De Wijn *et al.* 1971, Colpa *et al.* 1971, Schrama 1972, De Jongh 1972 b, c). We will collect and discuss these findings in § 4.2 and as a last reference mention the study of the fluorine N.M.R. linewidth in K_2MnF_4 by Maarschall (1970) and Bucci and Guidi (1970, 1974).

Values for the exchange constants may be obtained from various sources. From the fit of the high-temperature susceptibilities, shown in fig. 28, $J/k = -3{\cdot}76$ K and $-4{\cdot}20$ K are obtained for Rb_2MnF_4 and K_2MnF_4, respectively. The low-temperature determinations are in fair agreement with these values, since Breed (1969) finds $-3{\cdot}65$ K and $-4{\cdot}20$ K from the fit of the $\chi_\parallel$ curve to spin-wave theory, while De Wijn *et al.* (1973 b) obtain $-3{\cdot}69 \pm 0{\cdot}045$ and $-4{\cdot}205 \pm 0{\cdot}03$ K from the spin-wave analysis of the sublattice magnetization. We thus conclude to $J/k = -3{\cdot}73$ K and $-4{\cdot}20$ K, with no temperature dependence of the exchange for $T < 100$ K. The experimental values of $\chi_\perp(0)$ are moreover in good agreement with the spin-wave prediction (fig. 28), calculated with these exchange constants and the measured anisotropy parameters. From the spin-flop fields

† Our conclusions are corroborated by the recent neutron diffraction study on K_2MnF_4 by Birgeneau *et al.* (1973), who find no evidence for the second phase reported by Ikeda and Hirakawa. They obtain $T_c = 42{\cdot}14$ K, and further $J/k = -4{\cdot}23 \pm 0{\cdot}05$ K from the dispersion curve. Moreover, their magnetization curve differs considerably from that of Ikeda and Hirakawa in that the magnetization is higher over most of the critical region, and that no kink is found down to a temperature of $1 - T/T_c = 6 \times 10^{-4}$. For the β they report $\beta = 0{\cdot}15 \pm 0{\cdot}01$ for $6 \times 10^{-4} < 1 - T/T_c < 0{\cdot}3$. See also § 3.2.3.

Fig. 38

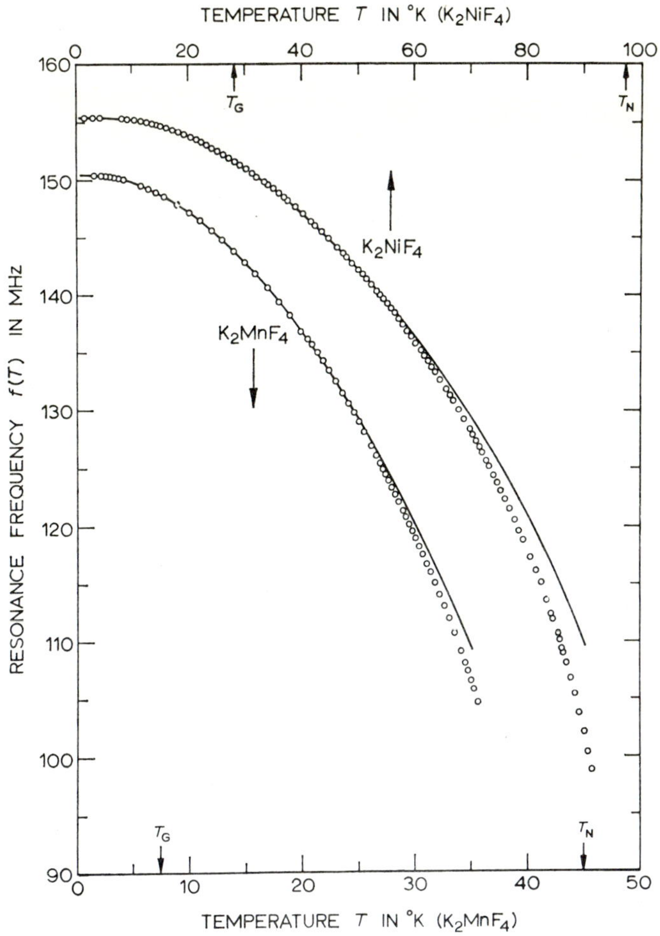

Temperature dependence of the sublattice magnetization in K_2MnF_4 and K_2NiF_4 as determined from the N.M.R. frequency of the fluorine nuclei by De Wijn *et al.* (1971). The circles are the measuring points, the solid curves have been calculated from spin-wave theory for a 2-d Heisenberg antiferromagnet with a small anisotropy.

(Breed 1967) $H_{SF} = 55{\cdot}1 \pm 1{\cdot}0$ and $50{\cdot}8 \pm 1{\cdot}3$ kOe (at $T = 4{\cdot}2$ K) one calculates $\alpha = 3{\cdot}8 \times 10^{-3}$ and $4{\cdot}0 \times 10^{-3}$ for K_2MnF_4 and Rb_2MnF_4, respectively. We may compare these anisotropy values with the dipolar anisotropies, that yield $\alpha = 4{\cdot}06 \times 10^{-3}$ and $4{\cdot}44 \times 10^{-3}$ for K_2MnF_4 and Rb_2MnF_4, respectively, in which calculation the effect of spin reduction has been included (Colpa, private communication). De Wijn *et al.* (1973 a) determined the spin-wave energy gaps at $T = 0$ K from AFMR measurements to be $7{\cdot}40 \pm 0{\cdot}05$ K and $7{\cdot}28 \pm 0{\cdot}05$ K for K_2MnF_4 and Rb_2MnF_4 respectively, which yield $\alpha = 3{\cdot}9 \times 10^{-3}$ and $4{\cdot}7 \times 10^{-3}$ for the potassium and the rubidium salt, respectively. Since the AFMR results are the most accurate, we have entered these in table 6. The listed values of T_c follow from the work of Breed *et al.* (1969), Birgeneau *et al.* (1970 b, 1973) and Ikeda and Hirakawa (1972).

$Mn(HCOO)_2 . 2H_2O$

Numerous experiments have been performed on this example of the 2-d Heisenberg antiferromagnet with $S = \frac{5}{2}$. Of great advantage is its low T_c value, which enables a reliable determination of the magnetic specific heat since the lattice contribution can be obtained with reasonable accuracy. As described above under the isomorphous Co salt, its magnetic structure consists of antiferromagnetic A sheets separated by paramagnetic B sheets. In some way this situation resembles that found in § 3.1.2 for the linear chain compound $CuSO_4 . 5H_2O$. Evidence for the peculiar magnetic structure can be found from various sources, for instance, the proton N.M.R. experiments of Abe and Matsuura (1964) and the susceptibility measurements of Abe and Torii (1965) and independently of Cohen *et al.* (1964). Apart from crystallographic considerations, strong support was also obtained from the heat capacity data of Pierce and Friedberg (1968), who found that even down to $T \simeq 0{\cdot}5\ T_c$, only about one-half of the expected entropy $R \ln (2S+1)$ was removed from the system. In this paper a MF calculation, based upon the above sketched division in differently behaving A and B sheets, was carried out and found to be consistent with the experimentally observed behaviour.

The overall magnetic properties are best described by considering the specific heat behaviour (fig. 39). At $T_c \simeq 3{\cdot}7$ K there occurs a transition to long-range antiferromagnetic order within the A sheets. The spike reflecting this cooperative phenomenon is superimposed on a broad anomaly, which is once more attributed to the short-range-order processes that are inherent on the ideal 2-d Heisenberg system. The fact that T_c lies nearly on top of this maximum, whereas for the 2-d copper compounds with $S = \frac{1}{2}$ (see below) $T_c \simeq 0{\cdot}5\ T(C_{max})$, is partly explained by the high spin value of $\frac{5}{2}$. The transition at $3{\cdot}7$ K is accompanied by the appearance of a weak ferromagnetic moment, directed along the c axis of the monoclinic structure ($a_0 = 8{\cdot}86$ Å, $b_0 = 7{\cdot}29$ Å, $c_0 = 9{\cdot}60$ Å ; $\beta = 97{\cdot}7°$). The extremely sharp peak observed at $T \simeq 0{\cdot}46\ T_c$ is associated with a spontaneous reorientation of the antiferromagnetic axis (first-order transition). Accordingly, below this temperature the direction of the weak ferromagnetic moment is found to be parallel to the b axis. Evidence for the weak ferromagnetic behaviour has been brought forward by Yamagata (1967) from torque measurements, by Matsuura *et al.* (1969) and Ajiro (1969) from susceptibility experiments (see also Ûryu 1965), and by Bertaut *et al.* (1969) who studied the magnetic structure as a function of temperature with neutron diffraction.

Fig. 39

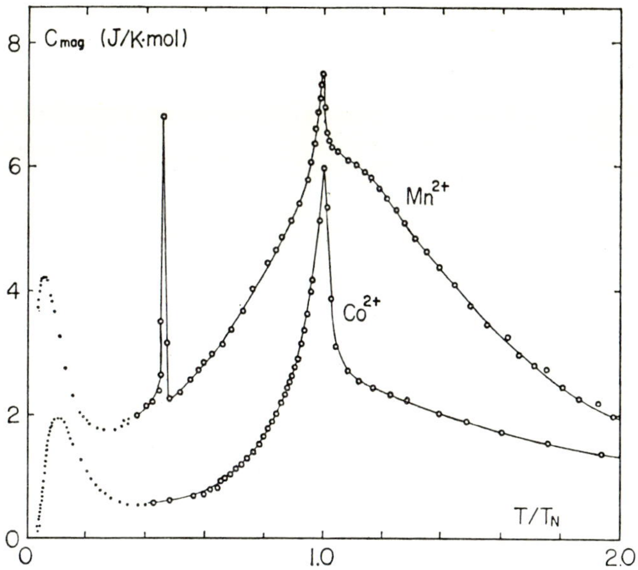

Specific heat of $Mn(HCOO)_2 . 2H_2O$ versus the relative temperature T/T_c. The measuring points have been taken by Pierce and Friedberg (1968) and by Matsuura *et al.* (1970 a). The heat capacity of $Co(HCOO)_2 . 2H_2O$ is included for comparison.

The third maximum observed in fig. 39 at $T \simeq 0{\cdot}2$ K represents the contribution of the paramagnetic B sheets (evidence for the hyperfine contribution has also been found at still lower temperatures). Below T_c the Mn^{2+} ions in the B sheets are gradually ordered as a consequence of the effective field exerted on them by the antiferromagnetically ordered spins in the A sheets (Pierce and Friedberg 1968, Burlet *et al.* 1969, Matsuura *et al.* 1970 a), resulting in a broad Schottky-type maximum.

The susceptibility (Matsuura *et al.* 1969) appears roughly as a paramagnetic curve, due to the contribution of the B spins, with superimposed on it two sharp peaks, one at T_c, the other at $T = 1{\cdot}72$ K $(= 0{\cdot}46\ T_c)$, both reflecting the weak ferromagnetic behaviour. Unfortunately, measurements in the region $T > 4{\cdot}2$ K are not available so that the positions of the maximum associated with the shortrange 2-d interactions has not been located. With the aid of table 4 and the J/k value to be derived below, this maximum may be expected to be found at about 6·5 K.

In order to obtain J/k one must therefore resort to the caloric data. From the heat capacity in the higher temperature region, Pierce and Friedberg (1968) found the intralayer exchange within the A sheets to be $J/k = -0{\cdot}37$ K. By analysing the Schottky anomaly at 0·2 K, Matsuura *et al.* (1970 a) and Takeda *et al.* (1971 a) were able to estimate the interlayer couplings, obtaining 7×10^{-2} J for the coupling between A and B sheets and 3×10^{-3} J for the interaction between the A sheets. The latter value is listed in table 6 as representing 'the' interlayer coupling.

As regards the value of the anisotropy a definite conclusion cannot be drawn, due to the fact that various mechanisms contribute to it and experimental determinations are still lacking. The dipolar anisotropy is of the order of 1×10^{-3} of the exchange, but, as pointed out by Bertaut *et al.* (1969), it cannot be the only source since for $1{\cdot}7 < T < 3{\cdot}7$ the spins are within the magnetic layer, an orientation that is not favoured by the dipolar anisotropy in case of an antiferromagnetic intralayer exchange. From the study of the weak ferromagnetic behaviour it transpires that the anisotropic part of the exchange interaction is less than 1% of the isotropic part. As a third factor we have to consider the single-ion anisotropy. Although this quantity is small for Mn^{2+} ions it may not be neglected because of the small value of the exchange. According to the E.S.R. experiments of Morigaki and Abe (1967), the single-ion anisotropy is about 0·04 K, giving $\alpha \simeq 3 \times 10^{-2}$.

Comparing the T_c/θ value of $Mn(HCOO)_2 . H_2O$ with other $S = \frac{5}{2}$ examples of table 6 of which an α value is known, one would conclude to a value of about 1×10^{-3}. This value is in accord with what one may derive from the magnetic phase diagram (Ajiro 1969, Schutter *et al.* 1972). If the lowest transition observed with the field parallel to the b axis is indeed due to spin-flopping, one may obtain $H_A \simeq 2{\cdot}5 \times 10^{-3}$ from the value of the spin-flop field $H_{SF} \simeq 4$ kOe. This would imply that the different sources of anisotropy somehow cancel one another. Since we have seen that the spin direction does not correlate with the dipolar anisotropy, this does not seem to be an unreasonable assumption. We might add that also in the case of the other Mn salts, the measured anisotropy is often lower than the calculated dipolar value.

Lastly, we point to the neutron scattering study of Skalyo *et al.* (1969 b), which provides proof of the 2-d character of the substance, since up to $T = 2T_c$ the existence of 2-d magnetic correlations could be established. These authors also studied the temperature dependence of the magnetic moments on the A and B sheets. For the critical index β of the sublattice magnetization of the A sheets they found $\beta = 0{\cdot}23 \pm 0{\cdot}01$ in the range $5{\cdot}5 \times 10^{-2} < 1 - T/T_c < 0{\cdot}55$. In a subsequent proton N.M.R. study, Ajiro *et al.* (1970) obtained the slightly different result $\beta = 0{\cdot}22 \pm 0{\cdot}02$ for $1{\cdot}5 \times 10^{-2} < 1 - T/T_c < 0{\cdot}47$. We will come back to these β values in § 3.2.3.

Rb_2MnCl_4 *and* Cs_2MnCl_4

The magnetic structure of these chlorine compounds has recently been investigated by Epstein *et al.* (1970) and Gurewitz *et al.* (1970) with neutron diffraction. They found it to be of the K_2NiF_4 type when the crystals were prepared from molten salts. On the other hand crystals of Rb_2MnCl_4 prepared from aqueous solutions were shown to possess the magnetic

Ca_2MnO_4 structure. In this case the two phases do not co-exist in the same sample, as in Rb_2MnF_4. Instead the authors observed that after annealing the crystals prepared from aqueous solutions possess the K_2NiF_4 structure too, which seems therefore to be the most stable phase. The transition temperature was found to be the same for both phases, being $T_c = 57$ K and 52 K for the Rb and the Cs salt, respectively. Spin correlations within the magnetic layers were observed to persist up to 200 K.

Magnetic measurements on Rb_2MnCl_4 have been carried out by De Jongh (unpublished). From susceptibility and spin-flop experiments, the values $J/k = -6{\cdot}2$ K, $T_c = 56{\cdot}5 \pm 1{\cdot}5$ K and $\alpha = H_A/H_E \simeq 1{\cdot}5 \times 10^{-3}$ could be derived, the latter being equal to the calculated dipolar anisotropy within the experimental error (Colpa, private communication).

$(C_nH_{2n+1}NH_3)_2MnCl_4$ ($n = 1, 2, 3, \ldots$)

As pointed out by Van Amstel and De Jongh (1972), the face-centred tetragonal structure of these compounds offers the possibility of finding the ' best ' 2-d antiferromagnets, in the sense that it allows in principle for the smallest $|J'/J|$ values that can possibly be reached. This arises from the fact that in this structure the symmetry argument leading to the absence of an interaction between neighbouring layers, that provides for the 2-d properties in the K_2NiF_4 structure, is combined with the mechanism of separating the layers by increasing n, which leads to the pronounced 2-d character of the Cu compounds $(C_nH_{2n+1}NH_3)_2CuCl_4$ (see § 2.4). For instance, the dipolar interlayer coupling in $(C_nH_{2n+1}NH_3)_2MnCl_4$ has been calculated by Colpa to be $2{\cdot}5 \times 10^{-9}$, $2{\cdot}2 \times 10^{-10}$ and $1{\cdot}3 \times 10^{-11}$ of the exchange, for $n = 1, 2, 3$, respectively (using the J/k value for the methyl compound derived below). The superexchange interlayer coupling is even for the $n = 1$ compound estimated to be a mere 10^{-10} of J, since it has to occur via two (CH_3NH_3) groups and two Cl anions. The lower bound to $|J'/J|$ will in this case no doubt be set by lattice imperfections or the presence of phonons, as has been pointed out above. We have tentatively entered the value 10^{-8} in table 6. Since the crystal structure allows for the replacement of Mn by other metal ions, for instance the isomorphous Fe compounds ($n = 1, 2, 3$) have been found to be also tetragonal (Mostafa and Willett 1971), it certainly offers many possibilities.

The temperature dependence of the E.S.R. linewidth in methyl and ethyl ammonium manganese chloride has been studied by Boesch *et al.* (1971). Indications were found that the position of the transition temperatures will probably be near 50 K. This was corroborated by the susceptibility measurements of Van Amstel and De Jongh (1972) on $(CH_3NH_3)_2MnCl_4$, who obtained $T_c = 47 \pm 3$ K. The exchange constant was determined as $J/k = -5{\cdot}0 \pm 0{\cdot}2$ K, while the anisotropy derived from the value of the spin-flop field was $\alpha = 1{\cdot}1 \times 10^{-3}$.

The transition temperatures of the Fe salts were found at 96 K and 90 K for $n=1$ and 2, respectively (Mostafa and Willett 1971). A sharp peak at T_c was observed in the (otherwise) antiferromagnetic susceptibility of the methyl compound. This will likely be due to the presence of a weak ferromagnetic moment. It would be interesting to analyse this susceptibility divergence in view of the large value of the susceptibility exponent γ that is expected for 2-d lattices ($\gamma=1{\cdot}75$–$2{\cdot}0$). These compounds have not been included in table 6, due to the limited amount of information known to date.

$BaMnF_4$

This manganese compound distinguishes itself by its very low value for the anisotropy parameter α, in spite of its rather low J/k (see table 6). Once more we may derive the exchange from the susceptibility curve, which has in this case been measured by Holmes *et al.* (1969). From the fit of the high-temperature susceptibility shown in fig. 28 we obtained $J/k=-2{\cdot}72$ K. The value following from $\chi_{\perp}(0)=3{\cdot}0\times10^{-2}$ cm^3/mole is in fair agreement, since taking $\Delta S(\alpha)=0{\cdot}19$ and $e(\alpha)=e(0)=0{\cdot}632$ and allowing for $\chi_{dia}=-0{\cdot}1\times10^{-3}$ cm^3/mole, one derives $J/k=-2{\cdot}78$ K. We conclude therefore to $J/k=-2{\cdot}75$ K, with no apparent temperature dependence below $T=100$ K. The anisotropy follows from the spin-flop field via the formula $H_{SF}{}^2=2H_AH_E(1-\chi_{\parallel}/\chi_{\perp})^{-1}$. With $H_{SF}=10{\cdot}4$ kOe and $\chi_{\parallel}/\chi_{\perp}=0{\cdot}047$ (Holmes *et al.* 1969) this yields $\alpha=3{\cdot}1\times10^{-4}$. The transition temperature $T_c\simeq 24$–25 K is indicated by the observed behaviour of the AFMR ($T<T_c$) and E.S.R. ($T>T_c$) modes, and by the parallel susceptibility curve.

$MnTiO_3$

Manganese titanate is the only example in table 6 of a 2-d antiferromagnet in which the 2-d network is not quadratic. The compound has the hexagonal ilmenite ($FeTiO_3$) structure and consists of magnetic Mn^{2+} layers, separated from each other by two oxygen and one Ti sheets. In the hexagonal layers, the Mn^{2+} ions have three nearest neighbours (honeycomb lattice). Susceptibility experiments on powdered specimens have been performed by Heller (1964) and Sawaoka *et al.* (1966), whereas Akimitsu *et al.* (1970) have studied a single crystal. The susceptibility shows the usual characteristics, with a broad maximum at about 100 K, whereas the values of T_c reported from various experiments lie in the range 60–65 K.

The early neutron diffraction investigation of Shirane *et al.* (1959) confirmed the antiferromagnetic structure within the layer. In later work, Akimitsu *et al.* (1970) observed the typical 2-d correlations, thereby confirming the 2-d character of the salt.

To derive a value of J/k is not so easy because of the limited amount of information available. In the high-temperature region we may use the series expansion of the susceptibility of the honeycomb lattice with $S=\frac{5}{2}$, of which six terms have been obtained by Rushbrooke and Wood (1958). We have analysed this expansion in the same way as for the quadratic lattice. The susceptibility calculated from six terms exhibits a maximum at about $kT/|J|S(S+1)\simeq 1{\cdot}5$, but this number is uncertain, due to the limited number of terms. For instance, with five terms the maximum is found at 1·15. Assuming that the maximum will be somewhere near 1·35, it follows that $J/k\simeq -8{\cdot}5$ K. We have tried to fit the data of Akimitsu *et al.* to the series prediction for $T>T_{max}$, but this could not be achieved properly. It seems as if other contributions to the susceptibility (apart from χ_{dia}) are present, which may be attributed to the presence of the Ti ions. An indication for this may also be the fact that different values for χ_{max} have been found, ranging from 8×10^{-5} cm³/g (Heller 1964) to $9{\cdot}8\times 10^{-5}$ cm³/g (Akimitsu *et al.* 1970).

Calculating J/k from $\chi_{\perp}(0)=9{\cdot}7\times 10^{-5}$ cm³/g (Akimitsu *et al.* 1970) is therefore also a risky matter, more so since quantitative estimates of the effects of zero-point spin deviations ($\Delta S(0)$ and $e(0)$) are missing for the honeycomb lattice. Nevertheless one may try to estimate $\Delta S(0)$ and $e(0)$ from the corresponding values for the quadratic lattice, by comparing the differences in these quantities found for the simple cubic ($z=6$) and the body-centred cubic ($z=8$) lattice. In this way we may guess that $\Delta S(0)\simeq 0{\cdot}25\, S$ and $e(0)\simeq 0{\cdot}63$. After correction for $\chi_{dia}\simeq -0{\cdot}05\times 10^{-3}$ cm³/mole, one then obtains from $\chi_{\perp}(0)$ $J/k\simeq -9{\cdot}4$, in reasonable agreement with the estimate on the basis of $T(\chi_{max})$. Without the corrections for zero-point spin deviations, the value would have been $-8{\cdot}2$ K.

With $J/k\simeq -9{\cdot}0$ K one may subsequently calculate the anisotropy from the zero-field AFMR frequency extrapolated to $T=0$, as determined by Stickler *et al.* (1967). From $\nu=156$ kMc/sec the value $\alpha=H_A/H_E\simeq 1{\cdot}2\times 10^{-3}$ is obtained. The number 5×10^{-5} quoted by De Jongh *et al.* (1972 a), is wrong, due to a calculational error, arising from the fact that ω_R and ν_R are interchanged in the paper of Stickler *et al.* (1967).

The ratio $|J'/J|$ is not known for this compound, but will be considerably larger than in the K_2NiF_4 structure, since within the hexagonal symmetry there exists no decoupling of nearest neighbouring layers. Accordingly, the sublattice magnetization in the region $1-T/T_c<0{\cdot}125$, was found to follow a power law with the ‘ 3-d ’ β value $0{\cdot}32\pm 0{\cdot}01$, pointing indeed to a substantial interlayer coupling.

We next proceed to discuss the ferromagnetic layer-type compounds, gathered in table 7. It is seen that only Cu and Cr compounds have been found to approximate this model. This arises because in the other magnetic ions with a fairly isotropic interaction, e.g. Mn^{2+} and Ni^{2+}, apparently the exchange nearly always has the antiferromagnetic sign.

The interaction between the layers is mostly antiferromagnetic, so that below T_c the ferromagnetic layers become ordered antiparallel with

Table 7. Properties of examples of the 2-d Heisenberg ferromagnetic model. The Cu compounds have $S=\frac{1}{2}$, the Cr compounds $S=\frac{3}{2}$. The listed quantities have been defined in the heading of table 5. In addition the distance d_1, between magnetic ions in the same layer and the ratio d_2/d_1 have been added, where d_2 is the distance separating the magnetic ions of nearest-neighbouring layers. In the Cu compounds the 2-d network is nearly quadratic, for the Cr salts it is the honeycomb or the triangular lattice. The values for the anisotropy parameters and $|J'/J|$ have mainly been derived from field dependent susceptibility, magnetization or magnetic torque measurements.

Compound	T_c (K)	J/k (K) (from χ)	J/k (K) (from C_m)	H_A/H_E	$\|J'/J\|$	kT_c/J	T_c/θ	d_1 (Å)	d_2/d_1
$Rb_2CuCl_2Br_2$	17·2	16·3		$\simeq 10^{-4}$ (H_A^{I}) $1{\cdot}4\times10^{-2}$ (H_A^{II})	$2{\cdot}6\times10^{-2}$	1·06	0·53	5·130	1·735
Rb_2CuCl_3Br	15·2	17·6		$\simeq 10^{-4}$ (H_A^{I}) $5{\cdot}0\times10^{-3}$ (H_A^{II})	$1{\cdot}6\times10^{-2}$	0·86	0·43	5·104	1·714
Rb_2CuCl_4	13·8	18·8		$\simeq 10^{-4}$ (H_A^{II}) $1{\cdot}9\times10^{-3}$ (H_A^{II})	8×10^{-3}	0·73	0·37	5·088	1·683
K_2CuF_4	6·5	11·2	8·8	4×10^{-5} (H_A^{I}) $1{\cdot}2\times10^{-2}$ (H_A^{II})	$\simeq 3\times10^{-3}$	0·58	0·28	4·155	1·68
$(CH_3NH_3)_2CuBr_4$	15·8		26			0·61	0·30	5·466	1·888
$(C_2H_5NH_3)_2CuBr_4$	10·85	19·0	22·2	3×10^{-3} (H_A^{I}) 4×10^{-3} (H_A^{II})	$1{\cdot}7\times10^{-3}$	0·57	0·29	5·541	2·061
$(C_3H_7NH_3)_2CuBr_4$	10·50	21·3	18·0	4×10^{-4} (H_A^{I}) 4×10^{-3} (H_A^{II})	$2{\cdot}0\times10^{-3}$	0·49	0·25	5·548	2·300
$(C_4H_9NH_3)_2CuBr_4$	11·33	21·9	21·1		$\simeq 10^{-3}$	0·52	0·26	5·530	2·68
$(C_5H_{11}NH_3)_2CuBr_4$	11·4	22·0	21·1		$\simeq 10^{-3}$	0·52	0·26		

$(NH_4)_2CuCl_4$	11·2	17			$3·2 \times 10^{-3}$	0·66	0·33	5·09	1·75
$(CH_3NH_3)_2CuCl_4$	8·91	19·2	17·8	$1·6 \times 10^{-5}$ $(H_A{}^{I})$ 3×10^{-3} $(H_A{}^{II})$	$5·5 \times 10^{-4}$	0·465	0·23	5·247	1·900
$(C_2H_5NH_3)_2CuCl_4$	10·20	18·6		$1·6 \times 10^{-4}$ $(H_A{}^{I})$ 3×10^{-3} $(H_A{}^{II})$	$8·5 \times 10^{-4}$	0·55	0·27	5·240	2·139
$(C_3H_7NH_3)_2CuCl_4$	7·65	16·0	14·3	1×10^{-4} $(H_A{}^{I})$ 3×10^{-3} $(H_A{}^{II})$	$\simeq 5 \times 10^{-5}$	0·48	0·24	5·297	2·428
$(C_4H_9NH_3)_2CuCl_4$	7·33	15·4	13·7		$\simeq 1 \times 10^{-4}$	0·48	0·24	5·208	3·040
$(C_5H_{11}NH_3)_2CuCl_4$	7·30	15·9	15·1		$10^{-5} - 10^{-4}$	0·46	0·23	5·265	3·379
$(C_6H_{13}NH_3)_2CuCl_4$	7·75	17·1	14·9		$10^{-5} - 10^{-4}$	0·45	0·23		
$(C_{10}H_{21}NH_3)_2CuCl_4$	7·91	17·9		6×10^{-5} $(H_A{}^{I})$ 3×10^{-3} $(H_A{}^{II})$	$\lessapprox 10^{-5}$	0·44	0·22	5·239	4·921
$AgCrSe_2$	50	$\simeq 6$			$\simeq 0·17$	$\simeq 8·3$	$\simeq 0·55$		$\simeq 2$
$NaCrS_2$	18	$\simeq 3·0$		$\simeq 5 \times 10^{-2}$	$\simeq 0·21$	$\simeq 6·0$	$\simeq 0·45$		$\simeq 2$
$NaCrSe_2$	40	$\simeq 7$			$\simeq 2 \times 10^{-2}$	$\simeq 5·7$	$\simeq 0·38$		$\simeq 2$
CrI_3	6·8	13·5		$3·3 \times 10^{-2}$	0·13	5·04	0·67		$\simeq 2$
$CrBr_3$	32·7	8·25		$1·6 \times 10^{-2}$	6×10^{-2}	3·95	0·53		$\simeq 2$
$CrCl_3$	16·8	5·25		$\simeq 3 \times 10^{-5}$ $(H_A{}^{I})$ $\simeq 1·5 \times 10^{-4}$ $(H_A{}^{II})$	$3·4 \times 10^{-3}$	3·20	0·43		$\simeq 2$

respect to one another. This situation is favoured by the dipolar interlayer coupling, although this is mostly too small to account for the observed value of J'. The only exceptions are K_2CuF_4 and $(CH_3NH_3)_2CuCl_4$, which behave ferromagnetically also below T_c. This points to the presence of various types of interlayer interactions, with different signs.

As already remarked above, the combination $J>0$, $J'<0$ has the advantage that quantitative estimates of J' are easily obtained from the value of the (antiferromagnetic) susceptibility at or below T_c ($\chi_\perp$), which is inversely proportional to $|J'|$. It is remarked that, due to the fact that $J \gg |J'|$, effects of zero-point spin deviations are negligible in this special type of antiferromagnets (De Jongh 1972 a), justifying an MF calculation of J' from $\chi_\perp$ via the relation $\chi_\perp = N_0 g^2 \mu_B{}^2/4z|J'|$. Since the actual $\chi_\perp$ will in most cases decrease below T_c by, amongst other things, the presence of anisotropy, one may conveniently use $\chi(T_c)$ in this formula, which equals $\chi_\perp(0)$ (for $H_A=0$) in both MF and spin-wave theory.

The anisotropy in these compounds is of orthorhombic symmetry, with $H_A{}^{II} \gg H_A{}^{I}$. This points to a planar Heisenberg character, were it not that also $H_A{}^{II} \ll H_E$. We prefer therefore to classify them as Heisenberg compounds, bearing in mind that there is a small planar type anisotropy superimposed upon the overall Heisenberg character. There certainly remains the possibility that, near enough to T_c, the critical behaviour will be more appropriately described by the planar model.

Values for the in-plane ($H_A{}^I$) and the out-of-plane ($H_A{}^{II}$) anisotropy may be obtained by torque or ferromagnetic resonance measurements, or in the case of an antiferromagnetic J' by measuring the magnetization or the susceptibility as a function of field in the different crystallographic directions (at $T \ll T_c$). From the values of the spin-flop field and the fields needed to saturate the sample, both $H_A{}^I$ and $H_A{}^{II}$ may then be deduced (cf. § 4.5).

The compounds of table 7 fall into three groups, those of the K_2NiF_4 type, the $(C_nH_{2n+1}NH_3)_2CuX_4$ series and the chromium compounds. We will start with the first category.

K_2CuF_4

The experiments performed on K_2CuF_4 include magnetization (Yamada 1970), N.M.R. (Yamada *et al.* 1971), neutron scattering (Hirakawa and Ikeda 1972) and specific heat measurements (Yamada 1972). In the latter experiment $T_c = 6{\cdot}25$ K is indicated. The exchange constant J/k was obtained as 11·2 K from the high-temperature susceptibility and as 8·8 K from the linear temperature dependence of the spin-wave contribution to the heat capacity (see below). Similar differences have been found in the case of the $(C_nH_{2n+1}NH_3)_2CuX_4$ series (see below). The values for the anisotropy ($H_A{}^I$ and $H_A{}^{II}$) were obtained from the magnetization measurements. The ferromagnetic coupling between the layers has been estimated to be about 0·03 K ($J'/J \simeq 3{\cdot}5 \times 10^{-3}$) by Hirakawa and Ikeda (1972).

In view of the rather large value of J'/J, as compared with those encountered in the 2-d antiferromagnets, one expects the critical behaviour of the magnetization to be 3-d in character. On the other hand the interlayer coupling is small enough for the spin-wave spectrum to be predominantly 2-d in nature (compare with the results obtained for $FeCl_2$ and $(C_nH_{2n+1}NH_3)_2CuX_4$). The apparent agreement of the temperature dependence of the magnetization in the region $0{\cdot}12 < kT/J < 0{\cdot}36$ with the $T^{3/2}$ law predicted by spin-wave theory for a 3-d ferromagnet, as reported by Kubo, may therefore be fortuitous. Although by taking $T_c = 6{\cdot}25$ K, Hirakawa and Ikeda deduced a β value of about 0·22 for $0{\cdot}01 < 1 - T_c/T < 0{\cdot}17$, this result does not seem to be conclusive. We would rather say that the 2-d behaviour in the region near T_c is spoiled by interference of J'. For instance, by taking $T_c = 6{\cdot}32$ K, which is the temperature at which the neutron intensity (which is proportional to M_s^2) actually falls of, the authors reported that the fit of the data to a power law did not yield a straight line, so that a unique value for β could not be derived.

Rb_2CuCl_4

The system Rb_2CuCl_4, Rb_2CuCl_3Br, $Rb_2CuCl_2Br_2$ is currently being investigated by Witteveen (1973). The transition temperatures have been determined from specific heat and susceptibility measurements, the J/k's from the analysis of the high-temperature susceptibility (see below). The anisotropy parameters follow from the measurement of the magnetization curves at temperatures far below T_c. (We are much indebted to H. T. Witteveen for providing these results prior to publication.)

It is quite interesting to observe how by the successive replacement of the Cl^- ions by Br^- the anisotropy H_A^{II} increases by an order of magnitude, whereas the anisotropy within the layer is hardly affected. This arises most probably because the Br^- ions fill in the out-of-plane positions in the octahedral environment of the Cu_2^+ ions. Similar indications for the presence of an anisotropy associated with the superexchange mechanism are frequently found when comparing otherwise isomorphous Cl and Br compounds (e.g. $CrCl_3$ and $CrBr_3$, $(NH_4)_2CuBr_4 . 2H_2O$ and $(NH_4)_2CuCl_4 . 2H_2O$, $(C_nH_{2n+1}NH_3)_2CuX_4$ ($n = 0, 1, 2, 3, \ldots, 10$; X = Cl *or* Br)

This series of compounds has been the object of extensive studies at our laboratory, including specific heat, E.S.R., susceptibility, magnetic torque and thermal conductivity measurements. Although the crystal structure is orthorhombic, the magnetic layers are very nearly quadratic since the difference between the lattice parameters within the Cu planes is only about 3%. We have already outlined in § 2.4 how the pronounced 2-d properties in these compounds arise from the separation mechanism, based upon the fact that the organic alkyl ammonium groups can be greatly enlarged. As may be inferred from table 7 the distance between the copper ions in neighbouring magnetic layers is increased by a factor 3 by varying n from 1 to 10, whereas the configuration within the layer is hardly affected†. As a consequence of this piling up of organic material between the magnetic layers, the estimates of the interlayer super-exchange interaction, as well as the interlayer dipolar coupling (in the case of a pure antiferromagnetic arrangement of the ferromagnetic layers), amount to $|J'/J| < 10^{-5}$ for $n > 3$. In the case of $(NH_4)_2CuCl_4$ and $(C_2H_5NH_3)_2CuCl_4$ the directions of the moments of neighbouring layers are indeed fully antiparallel (in zero external field). In $(CH_3NH_3)_2CuCl_4$ the interlayer interaction is ferromagnetic, whereas in other compounds weak ferromagnetic moments have been observed, the direction of this moment being within the layers for the chlorine and perpendicular to it for the bromine compounds. Consequently, the magnetic behaviour found below T_c is often quite complex and difficult to analyse. It is remarked that with these small J' values also the Earth's magnetic field comes into play, that is about 10^{-6} of J, and was not compensated for in the experiments. In view of the above we may expect that the fields acting on the sample, other than arising from the anisotropy, will be of the order of 10^{-5} H_E or less for the compounds with $n > 3$ (cf. the observed values for $n \leqslant 3$ in the Cl series).

The anisotropy within the layer H_A^{I} is seen to vary between 10^{-5} and 10^{-4} of H_E, whereas the out-of-plane anisotropy H_A^{II} is typically 10^{-3} H_E for all compounds. As mentioned above, one intuitively expects that H_A^{I} will be the quantity that must be taken into account in discussions concerning the occurrence of long-range order, since H_A^{II} merely introduces a planar Heisenberg character into the system.

The above discussion also indicates the difficulties encountered in the interpretation of magnetic measurements on these compounds. As an illustration we point out that in the case of an antiferromagnetic interlayer coupling J' as small as 5×10^{-5} of J, the susceptibility at T_c reaches a value about equal to that expected for a ferromagnetic platelet-shaped sample (J and J' both > 0), which is determined by the demagnetizing factor. With these extremely small J' values, it is all but impossible to decide experimentally whether the interlayer coupling is ferro or antiferromagnetic if further information is lacking. This may also be understood from the fact that for $J' < 0$, the field needed to saturate the sample is roughly given by $4z'|J'|S/g\mu_B$, which for $|J'/J| \simeq 10^{-5}$ amounts to 10 Oe, only.

It can be seen from table 7 that the compounds with a positive J', K_2CuF_4 and $(CH_3NH_3)_2CuCl_4$, form an exception in the sense that they have considerably smaller $|J'/J|$ values than their nearest neighbours in

† Recently Kitamura and Tsujikawa have extended the series to $n = 18$, in which case $d_2/d_1 \simeq 8$ (private communication, to be published).

the row of compounds (compare the T_c/θ values). The value of $|J'/J|$ of $(CH_3NH_3)_2CuCl_4$ has been determined by Yamazaki (1973) from E.S.R. measurements and by Bloembergen (private communication) from torque measurements. The only explanation for this peculiarity that we can give is that in these compounds there is a competition between interlayer couplings of different sign. Assuming that in the other salts the various interlayer interactions would all prefer an antiferromagnetic arrangement of the layers, the argument would be that they add up, except in K_2CuF_4 and $(CH_3NH_3)_2CuCl_4$, where they would partly cancel one another.

In order to obtain more quantitative estimates of the anisotropy and of J', Bloembergen *et al.* (1972, 1973) have started a programme of magnetic torque measurements. Although the analysis of these measurements is still under way, the preliminary results do confirm the picture just outlined.

After this general description we will present some of the experimental results of importance within the present context. The fact that a ferromagnetic interaction within the Cu layers is the predominant exchange has been proven by susceptibility experiments in the paramagnetic region $T \gg T_c$ (De Jongh *et al.* 1969, De Jongh and Van Amstel 1970), and by measurements of the spin-wave contribution to the heat capacity at $T \ll T_c$ (Colpa 1972 b). In the latter experiment, a linear dependence of the specific heat on temperature was clearly indicated, as predicted by simple spin-wave theory for a 2-d Heisenberg ferromagnet (see fig. 67 and the discussion in § 4.2). The high-temperature susceptibility yields positive Curie–Weiss temperatures, moreover the data could be adequately analysed in terms of the series expansion for this model. For $S=\frac{1}{2}$, ten terms are known in this expansion (Baker *et al.* 1967 a), which give a prediction for the susceptibility that may be trusted down to $T \simeq 1 \cdot 5\, J/k$. As a typical example we give in fig. 40 the fit of the (powder) suceptibility of $(C_2H_5NH_3)_2CuCl_4$ to the series expansion result (curve 1). Here $C/\chi T$ is plotted versus θ/T, so that the Curie–Weiss law appears as the straight line $C/\chi T = 1 - \theta/T$ (curve 3). The exchange constant J/k is the only unknown needed to scale the experimental points upon the theoretical curve. Note the huge deviation of the 2-d ferromagnetic susceptibility from the Curie–Weiss law, even at $T = 2\theta$. For comparison the deviation of the series result for the ferromagnetic b.c.c. lattice has also been included (curve 2). It is seen that even for a 3-d ferromagnet, estimates of J/k from a value of the Curie–Weiss θ determined in the range $T \simeq 4\theta \simeq 6T_c$, may result in serious errors.

For $T < J/k$ $(=\frac{1}{2}\theta)$ the susceptibility becomes field dependent in fields of a few kOe. In fig. 41 it is shown how the (initial) susceptibilities of eleven Cu compounds of the series in the region up to $T \simeq J/k$ scatter evenly around a common curve, in spite of the fact that the various compounds differ in magnitude and sign of J' and in anisotropy. For each compound the value of J was determined by the fit to the series expansion result for $T > 1 \cdot 5\, J/k$. Since there is no apparent difference for the various compounds for $T > 0 \cdot 9\, J/k$, one may consider the curve

Fig. 40

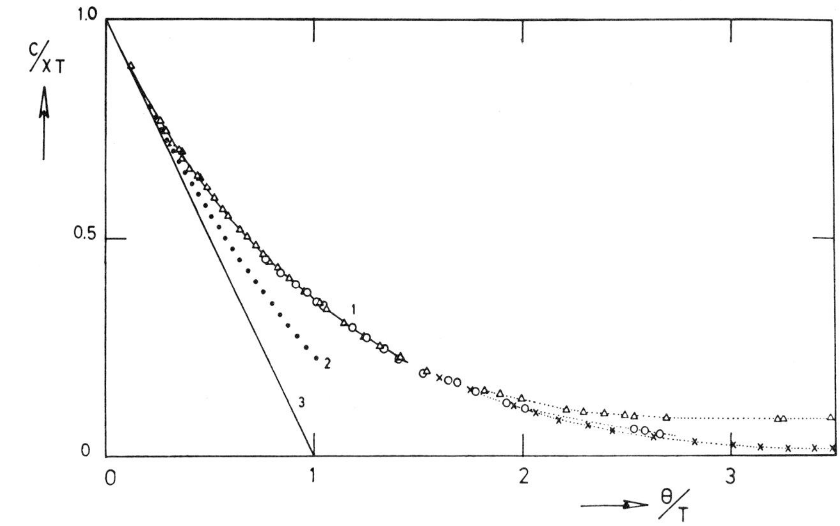

The susceptibility of the ferromagnetic layer compound $(C_2H_5NH_3)_2CuCl_4$ in the high-temperature region ($T \gg T_c$; $\theta/T_c \simeq 3{\cdot}6$). The full curve 1 drawn for $\theta/T < 1{\cdot}4$ has been calculated from the high-temperature series expansion for the quadratic Heisenberg ferromagnet with $S=\frac{1}{2}$. The exchange constant J/k was obtained by fitting the data to this prediction. The dotted curve 2 represents the series expansion result for the b.c.c. Heisenberg ferromagnet. The straight line 3 is the MF prediction $C/\chi T = 1 - \theta/T$ for the quadratic ferromagnet. (After De Jongh *et al.* 1972 b.) △ : $H = 10$ kOe ; ○ : $H = 4$ kOe ; × : $H = 0$ (a.c. susceptibility measurements).

drawn through the data of fig. 41 as an experimental continuation of the series expansion prediction with ten terms into a region wherein it is no longer trustworthy.

For $T < 0{\cdot}9\,J/k$, the influence of the interlayer interaction and the anisotropy gradually become manifest. This is depicted in fig. 42 where the parallel susceptibility of $(C_2H_5NH_3)_2CuCl_4$, in which the ferromagnetic layers order antiferromagnetically at T_c, is compared with the suceptibility of $(CH_3NH_3)_2CuCl_4$ and $(C_{10}H_{21}NH_3)_2CuCl_4$ (De Jongh *et al.* 1972 b). The χ of the methyl compound diverges at T_c, since $J' > 0$. In practice this means of course that $\chi(T)$ becomes equal to $1/D$, where D is the demagnetizing factor of the sample. The same situation was found in the decyl compound, which led the authors to believe initially that in this case J' is positive too. However, as stated above, we are at the moment inclined to think that for all compounds with $n > 1$, J' is negative, attributing the high values of χ reached at T_c in $(C_{10}H_{21}NH_3)_2CuCl_4$ to the extremely low value of J'. Be this as it may, from the much lower values of $|J'/J|$ in the methyl and decyl compound (compare the T_c/θ values) and from the fact that their susceptibilities coincide, it can be

Fig. 41

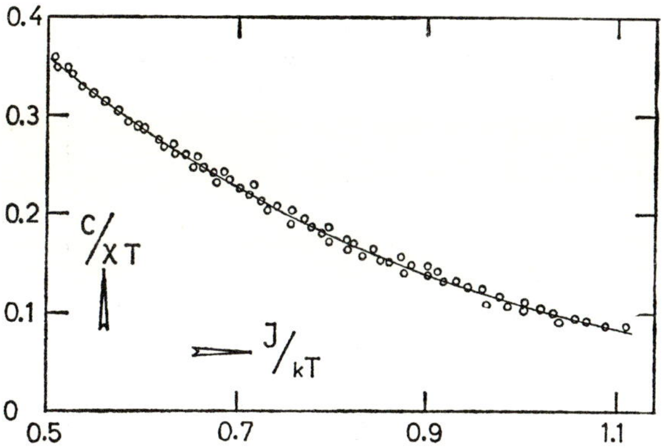

Susceptibility data of eleven different members of the series $(C_nH_{2n+1}NH_3)_2CuX_4$ in the region $0{\cdot}5 < J/kT < 1{\cdot}1$ $(J/k = \theta/2)$. Since the results for the various compounds coincide, although they differ in strength and type of the inter-layer interaction as well as in anisotropy, the straight line through the data may be considered to represent the χ of the ideal quadratic Heisenberg ferromagnet (tabulated values have been published by De Jongh and Van Amstel 1970). As such this result may be seen as an extension of the series expansion prediction, which is trustworthy up to $J/kT < 0{\cdot}6$ only.

inferred that in the temperature region shown, the upper curve in fig. 42 may be regarded as representing the susceptibility of a system of isolated ferromagnetic layers. The lower curve of $(C_2H_5NH_3)_2CuCl_4$ is then obtained by ' switching on ' the weak antiferromagnetic interaction between these layers. We point out that the value of $\chi T_c/C$ in this unusual type of antiferromagnet is extremely high ($\simeq 85$), as a consequence of the small value of $|J'/J|$. In ' normal ' antiferromagnets, in which a similar division of the spins in (ferromagnetic) sublattices can be made, $\chi T/C$ is usually smaller than unity, since the ferromagnetic intra-sublattice interaction, corresponding to J, is mostly much smaller than the antiferromagnetic intersublattice interaction (J').

In table 7 the values of the exchange constants J/k, as obtained from the analysis of the high-temperature susceptibility with the aid of the series expansion, have been compared with those derived from the coefficient of the linear spin-wave term in the heat capacity. It is seen that there is a systematic difference between the two results. Taking into account the errors of a few per cent involved in both determinations, we may say that the mean difference is about 10%. It is noteworthy that the same discrepancy has been obtained in the case of K_2CuF_4 (Yamada 1972).

An obvious way out of the problem would be to postulate a temperature dependence of the exchange, the susceptibility analysis being performed in the region $T > 1{\cdot}5\,J/k$, whereas the specific heat data considered were

Fig. 42

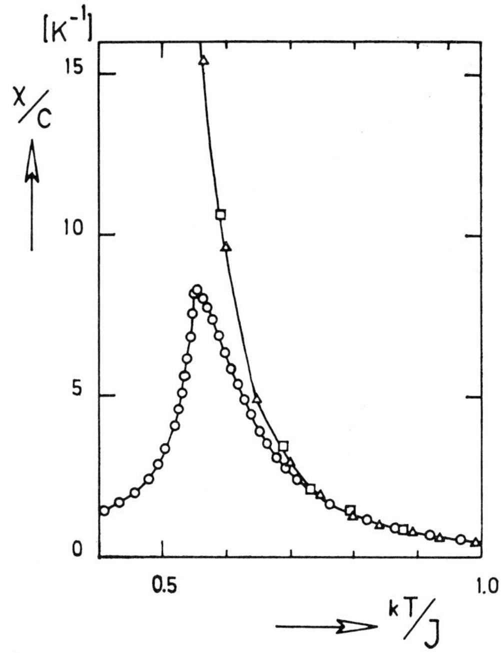

Comparison of the parallel susceptibilities of three Cu compounds. ○ : $(C_2H_5NH_3)_2CuCl_4$; △ : $(CH_3NH_3)_2CuCl_4$; □ : $(C_{10}H_{21}NH_3)_2CuCl_4$. In the temperature region shown, the curve common to the methyl and decyl compound represents the (diverging) susceptibility of a system of isolated ferromagnetic layers, with an intralayer exchange J. The curve for $(C_2H_5NH_3)_2CuCl_4$ can be thought of as being obtained from this by including a weak antiferromagnetic coupling $J'(|J'|/J \simeq 8 \times 10^{-4})$ between the layers. (After De Jongh *et al.* 1972 b)

for $T < 0{\cdot}05\,J/k$. However, the χ data show no apparent sign of a temperature dependence of J in the range $T < 150$ K. Moreover, one would expect an increase of J as the temperature is lowered, as in fact is mostly found experimentally (also for Cu compounds), whereas in table 7 the reverse is seen to be the case. The possible effect of J' on the anisotropy and of an interaction with second neighbours within the layer have been considered in detail by Colpa (1972 b) with negative results. Taking into account these by-effects will either increase the discrepancies or their influence is too small to be perceived.

Since the series expansion for the susceptibility is an exact result, at least as far as the temperature is high enough not to invalidate the calculation because of the limited number of terms, a possible explanation would be to assume the spin-wave prediction to be quantitatively in error. Indeed, such a situation is encountered in the magnetic chain systems, where spin-wave theory does predict the right variation with temperature of the ferro and antiferromagnetic specific heats at low temperatures, but quantitatively gives quite incorrect answers regarding the amplitude of these temperature dependences (Bonner and Fisher 1964). We will return to this matter in § 4.2. In any case it is clear that the determination of J/k on the basis of the spin-wave theory stands on a less sure footing than that from the series expansion for χ, so we have used the latter in the subsequent calculations.

We remark that the parameters of $(NH_4)_2CuCl_4$ listed in table 7 have been derived by Lécuyer *et al.* (1972) from powder susceptibility measurements that were analysed in the same manner as described above. Their result provides a welcome addition to the work on the other Cu compounds.

With the aid of the so-derived J/k values, Bloembergen *et al.* (1970 and to be published) have succeeded in separating the lattice contribution from the total measured heat capacity of these compounds in the following way. Assuming the magnetic part in the high-temperature region to be given by the series expansion prediction for C_m calculated with the known J/k's, the lattice contribution is known for say $T \simeq 2J/k$ and an extrapolation down to $T=0$ can be made, using different procedures of a varying degree of sophistication. As consistency checks, one has available the requirements that the total energy and entropy derived from the subsequently resulting magnetic specific heat must be equal to the expected values for $S=\frac{1}{2}$. In addition the lattice specific heat thus obtained can be compared with the measured heat capacity of the isomorphous non-magnetic compounds, which are obtained by replacing the Cu atoms by Cd or Zn. Moreover the specific heat in the spin-wave region is known, since at these low temperatures the lattice part is either negligible or can be more easily accounted for.

Proceeding in this way, estimates of C_m for a large number of Cu compounds have been obtained that more or less fall on the same curve. From these one may then construct a ' mean ' curve, which should represent the behaviour of the ideal model within an accuracy of a few per cent. In fig. 43 the resulting prediction is shown, with superimposed the small spike (2) due to the occurrence of long-range order at T_c, as found in $(C_2H_5NH_3)_2CuCl_4$, which has the highest $|J'/J|$ value of the members of the Cl series on which measurements have been made. In the other compounds similar but smaller anomalies have been observed, the interesting feature being that the heat content of these peaks (i.e. the area under the spike) diminishes with increasing n. This is what may be anticipated, because by increasing n the ideal 2-d model is better approximated (as can be inferred from the T_c/θ values) and since one expects the size of the anomaly at T_c to reflect the strength of the deviations from ideality that are present. Accordingly, for the chlorine series the highest peak was found in case of the ethyl compound. As can be seen from fig. 43, even in this case the peak at T_c is so small that one may easily accept the interpretation that it arises from spurious by-effects and is not an intrinsic property of the ideal system. To illustrate this point we have also

Fig. 43

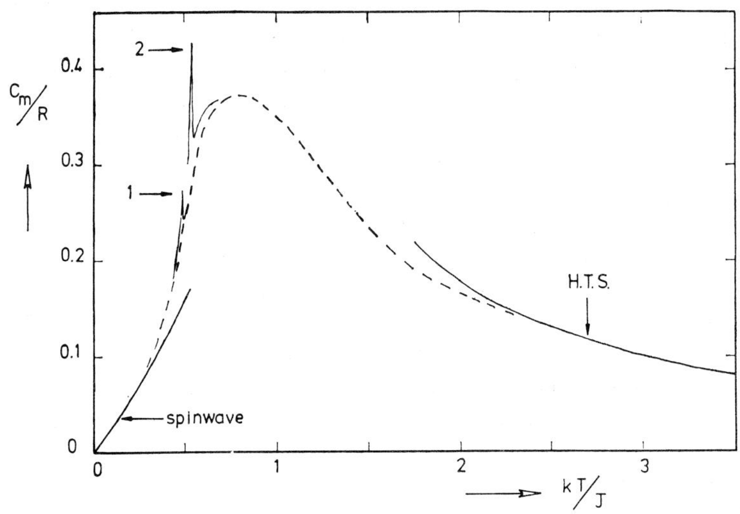

The dashed curve in this figure represents the heat capacity of the ideal quadratic Heisenberg ferromagnet as predicted on basis of the experimental findings (P. Bloembergen, private communication). It is the mean of the results obtained for various members of the series $(C_nH_{2n+1}NH_3)_2CuX_4$. Except near to T_c, the data of the individual compounds scatter evenly within a few per cent around this common curve. As in the magnetic chains there is a small spike observed at T_c in each compound, superimposed upon the broad maximum. The arrows 1 and 2 indicate the peaks observed in $(CH_3NH_3)_2CuCl_4$ and $(C_2H_5NH_3)_2CuCl_4$, respectively, and have been included to show the behaviour observed near the transition point, arising from the existing deviations from the ideal model, which are most pronounced in the ethyl compound. The full curves are the series expansion prediction at the high-temperature side, and the spin-wave contribution at low temperatures, as measured and calculated by Colpa (1972).

included in fig. 43 the spike observed in $(CH_3NH_3)_2CuCl_4$ for comparison (1). This compound has a much lower T_c/θ value, because of its smaller interlayer interaction. As discussed above this is probably due to an accidental partial cancellation of interlayer interactions of different sign.

The advantage of having the non-magnetic isomorphs has also been exploited in measurements of the heat conduction of these compounds. In fig. 44 the thermal conductivities of $(CH_3NH_3)_2CuCl_4$ and $(CH_3NH_3)_2CdCl_4$ are compared, from which it is seen that the lattice contribution is rather well represented by the behaviour of the non-magnetic compound. The small shifts needed to let the curves coincide for temperatures exceeding the T_c of the Cu compound may be easily attributed to small differences in the Debye temperature (horizontal scale) and in the dimensions of the two samples (vertical scale). Below T_c the curve of $(CH_3NH_3)_2CuCl_4$

clearly shows a huge contribution of the magnetic excitations to the heat conductivity. That the surplus conductivity is indeed of magnetic origin could be confirmed by subsequent measurements in magnetic fields of varying strength.

In concluding the discussion of the Cu compounds of table 7 we remark that in trying to correlate the observed decrease of T_c/θ with either $|J'/J|$ or $H_A{}^{I}$ or $H_A{}^{II}$, the following features become apparent. Firstly in comparing Rb_2CuCl_4, $(CH_3NH_3)_2CuCl_4$ and $(C_2H_5NH_3)_2CuCl_4$ one concludes that for these compounds it is the quantity $|J'/J|$ that determines

Fig. 44

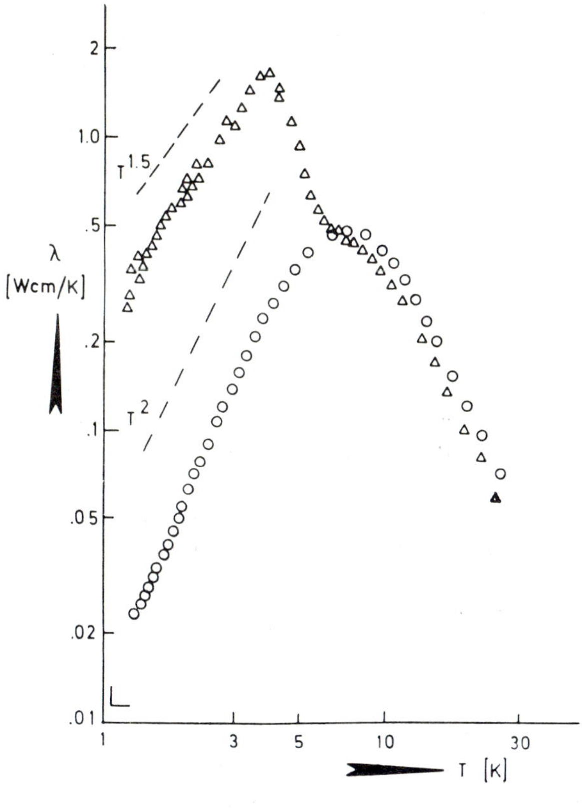

The thermal conductivity (in zero field) of $(CH_3NH_3)_2CuCl_4$ as compared with that of the isomorphous, non-magnetic Cd compound. The small shifts needed to let the two curves coincide above the transition temperature, $T_c = 8{\cdot}91$ K, of the Cu salt have been indicated in the left-hand corner below. The Cd data can be seen as representing the lattice part, the huge surplus conductivity observed below T_c in the Cu compound may be attributed to the contribution of the magnetic excitations. The $T^{3/2}$ dependence at low temperatures is expected for the magnon conductivity of a 2-d ferromagnet from simple spin wave theory. (After Gorter *et al.*, 1969.)

for the most part the position of the transition temperature. Also, from the fact that T_c/θ decreases in going from Rb_2CuCl_4 to K_2CuF_4, whereas the latter has a ten times larger H_A^{II} value, it seems that the out-of-plane anisotropy is indeed ineffective in establishing the long-range order, in accordance with our assumption that H_A^{II} merely introduces a planar Heisenberg character. For the $(C_nH_{2n+1}NH_3)_2CuCl_4$ series with $n > 3$ the quantity $|J'/J|$ becomes of the same order or smaller than H_A^I so that we expect that for these salts the actual value of T_c/θ is determined by a mixture of both these deviations from the ideal model. Furthermore, we point out that with the aid of table 7 one may estimate the $|J'/J|$ values of DPAN and $CuF_2 . 2H_2O$ to be 4×10^{-2} and $1{\cdot}5 \times 10^{-2}$, respectively, from the T_c/θ values of these compounds.

Concerning the temperature dependence of the sublattice magnetization, the only result obtained so far is an indirect determination following from the behaviour of the critical saturation field in $(C_2H_5NH_3)_2CuCl_4$ with temperature (De Jongh *et al.* 1972 b). According to both spin-wave and MF theory this should reflect the temperature dependence of the magnetization (cf. § 4.5). A 3-d value of β was found ($\simeq \frac{1}{3}$) which is to be expected, since the interlayer interaction is fairly large ($|J'/J| \simeq 10^{-3}$; see also § 3.2.3) in comparison with K_2NiF_4.

$AgCrSe_2$, $NaCrS_2$ *and* $NaCrSe_2$

In the hexagonal structure of these compounds the Cr^{3+} ions form ferromagnetic layers in which each magnetic ion is surrounded by six nearest neighbours (triangular lattice). These layers are antiferromagnetically coupled and are separated by three non-magnetic sheets, the ratio d_2/d_1 being about 2. Unfortunately they have not been greatly studied so far. Bongers *et al.* (1968) have measured the susceptibility of powdered specimens, whereas for the case of $NaCrS_2$ additional single-crystal data have been obtained by Blazey and Rohrer (1969). These authors also derived the antiferromagnetic phase diagram of this salt from magnetization and differential susceptibility measurements. Thus for $NaCrS_2$, values for the antiferromagnetic J' and H_A could be obtained from the values of the antiferromagnetic susceptibility at T_c, from the spin-flop field and from the critical field needed to saturate the sample, yielding $J_{af}/k \simeq -0{\cdot}6$ K and $H_A^I \simeq 3$ kOe. Apparently the value of H_A^{II} is only slightly larger than H_A^I. These results may however suffer from substantial errors, since Blazey and Rohrer did not correct their data for demagnetizing effects.

Since the high-temperature susceptibility was not analysed in terms of a series expansion, we could only derive a crude estimate of the ferromagnetic intralayer exchange from the reported value of the Curie–Weiss temperature. The same can be said for $NaCrSe_2$ and $AgCrSe_2$. In the latter case J' was calculated from the value of the susceptibility at T_c. We point out that $NaCrSe_2$, with its low $|J'/J|$ and T_c/θ values, presents an interesting object of further studies, since the experimental information on ferromagnetic layer compounds with $S > \frac{1}{2}$ is still limited.

CrI_3, $CrBr_3$ *and* $CrCl_3$

The hexagonal structure of the chromium tri-halides consists of ferro magnetic honeycomb layers ($z = 3$) of Cr^{3+} ions, separated from one another by two layers of halogen ions. Again the ratio d_2/d_1 is about 2. The large variation in the quantity $|J'/J|$ for these compounds, as observed from table 7, may be attributed to the existence of different interlayer interactions of opposite signs. Thus in $CrBr_3$, Samuelsen *et al.* (1971) reported that of the main three interlayer interactions two were ferromagnetic and one antiferromagnetic. The fact that in $CrCl_3$ the ferromagnetic layers are coupled antiferromagnetically with respect to one another, whereas in $CrBr_3$ and CrI_3 the effective J' is positive, also points to a variation in sign and magnitude of these three interlayer interactions. The low value of $|J'/J|$ in $CrCl_3$ would then be the result of a partial cancellation of the individual interactions.

The exchange constants listed in table 7 are the result of spin-wave theoretical calculations of the magnetization, in which a simplified two-parameter model was used, the various inter and intralayer interactions being substituted by an effective J' and J, respectively. In this way the results obtained for the different halides can be better compared. It should be pointed out that a more extensive analysis on the basis of a many-parameter model, which has been carried out for $CrBr_3$ (Samuelsen *et al.* 1971), leads to slightly different results. In this case an inelastic neutron scattering experiment was concerned in which the dispersion in the different crystallographic direction was measured. Quite similarly, as in the case of other 2-d magnets (cf. figs. 24 and 37), the spectrum was found to be very flat, with little dispersion in the c direction.

When likewise analysed in terms of a two-parameter model, however, these neutron data yield quite similar results for J and J', as given in table 7, where the listed parameters for $CrBr_3$ follow from the work of Davis and Narath (1964), who investigated the temperature dependence of the magnetization with an N.M.R. technique and fitted their measurements to a renormalized spin-wave theory on the basis of the two-parameter model. In the same way, Narath and Davis (1965) and Narath (1965) obtained the exchange constants of $CrCl_3$ and CrI_3. We remark that the values for the intralayer exchange J thus derived are considerably lower than those calculated from the measured Curie–Weiss temperatures (Hansen and Griffel 1959), in accordance with the picture given in the discussion of fig. 40.

For $CrBr_3$ slightly different transition temperatures have been found for different samples (Samuelsen *et al.* 1971, Ho and Litster 1970, Senturia and Benedek 1966, Jennings and Hansen 1965). A mean value is given in table 7. The T_c of $CrCl_3$ has been determined in a heat capacity experiment by Hansen and Griffel (1958), whereas the value for CrI_3 follows from the work of Williams and Sherwood (see Dillon and Olson 1965). In the heat capacity measurements (Jennings and Hansen 1965) large short-range-order contributions have been observed, the amount of entropy lost above T_c being 45% and 63% of the total $R \ln (2S+1)$ for $CrBr_3$ and $CrCl_3$, respectively.

The observed anisotropy is uniaxial in CrI_3 and $CrBr_3$, favouring the hexagonal *c* axis. Quantitative values have been determined by Dillon (1964) and Samuelsen *et al.* (1971) for $CrBr_3$ and by Dillon and Olson (1965) for CrI_3. Contrastingly, $CrCl_3$ has a (much smaller) anisotropy of orthorhombic symmetry. In this case $H_A{}^{I}$ and $H_A{}^{II}$ could be estimated from the spin-flop field and the saturation fields, respectively (Narath and Davis 1965). The fact that in $CrCl_3$ the direction of the moments is within the layer, whereas it is perpendicular to it in $CrBr_3$ and CrI_3, points to the presence of an anisotropy in the superexchange mechanism via the halogen ion, as mentioned above. In the case of ferromagnetic layers, the dipolar anisotropy favours an orientation within the layer.

In view of the large $|J'/J|$ value of $CrBr_3$ it is not surprising that the critical behaviour of this salt has been found to be fully 3-d in character (Ho and Litster 1970, Senturia and Benedek 1966). More about the critical parameters found for this salt will be said in § 4.4.

3.2.3. *Concluding remarks*

After having taken stock of the wealth of available experimental information, we are now in a position to try to tackle some of the fundamental questions, left by theory, concerning the thermodynamic behaviour of the isotropic 2-d systems (see § 3.2.1). As regards the 2-d Ising system, the situation is fairly well established theoretically and, as we have seen, the experimental work nicely confirms the picture given by theory.

Let us then focus attention on the examples of the 2-d Heisenberg model, and first of all make some remarks about the high degree of approximation that has already been reached. Regarding the interlayer interaction, J', we have learned that the estimate $|J'/J| \simeq 10^{-6}$ or smaller for the members of the K_2NiF_4 group is most probably correct. Although no exact determination of J' has hitherto been accomplished, there are indeed indications (see discussion of Rb_2MnF_4) that it is of the order of the dipolar coupling, which has been calculated to be about 10^{-8}–10^{-7} of J. Furthermore, we have found that there is probably no lower limit to the value of $|J'/J|$ that can be reached experimentally, other than that set by lattice imperfection or other sources, e.g. phonons (see under $(C_nH_{2n+1}NH_3)_2MnCl_4$).

Accordingly, it is the anisotropy that is the prime deviation from the ideal system that we are left with. In fact, when comparing the $|J'/J|$ and $\alpha = H_A/H_E$ values obtained, the conclusion is that in most examples α is at least one order of magnitude larger than $|J'/J|$. Unfortunately values of α smaller than about 10^{-3} are not easily reached, since even for a fully isolated (but finite) layer one always has the intralayer dipolar anisotropy to deal with. The only way out is in having competing anisotropy mechanisms, with different preferential directions, so that the various sources more or less cancel one another. This however is a rather unpredictable occurrence, so that it becomes a matter of trial and error to find such a compound.

The most prominent questions to answer are those concerning the ordering problem. Do the experiments confirm the theoretical proofs excluding the establishment of long-range order at a finite temperature? In our opinion the answer is definitely yes, based upon the observed specific heat behaviour. Especially in the case of the 2-d ferromagnetic Cu compounds, we have encountered overwhelming evidence for the fact that the specific heat of the ideal model will be a smooth non-anomalous curve, quite similar to those found for the chain structures†. Less conclusive but nevertheless clear indications to this end are the heat capacities of the quasi 2-d antiferromagnets $CuF_2 \cdot 2H_2O$ and $Mn(HCOO)_2 \cdot 2H_2O$. We remark that from the experimental findings predictions for the height and the position of the specific heat maximum for the 2-d Heisenberg systems may be obtained. Estimates for various S have been compiled in table 8, where we have used the heat capacities of $(C_nH_{2n+1}NH_3)_2CuX_4$, $CrBr_3$, $CuF_2 \cdot 2H_2O$, and of Ni and Mn formate. In comparing these results with the values of these parameters in the Heisenberg chain model (table 1), a number of interesting conclusions may be drawn. Firstly, one observes a similar dependence on S and also the fact that the maximum occurs at a higher temperature in the case of an antiferromagnetic interaction. Secondly, there is the striking feature that the predictions for C_{max}/R for different S of the 2-d antiferromagnet are equal to those given for the antiferromagnetic chain in table 1. In contrast with this there is no such correlation for the ferromagnetic case. Whereas the ferromagnetic chain maxima are considerably lower than their antiferromagnetic counterparts, in two dimensions the heights of the antiferromagnetic and ferromagnetic specific heat maxima seem to have about the same value.

The second question is: Do finite transition temperatures exist in these systems, at which the ferromagnetic (or the staggered antiferromagnetic) susceptibility diverges, in spite of the fact that there cannot be long-range order? As argued in § 3.2.1, in order to answer this, one must try to establish whether the experimentally observed T_c's are upwardly shifted (by the existing deviations of ideality) with respect to $T=0$ or with

† Although the experiments clearly indicate the absence of a diverging specific heat for the ideal 2-d Heisenberg model, the possibility of a weaker singularity, e.g. a diverging temperature derivative of C_m at the point where the susceptibility becomes infinite, cannot be excluded. Quite recently, Betts *et al.* (1973) have found evidence for the C_m of the 2-d, $S=\frac{1}{2}$, XY model to be qualitatively of the same form as the curve shown in fig. 43.

Table 8. Values for the height and the temperature of the specific heat maximum of the quadratic Heisenberg layer as predicted by the experimental findings. The observed dependence on the spin value S may be compared with that encountered for the magnetic chains, as given in table 1.

S	Ferromagnetic		Antiferromagnetic	
	C_{max}/R	$kT(C_{max})/J$	C_{max}/R	$kT(C_{max})/\|J\|$
$\frac{1}{2}$	$\simeq 0{\cdot}37$–$0{\cdot}38$	$\simeq 0{\cdot}8$	$\simeq 0{\cdot}35$	$\simeq 1{\cdot}4$
1	—	—	$\simeq 0{\cdot}52$	$\simeq 2{\cdot}1$
$\frac{3}{2}$	$\simeq 0{\cdot}65$	$\simeq 4$	—	—
$\frac{5}{2}$	—	—	$\simeq 0{\cdot}71$	$\simeq 10{\cdot}6$

respect to a finite (Stanley–Kaplan) temperature T_{SK}. In other words, do the experimental T_c's, when studied as a function of the strength of these deviations, extrapolate to zero or rather to a finite value? Evidently, making such extrapolations to the ideal case is the only way open to the experimentalist to contribute to the solution of the problem.

Bloembergen *et al.* (1970) were the first to make such an inquiry, with the aid of the series of ferromagnetic Cu compounds. They argued that

Fig. 45

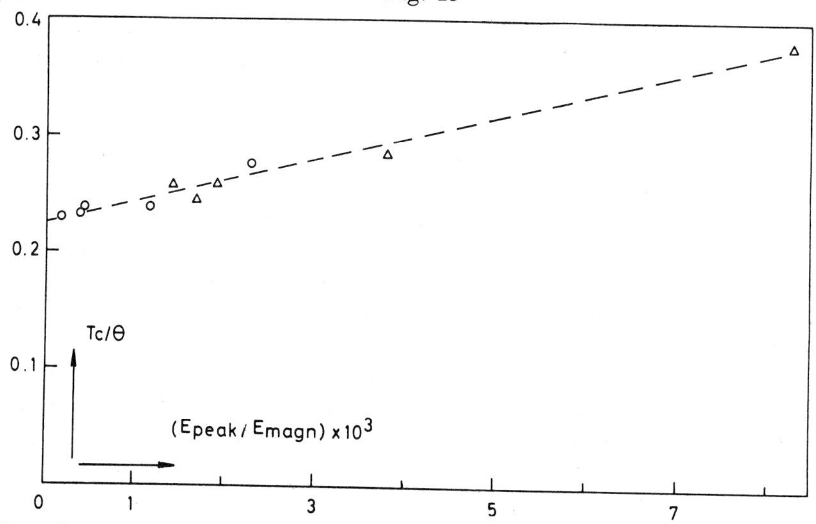

The relative transition temperatures T_c/θ of ten Cu compounds of the series $(C_nH_{2n+1}NH_3)_2CuX_4$ plotted versus the energy content of the small spikes which are found at T_c in the heat capacity, superimposed upon the ideal behaviour. △ : X=Br, $n=1, \ldots, 5$; ○ : X=Cl, $n=1, \ldots, 5$. The direction of increasing n is from the right to the left. It is seen that for most of the materials the energy content of the peak is only about 10^{-3} or less of the total energy involved in the magnetic ordering. (After Bloembergen *et al.* 1970.)

the energy content of the small spikes, found superimposed upon the flanks of the broad maxima, may be regarded as reflecting the strengths of the various deviations from the ideal system (for the latter the spike obviously will be absent).† Evaluating the area under these small anomalies for ten different compounds of the series, they studied the way in which the relative transition temperature T_c/θ depends on this minute amount of energy. Their findings have been reproduced in fig. 45, from which it may be seen that, when the peak energy is plotted on a linear scale, the extrapolation to the ideal case ($E_{peak}=0$) does indeed seem to yield a finite transition temperature. Note that for the compounds with large n the energy content of the specific heat spike diminishes to values of the order of 10^{-4} of the total magnetic energy.

Fig. 46

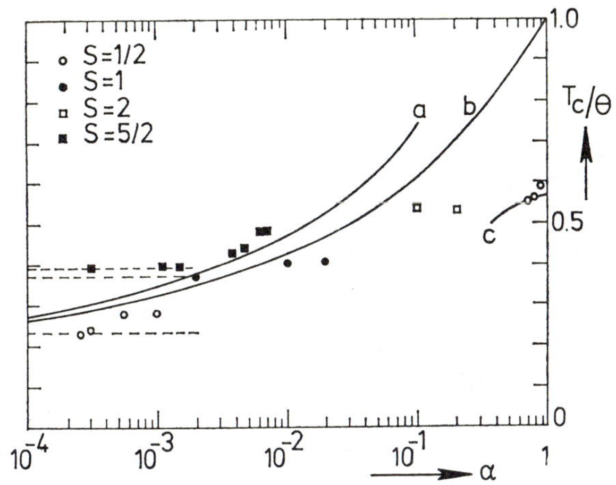

Relative transition temperatures T_c/θ as a function of the anisotropy parameter α, for the quadratic layer-type compounds of different spin value. The symbols refer to the examples of the 2-d antiferromagnet for which the anisotropy is expected to be at least one order of magnitude larger than the interlayer interaction, both as compared to the intralayer exchange. Only for the $S=\frac{1}{2}$ case have some ferromagnetic layer compounds been included which satisfy the same requirement. The various materials have been collected in tables 6 and 7 and the references to the data are given in the text. The curves a, b and c are theoretical results discussed in the text. The three dashed curves represent the limiting values to which the experimental data for $S=\frac{1}{2}$, 1 and $\frac{5}{2}$ seem to converge. (After De Jongh *et al.* 1972 a)

† It has recently been realized by the authors that the argument is not entirely correct if, as we expect, the planar part of the anisotropy is indeed ineffective in establishing the long-range order, since it would then not contribute to E_{peak}. Because H_A^{II}/H_E has about the same value ($\simeq 10^{-3}$) for all the compounds, its effect cannot be eliminated by the extrapolation, so that the obtained result for T_c/θ might in principle be affected.

As pointed out by De Jongh *et al.* (1972 a), another extrapolation can be made by taking advantage of the large number of 2-d antiferromagnets of widely varying anisotropy that has been discovered. Since in these compounds the interlayer interaction is usually considerably less than the anisotropy (both compared to J), one may assume that the anisotropy is the predominant deviation from the ideal system one has to reckon with. As a happy coincidence the anisotropy in antiferromagnetic substances can be determined relatively easy from spin-flop or AFMR measurements, so that in most cases a quantitative estimate of α is available. A plot of T_c/θ versus $\alpha = H_A/H_E$ for these substances will indeed yield the anisotropy dependence of the transition temperature. In fig. 46 this has been done for the materials discussed in the preceding section that receive consideration. As could already be inferred from table 6, the T_c/θ values do seem to converge to finite values, indicated by the horizontal broken lines. These limiting values are dependent on the spin value, as is to be expected, since the deviations from MF theory become larger with decreasing S. For comparison three theoretical predictions concerning the dependence of T_c/θ on α have been included. Curve *a* has been calculated by Lines (1970), who considered the quadratic Heisenberg antiferromagnet, introducing the anisotropy in the form of a field in the Hamiltonian. Curve *b* has been obtained by Dalton and Wood (1967) for the ferromagnetic quadratic lattice with an anisotropic exchange. For small α the results are seen to be very nearly the same, whereas for $\alpha > 10^{-3}$ the difference between the two approaches clearly manifests itself. In both cases the Green-function method in the random-phase approximation has been used. Since in this method the occurrence of a transition is associated with the onset of long-range order, T_c is predicted to decay to zero for $\alpha \to 0$. Furthermore, this approximate theory is expected to give best results for the high spin values. Lastly, curve *c* represents the dependence of T_c on α obtained by Dalton and Wood from series expansions of the susceptibility of the quadratic $S = \frac{1}{2}$ Heisenberg ferromagnet with an anisotropic exchange. For $\alpha = 1$ the result correctly coincides with the prediction for the $S = \frac{1}{2}$ Ising model.

From inspection of fig. 46 one may observe that especially for the higher spin value $S = \frac{5}{2}$, for which deviations from the curves *a* and *b* will already occurr at relatively high α values since T_c/θ increases with S, there is a clear departure from these predictions, the experimental points in the range 10^{-4}–10^{-3} lying 20–30% higher.

In table 9 we have listed the limiting temperatures, derived from the plots of figs. 45 and 46, and compared them with the T_{SK}'s obtained by Stanley and Kaplan from their analysis of the susceptibility series. These were found to agree to within a few per cent with the formula (Stanley and Kaplan 1966)

$$T_{SK} \simeq \tfrac{1}{5}(z-1)[2S(S+1)-1]J/k, \tag{3.8}$$

The $S = \frac{1}{2}$ case is excluded, since for this spin value the evidence from the

Table 9. Limiting values of T_c/θ as a function of S derived from the plots of figs. 45 and 46. These results are compared with the predicted transition temperatures T_{SK}/θ to a state of infinite susceptibility, as obtained by Stanley and Kaplan from their analysis of the high-temperature susceptibility series expansions for the quadratic lattice (eqn. (3.8)). One should also take into account the value $T_c/\theta \simeq 0{\cdot}40$ obtained from the series for the classical model ($S=\infty$), which differs from that calculated from eqn. (38), as well as the result $T_c/\theta \simeq 0{\cdot}35$ for the quadratic planar model with $S=\infty$ (Stanley 1968 a).

S	$\frac{1}{2}$	1	$\frac{5}{2}$	∞
T_c/θ	0·22	0·36	0·39	—
T_{SK}/θ	—	0·34	0·42	0·45

series was not conclusive. Dividing by the Curie–Weiss temperature, with $z=4$, gives

$$T_{SK}/\theta \simeq 0{\cdot}225[2-1/S(S+1)] \tag{3.9}$$

from which the values given in table 9 may be calculated. We remark that the uncertainties in these predictions are fairly large. As an indication of the errors involved we point to the result $T_{SK}/\theta \simeq 0{\cdot}30$–$0{\cdot}35$ for the $S=\infty$ case that was subsequently obtained by Stanley (1967) from the series for the classical Heisenberg model, much lower than the value calculated from eqn. (3.9). On the other hand, Lines (1971) derived $T_{SK}/\theta = 0{\cdot}40$ from the same $S=\infty$ series (see also Ritchie and Fisher 1973). One is therefore inclined to take $0{\cdot}40 \pm 0{\cdot}05$ as the estimate for $S=\infty$, which is indeed close to the experimentally found value for $S=\frac{5}{2}$. Since the errors involved in the experimental T_c/θ values are also considerable, one should not put too much weight on the quantitative agreement. In qualitative respect, however, experiment and theory do yield the same picture.

In summing up we state that the experimental evidence obtained so far, favours the existence of finite T_{SK}'s, although the precise values remain rather uncertain. In § 3.2.1 it has been explained how the occurrence of such a temperature at which the susceptibility diverges will lead to the establishment of long-range order under experimental conditions, although this could not happen in the ideal model.

We will conclude the discussion of 2-d magnets with a few remarks concerning the temperature dependence of the spontaneous magnetization is these systems. To this end we have reproduced in figs. 47 and 48 the magnetization curves of a number of isotropic and anisotropic compounds, respectively, available from the literature. For comparison the result for the quadratic Ising lattice ($S=\frac{1}{2}$) has been included in both figures, whereas in fig. 47 the behaviour of a typical 3-d isotropic salt (MnF_2) has also been drawn.

Firstly we observe that, since the 2-d Heisenberg lattice cannot sustain long-range order, the conclusion must be that any spontaneous magnetization found in an experimental example cannot be an intrinsic property of the ideal system but is the result of the deviations from ideality, i.e. the anisotropy and/or the interlayer interaction. In § 3.2.1 we made the assumption that the behaviour of the spontaneous magnetization will be of a 2-d anisotropic character or of a 3-d (Ising or Heisenberg) nature, according to whether $g\mu_B H_A \gg |J'|$ or conversely. Furthermore, in the former case one expects that the effect of J' will still become manifest, if only the transition temperature is approached closely enough, the width of this range around T_c being dependent on the strength of J'.

Fig. 47

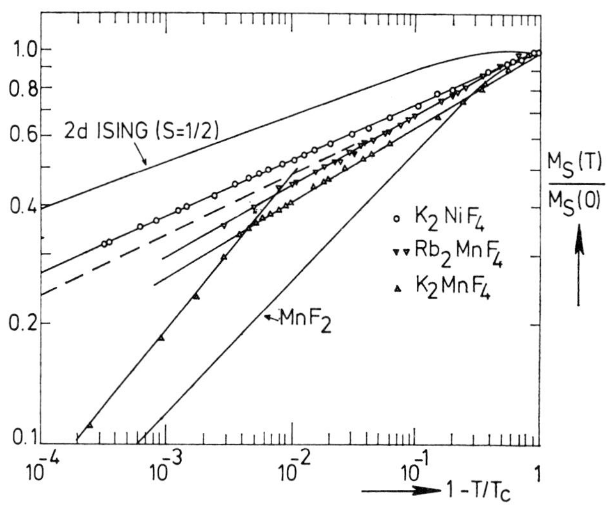

Magnetization curves of three examples of the quadratic Heisenberg antiferromagnet in the critical region, plotted on a double logarithmic scale. The prediction for the quadratic $S=\frac{1}{2}$ Ising lattice and the result obtained for MnF_2 experimentally by Heller (1966) are also shown. For references to the data, see the text.

Arguing in this way, one may define a ' changeover ' temperature $T^*(J')$, below which the system behaves two-dimensionally, whereas in the region $T_c - T^*$ the behaviour is of a 3-d character. It is not obvious *a priori* that the changeover will indeed occur abruptly at a certain definite temperature, or that the transition will be more or less smeared out. As seen from figs. 45 and 46 two examples have been found that seem to behave according to the picture just given, and the changeover does take place rather suddenly, in particular in K_2MnF_4. In Rb_2FeF_4 the temperature T^* is about 0·97 T_c, whereas in K_2MnF_4 it is about 0·996 T_c. The explanation for the fact that in the other compounds no such effects have been observed would be that in those cases the transition

Fig. 48

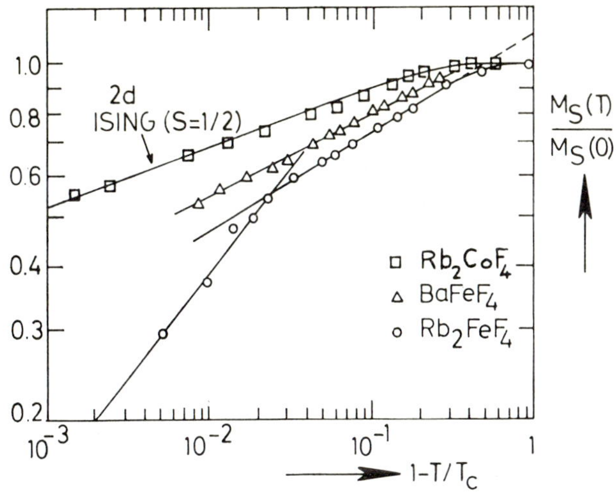

A similar plot as in fig. 47, but now for three anisotropic examples of 2-d antiferromagnetism. The origin of the data has been cited in the text.

temperature T_c has not been approached closely enough. Thus in K_2NiF_4, T^* would be nearer to T_c than a hundredth of a per cent! Since the difference $T_c - T^*$ is related to $|J'/J|$ it would follow that for some reason or the other this quantity is much larger in Rb_2FeF_4 than in K_2NiF_4, the case of K_2MnF_4 being intermediate between these two. On the other hand, the situation is far from being solved, since the apparent cross-over behaviour may very well be caused by a distribution of transition temperatures throughout the crystal (Birgeneau *et al.* 1973). Clear indications for this have been found in the case of Rb_2FeF_4 (see above). The effect of such a distribution on the data analysis has been discussed both qualitatively and quantitatively by Birgeneau *et al.*, explaining the apparent cross-over observed by Hirakawa and Ikeda in K_2MnF_4 by a spread in T_c of the order of 10^{-3} T_c in their sample. Their own data, in which no kink could be observed, are indicated in fig. 47 by the broken line. The conclusion therefore must be that in these two compounds the apparent cross-over is most probably a spurious effect. An example of a changeover of an exponent that is more likely to be real will be given in § 4.4.

In any case one would expect that interlayer interactions of the order of 10^{-4}–10^{-3} will be sufficient to bring the region $T^* - T_c$ down to about 0·99 to 0·90 of T_c, so that log–log plots of $M_s(T)/M_s(0)$ versus $\epsilon = 1 - T_c/T$ will then yield β values which are either near to the 3-d value $\beta = \frac{1}{3}$ or lie in between $\frac{1}{3}$ and the 2-d Ising value $\frac{1}{8}$ if the changeover temperature T^* is smeared out. In order to test this we have gathered in table 10 the observed β's in the 2-d compounds of the preceding section, together with their anisotropy parameters and interlayer interaction constants. Also listed are the amplitudes of the singularity B, as defined by eqn.

Table 10. Critical behaviour of the spontaneous magnetization as observed in the quasi 2-d compounds. Listed are the anisotropy parameter α, the ratio $|J'/J|$, the critical exponent β, the amplitude B (eqn. (3.7)), the range of the relative temperature $\epsilon = 1 - T/T_c$ in which β has been derived and the method of measurement used. The results for β and B for the 2-d Ising and the 3-d Ising and Heisenberg models are also given.

Compound	α	$\|J'/J\|$	β	B	Range of $\epsilon = 1 - T/T_c$	Method of measurement
Rb_2CoF_4	$\simeq 0{\cdot}8$	$\simeq 10^{-6}$	$0{\cdot}122 \pm 0{\cdot}008$	$1{\cdot}18$	$0{\cdot}0005 \leqslant \epsilon \leqslant 0{\cdot}02$	Neutrons
$BaFe_4$	$\simeq 0{\cdot}2$	$\simeq 10^{-6}$	$0{\cdot}168 \pm 0{\cdot}005$	$1{\cdot}18 \pm 0{\cdot}01$	$0{\cdot}015 \leqslant \epsilon \leqslant 0{\cdot}20$	Mössbauer
$CoCl_2 . 6H_2O$	$\begin{cases} 4 \times 10^{-2}\ (H_A^{I}) \\ \simeq 0{\cdot}7\ (H_A^{II}) \end{cases}$	?	$\simeq 0{\cdot}18$		$0{\cdot}04 \leqslant \epsilon \leqslant 0{\cdot}4$	N.M.R.
Rb_2FeF_4	$\simeq 0{\cdot}1$	?(10^{-4}–10^{-6})	$\simeq 0{\cdot}20$	$\simeq 1{\cdot}2$	$0{\cdot}03 < \epsilon < 0{\cdot}3$	
			$\simeq 0{\cdot}3$		$0{\cdot}005 < \epsilon < 0{\cdot}03$	Neutrons
$FeCl_2$	$\simeq 1$	$7{\cdot}5 \times 10^{-2}$	$0{\cdot}29 \pm 0{\cdot}01$	$1{\cdot}47$	$0{\cdot}0001 \leqslant \epsilon \leqslant 0{\cdot}1$	Neutrons
K_2NiF_4	2×10^{-3}	$\simeq 10^{-6}$	$0{\cdot}138 \pm 0{\cdot}004$	$0{\cdot}973$	$0{\cdot}0003 < \epsilon < 0{\cdot}2$	Neutrons
Rb_2MnF_4	$4{\cdot}7 \times 10^{-3}$	$\simeq 10^{-6}$	$0{\cdot}16 \pm 0{\cdot}02$	$0{\cdot}96$	$0{\cdot}003 < \epsilon < 0{\cdot}1$	Neutrons
			$0{\cdot}18 \pm 0{\cdot}02$	$1{\cdot}02$	$0{\cdot}003 < \epsilon < 0{\cdot}9$	
$KFeF_4$	$6{\cdot}5 \times 10^{-3}$	$\simeq 10^{-4}$	$0{\cdot}185 \pm 0{\cdot}005$	$0{\cdot}99 \pm 0{\cdot}01$	$0{\cdot}01 < \epsilon < 0{\cdot}28$	Mössbauer
K_2MnF_4	$3{\cdot}9 \times 10^{-3}$	$\simeq 10^{-6}$(?)	$0{\cdot}188 \pm 0{\cdot}01$	$1{\cdot}0$	$0{\cdot}004 \leqslant \epsilon < 0{\cdot}6$	Neutrons
			$\simeq 0{\cdot}3$		$0{\cdot}002 \leqslant \epsilon < 0{\cdot}004$	
			$0{\cdot}15 \pm 0{\cdot}01$		$0{\cdot}0006 < \epsilon < 0{\cdot}3$	Neutrons
$RbFeF_4$	$6{\cdot}5 \times 10^{-3}$	$\simeq 10^{-3}$	$0{\cdot}245 \pm 0{\cdot}005$	$1{\cdot}18 \pm 0{\cdot}01$	$0{\cdot}01 < \epsilon < 0{\cdot}4$	Mössbauer
$Mn(HCOO)_2 . 2H_2O$	$\simeq 10^{-3}$	3×10^{-3}	$0{\cdot}23 \pm 0{\cdot}01$		$0{\cdot}04 < \epsilon < 0{\cdot}44$	Neutrons
			$0{\cdot}22 \pm 0{\cdot}002$		$0{\cdot}015 < \epsilon < 0{\cdot}47$	N.M.R.
$MnTiO_3$	$1{\cdot}2 \times 10^{-3}$	?	$0{\cdot}32 \pm 0{\cdot}01$		$\epsilon < 0{\cdot}125$	Neutrons
$(C_2H_5NH_3)_2CuCl_4$	$\begin{cases} 1{\cdot}6 \times 10^{-4}\ (H_A^{I}) \\ 1{\cdot}6 \times 10^{-3}\ (H_A^{II}) \end{cases}$	$8{\cdot}5 \times 10^{-4}$	$0{\cdot}32 \pm 0{\cdot}02$	$1{\cdot}0$	$0{\cdot}01 < \epsilon < 1$	Antiferromagnetic phase diagram
K_2CuF_4	$\begin{cases} 4 \times 10^{-5}\ (H_A^{I}) \\ 1{\cdot}2 \times 10^{-5}\ (H_A^{II}) \end{cases}$	$3{\cdot}5 \times 10^{-3}$	$0{\cdot}22$		$0{\cdot}001 < \epsilon < 0{\cdot}15$	Neutrons
			3-d behaviour		low temperatures	N.M.R.
$CrBr_3$	$1{\cdot}6 \times 10^{-2}$	6×10^{-2}	$0{\cdot}365 \pm 0{\cdot}015$	$1{\cdot}32 \pm 0{\cdot}07$	$0{\cdot}007 < \epsilon < 0{\cdot}05$	N.M.R.
			$0{\cdot}368 \pm 0{\cdot}005$	$1{\cdot}20 \pm 0{\cdot}04$	$0{\cdot}0004 < \epsilon < 0{\cdot}09$	Faraday effect
2-d Ising ($S=\frac{1}{2}$)			$0{\cdot}125$	$\simeq 1{\cdot}24$	$\epsilon < 0{\cdot}02$	
3-d Ising ($S=\frac{1}{2}$)			$0{\cdot}312$	$\simeq 1{\cdot}52$	$\epsilon < 0{\cdot}02$	
3-d Heis. ($S=\frac{1}{2}$)			$\simeq 0{\cdot}36$	$\simeq 1{\cdot}25$	?	

(3.7). For comparison the values of β and B for the 2-d quadratic Ising model and for the 3-d Heisenberg and Ising lattices are given, together with the range of ϵ in which the magnetization of these models does follow the limiting behaviour given by the power-law of eqn. (3.7) (Wielinga 1971).

By studying table 10 one may indeed observe a correlation between the β value and the interlayer coupling. $FeCl_2$, $CrBr_3$ and $(C_2H_5NH_3)_2CuCl_4$ clearly show 3-d behaviour. Note in particular that the parameters of the anisotropic compound $FeCl_2$ are very close to the 3-d Ising predictions, whereas for the reasonably isotropic salt $CrBr_3$ the results are in better agreement with those of the 3-d Heisenberg model. Also for $MnTiO_3$ a 3-d β value is found, which finds its explanation in the fact that in this hexagonal compound there is a direct coupling between the neighbouring antiferromagnetic layers, in contrast with the situation in the K_2NiF_4 structure. The ratio $|J'/J|$ will therefore probably also be of the order of 10^{-3}–10^{-2}, as in the Cr compounds of table 7, which have a rather similar structure. Such a direct coupling also exists in the structure of $RbFeF_4$, that accordingly has a higher β (within the same ϵ range) than $KFeF_4$, in which the interaction between the nearest layers is cancelled because of symmetry. In spite of this the ratio of $|J'/J|$ in $KFeF_4$ will likely be larger than in the K_2NiF_4 structure because the exchange paths connecting the next-nearest layers involve three F ions, instead of the two F and two K ions in K_2NiF_4.

The 'intermediate' β values of about 0·22 of $RbFeF_4$, $Mn(HCOO)_2 . 2H_2O$ and K_2CuF_4 may be understood in terms of a smeared-out transition range, taking into account that they have been determined in a relatively high range of the relative temperature ($\epsilon > 0{\cdot}01$). Although for K_2CuF_4 an additional lower decade in ϵ was explored, the uncertainty in the choice of T_c in this case enables a fairly wide range of possible β values. The same remark also applies to some of the other compounds of table 10, since one must keep in mind that for $\epsilon < 10^{-2}$ a knowledge of the T_c of the investigated sample better than 0·1% is necessary to obtain a reliable result for the β.

From the fact that in the 2-d Ising model the spontaneous magnetization retains near-saturation values up to a much higher relative temperature than in the 3-d models, it is explained why the log–log plots of the quasi 2-d salts in figs. 47 and 48 lie in between the 2-d Ising curve and the MnF_2 result, and also why the measured B values of the isotropic 2-d compounds are much lower than the Ising prediction. The particular value of $M_s(T)/M_s(0)$ at a certain relative temperature will be dependent on α, $|J'/J|$ and on S. Concerning the dependence on S, for instance, it is known from the 3-d models and the 2-d Ising model that the value of B will decrease by increasing S, although the exponent β will to all probability retain the same value. For compounds of similar $|J'/J|$ and α, the magnetization at a particular ϵ value will be lower the lower is S. Thus K_2NiF_4 is nearer to the Ising result for $S = \frac{1}{2}$ than Rb_2MnF_4, although it is more isotropic. Far away from T_c, for $\epsilon > 0{\cdot}1$, one may observe that the Fe compounds are even closer to the 2-d Ising prediction than K_2NiF_4, which can be explained from their highly anisotropic properties. Not surprisingly, the magnetization curve of one of the 2-d Co salts that have $S=\frac{1}{2}$ is extremely close to the Ising prediction (preliminary measurements of Samuelsen on Rb_2CoF_4, see Samuelsen 1973). In view of the above discussion we may add that only the experiments on extremely anisotropic compounds may provide essential information regarding the critical behaviour of the spontaneous magnetization in 2-d systems. By this we mean that the measurements of the long-range order in the isotropic salts, although extremely useful for other purposes, are concerned with a property that is not intrinsic to the ideal system. The purpose of such work should therefore lie elsewhere, for instance, to provide more detailed studies about the nature of the changeover effect.

In concluding this section we would like to add that a slightly different approach to the changeover mechanism may be taken. Quite generally one might assign a transition temperature T_c to an ideal model system, and then assume that the deviations from this model will result in a shift of T_c to a T_c', which would be the experimental transition temperature. For temperatures for which $|T-T_c|$ is large as compared with the difference $|T_c-T_c'|$, the critical behaviour will then be in accordance with the ideal model ; for instance, the susceptibility behaviour will yield the right exponent. But as T_c' is approached, the system realizes that its actual transition temperature is not T_c but a shifted T_c', and the character of the susceptibility plot is changed. One may thus define an inner region $|T_c'-T^*|$ in which the presence of the deviations are felt, the susceptibility in this range behaving as diverging at T_c', with a changed critical exponent.

In the 2-d quasi-isotropic magnets one may identify the ideal T_c with the Stanley–Kaplan temperature T_{SK}, whereas the shifted T_c' would be the experimentally observed transition temperature. It is of importance to note that for these systems the cross-over can be different for both sides of T_c. In the case of the spontaneous magnetization observed below T_c the cross-over will be from 2-d Ising to 3-d Ising-like behaviour, and is due solely to J'. Above T_c the susceptibility, for instance, may first change from 2-d Heisenberg to 2-d Ising and thereafter from 2-d Ising to 3-d Ising behaviour if, as in most cases, the anisotropy is considerably larger than the interlayer coupling. We will take up this matter again in § 4.4.

3.3. *Three-dimensional magnetic systems*

3.3.1. *Introductory remarks*

To his likely surprise, the reader will find the number of experimental examples treated in this section to be less than in the previous ones on 1-d and 2-d magnets. There are a number of reasons for this. Firstly, since in this paper 1-d and 2-d magnetism is reviewed for the first time, the authors thought it appropriate to give an extensive treatment of the lower dimensional systems, considering also the less ideal examples. Evidently, the fact that our own research has been in this field for several years plays a role too. Secondly, the number of 3-d magnetic compounds being so vast, it would be an almost impossible task to discuss them all, so that we decided to regard only the very best examples of either the Ising or the Heisenberg 3-d models (a good example of the 3-d planar Heisenberg model is yet to be found). It turns out that in this way one is left with few compounds. In particular the number of ferromagnetic insulators approximating the nearest-neighbour-only Ising or Heisenberg model to a sufficient degree turns out to be disappointingly small.

Before embarking on a discussion of the experimental examples we shall, as before, briefly mention some of the theoretical results that are available. Here again such a summary of necessity has to be of a very brief and general nature, due to the enormous amount of papers bearing on the subject. For more extensive information the reader is referred to the review papers mentioned in § 1.1.

As a first remark we recall that the MF theory is more closely approximated in 3-d systems, at least in comparison with the complete inapplicability of this theory in 1-d and 2-d cases. Referring to the discussion in § 1.2, the explanation is the relative weakness of the effects of short-range order in 3-d systems. In fact, if one is not interested in details (e.g. critical behaviour), effective field theories will give a satisfactory account of the overall behaviour found in 3-d magnets.

However, near to the transition point (and in the case of the isotropic magnets in the spin-wave region) the failure of the effective field concept becomes apparent and one has to take recourse to more sophisticated theories. Since no exact treatments of the 3-d Ising and Heisenberg models are available, the information about the thermodynamics has been supplied by approximate solutions. In the low-temperature region, $T \ll T_c$, one expects spin-wave theory to give a reasonable description of the thermodynamic behaviour of the Heisenberg model, as is indeed confirmed by the experimental findings (see below). On the other hand, in the Ising model the extreme anisotropy makes it possible to calculate the limiting ($T \to 0$) low-temperature behaviour of the thermodynamic functions, which is of exponential form. Combined with results obtained from low-temperature expansions this yields a thrustworthy prediction in the region $T \ll T_c$, in some cases even quite close to T_c. In the critical region, high ($T > T_c$) and low ($T < T_c$) temperature series expansions have been used in both models. By carefully analysing these series with various techniques one has obtained estimates of the critical behaviour of the specific heat, the ferromagnetic susceptibility and the magnetization, that compare favourably with the obtained experimental results (see § 4.3). Thus, in spite of the non-availability of exact solutions, one often has come quite far by linking together the results deduced by the approximate methods in various temperature regions (evidently the MF theory constitutes the high-temperature approximation). In that way a prediction valid over quite a wide range of temperatures relative to T_c can be constructed in practice.

Since the quantitative theoretical information obtained for the critical exponents (series expansions) and for the low-temperature behaviour (spin-wave theory) will be reviewed in §§ 4.2–4.4, together with the experimental findings, we shall not give a survey of these predictions here. They will, however, be mentioned when they arise in the discussion of the experimental examples below.

3.3.2. *Survey of experimental results*

As in the preceding section we will commence with the anisotropic compounds. In table 11 we have collected what we consider to be the best examples of the 3-d Ising model known at present, together with their critical parameters, as derived from the heat capacity data. These will serve to compare the experiments with the theoretical predictions included in the table.

$CoCs_3Cl_5$ *and* $CoRb_3Cl_5$

The origin of the anisotropic properties of these compounds has already been mentioned in § 2.2. They are isomorphous to $CoCs_3Br_5$, which has been discussed in the preceding section. The magnetic ions form a simple tetragonal lattice, with a c_0/a_0 ratio of about 1·1 so that the structure is approximately simple cubic. We have seen that in the case of the bromine compound there is apparently an accidental cancellation of the interaction along the c axis, giving the substance a pronounced 2-d magnetic character. In the two chlorine compounds, on the other hand, the interactions along the c axis and in the a–a plane are nearly equal. From E.S.R. measurements in $ZnCs_3Cl_5$ doped with Co^{2+}, Van Stapele *et al.* (1966) found that the exchange within the a–a plane is antiferromagnetic, whereas the interaction along the c axis is of ferromagnetic sign and about 25% smaller in magnitude. For $CoRb_3Cl_5$, on the other hand, Blöte (1972) deduces the coupling in the c direction to be also antiferromagnetic, by comparing the total energy involved in the magnetic ordering (derived from the heat capacity) and the measured Curie–Weiss θ. The occurrence of both ferro and antiferromagnetic interactions along the c axis in this series of compounds may provide an explanation for the apparent cancellation of this interaction in the case of $CoCs_3Br_5$.

Since in the Ising model the sign of the exchange constant is irrelevant as concerns the heat capacity, the experimental data on both $CoCs_3Cl_5$ and $CoRb_3Cl_5$ may be compared with the theoretical curve for the simple cubic Ising model. Basing themselves upon the work of Baker (1963) and Sykes *et al.* (1972), Blöte and Huiskamp (1969) obtained a prediction for this model that is shown as the full curve in fig. 49. It can be seen from this figure and table 11 that their data on $CoRb_3Cl_5$ fit the theory excellently. For $CoCs_3Cl_5$, which was studied by Wielinga *et al.* (1967)

Table 11. Critical entropy and energy parameters of theoretical 3-d Ising models and their experimental approximants. The theoretical values are from Sykes *et al.* (1972), Fisher (1967) and Domb and Miedema (1964). For references to the experimental data, see the text. The numbers in parentheses refer to the number of nearest magnetic neighbours.

Compound or model	S	J/k (K)	T_c (K)	T_c/θ	S_c/R	$\frac{(S_\infty - S_c)}{R}$	$\frac{(S_\infty - S_c)}{S_c}$	$\frac{-E_0}{RT_c}$	$\frac{-E_c}{RT_c}$	$\frac{(E_c - E_0)}{RT_c}$	$\frac{-E_c}{(E_c - E_0)}$
Ising, diamond (4)	$\frac{1}{2}$			0·6760	0·511	0·182	0·356	0·740	0·320	0·420	0·761
Ising, s.c. (6)	$\frac{1}{2}$			0·7518	0·5579	0·1352	0·2424	0·6651	0·2200	0·4451	0·4942
Ising, b.c.c. (8)	$\frac{1}{2}$			0·7942	0·5820	0·1111	0·1909	0·6296	0·1720	0·4576	0·3759
Ising, f.c.c. (12)	$\frac{1}{2}$			0·8163	0·5902	0·1029	0·1744	0·6126	0·1516	0·4610	0·3287
$DyPO_4$	$\frac{1}{2}$	−2·50	3·390	0·678	0·505	0·185	0·37			0·408	
$Dy_3Al_5O_{12}$	$\frac{1}{2}$	−1·85	2·54	0·68	0·489	0·204	0·42	0·77	0·38	0·392	0·97
$DyAlO_3$	$\frac{1}{2}$	$\simeq -2$	3·52	0·62	0·521	0·172	0·33	0·805	0·308	0·497	0·62
$CoRb_3Cl_5$	$\frac{1}{2}$	−0·511	1·14	0·74	0·563	0·137	0·24	0·673	0·226	0·447	0·51
$CoCs_3Cl_5$	$\frac{1}{2}$	−0·222	0·52	0·79	0·593	0·106	0·18	0·632	0·173	0·459	0·38
Ising, b.c.c. (8)	2			0·841	1·480	0·129	0·087		0·188		
Ising, f.c.c. (12)	2			0·864	1·486	0·123	0·083	1·16	0·167	0·990	
FeF_2	2	−2·69	78·3	0·91	1·40	0·21	0·15				

the agreement is slightly less. As mentioned by Wielinga *et al.* (1967) and Blöte and Huiskamp (1969), this may be attributed to the lower T_c of the chlorine compound, as a consequence of which dipolar interactions will be relatively more important. These long-ranged interactions can enhance the asymmetry of the specific heat curve by the fact that they may tend to increase the effective number of nearest neighbours. In fact it is observed from table 11 that the caloric data on $CoCs_3Cl_5$ are in better agreement with the predictions for the b.c.c. Ising model. Thus, although on the basis of the crystal structure one would choose a s.c. magnetic structure, the specific heat is best described by an Ising model with a higher (effective) coordination number.

Fig. 49

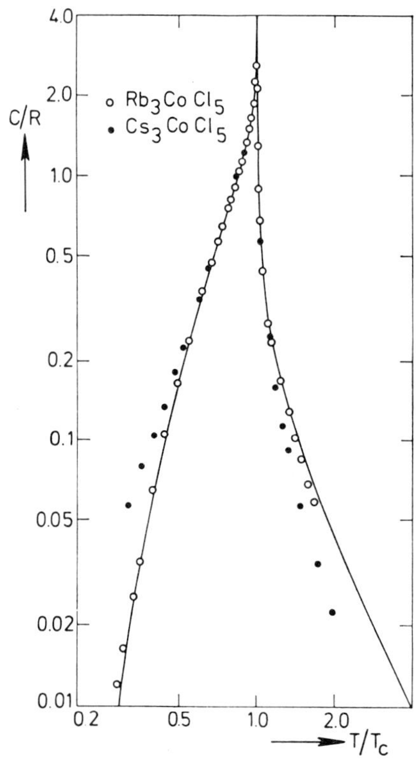

Heat capacities of $CoRb_3Cl_5$ and $CoCs_3Cl_5$ compared with the theoreticai prediction for the simple cubic Ising model. (After Blöte and Huiskamp 1969).

The transition temperatures and exchange constants listed in table 11 have been derived from the heat capacity data. In the case of $CoCs_3Cl_5$ the J/k so obtained was in reasonable agreement with the average value deduced from the E.S.R. experiment. It must be noted that the listed exchange constants include the dipolar contributions.

Magnetic experiments on $CoCs_3Cl_5$ have been performed by Mess *et al.* (1967). However, the existence of a ferromagnetic interaction in this compound makes a comparison with the antiferromagnetic Ising model irrelevant. Not surprisingly, the measured antiferromagnetic susceptibility deviates strongly from the prediction for the s.c. Ising antiferromagnet. Unfortunately, apart from the powder susceptibility, no magnetic measurements on $CoRb_3Cl_5$ have been performed up until the present.

The interactions in the three Dy compounds in table 11 are predominantly of the dipolar kind, but they are well approximated by an Ising $S=\frac{1}{2}$ model. The highly anisotropic properties of the Dy^{3+} ion have been discussed in § 2.2. The energy separation of the lowest Kramers doublet and the first excited state is about 70 cm^{-1} in $DyPO_4$ and DAG, and about 55 cm^{-1} in $DyAlO_3$. Since the ordering temperatures are in the liquid helium range, the population of the excited level is negligible, leaving a ground state with a nearly perfect uniaxial magnetic character ($g_{\parallel} \simeq 18$; $g_{\perp} \simeq 0{\cdot}5$).

$DyAlO_3$

The crystal structure of $DyAlO_3$ is a distorted perovskite in which the Dy ions occupy two magnetically inequivalent sites. The magnetic structure has been determined by neutron diffraction by Bidaux and Mériel (1968) and can in the antiferromagnetic state be described in terms of a four-sublattice model, with the principal axes within the *a–b* plane. There are six nearest neighbours, four in the *a–b* plane and two along the *c* axis. From their optical work, Schuchert *et al.* (1969) found the interactions to be mainly of dipolar origin with substantial further neighbour contributions. Furthermore, they deduced the exchange interactions within the *a–b* plane to be much smaller than along the *c* axis. Consequently, the total magnetic interaction along the *c* axis is predominant, which may explain why the critical parameters given in table 11, as obtained by Cashion *et al.* (1968) from the specific heat, are in better agreement with an Ising model having a lower coordination number than s.c. (compare with the diamond lattice, $z=4$). In table 11, T_c/θ has been deduced from the total magnetic energy (Cashion *et al.* 1968). By fitting the measured parallel susceptibility to the Curie–Weiss law in the region $6<T<20$ K Schuchert *et al.* (1969) obtained $\theta=18 \pm 3$ K, giving $T_c/\theta \simeq 0{\cdot}20$! This again illustrates the errors that may arise when the MF theory is applied, even in the temperature range $2T_c<T<5T_c$, in particular for $S=\frac{1}{2}$.

From their measurement of the magnetoelectric susceptibility, Holmes *et al.* (1971) were able to determine the critical behaviour of the sublattice magnetization. The power-law fit yielded $\beta=0{\cdot}311 \pm 0{\cdot}005$, $B=1{\cdot}51 \pm 0{\cdot}03$ and $T_c=3{\cdot}525 \pm 0{\cdot}001$ K. The latter value is in good agreement with $T_c=3{\cdot}52$ K, as derived from the specific heat. The β value is within the uncertainty equal to the theoretical prediction for

3-d Ising lattices (β is thought to be independent of the precise lattice structure). On the other hand, B is predicted to decrease slowly with coordination number, being 1·66, 1·57 and 1·49 for the diamond, s.c. and b.c.c. lattice, respectively (see, e.g., Fisher 1967).

$Dy_3Al_5O_{12}$

Dysprosium aluminium garnet (DAG) has the cubic garnet structure containing six magnetically inequivalent Dy ions. The local z axis is different for each of the six sites, with an equal number of moments pointing along the $\pm a$, $\pm b$, and $\pm c$ axes. As a consequence, the ordered antiferromagnetic state must involve at least six different sublattices. However, in a magnetic field along a (111) axis, the a, b, and c axes become equivalent in threes, so that the substance can be described within a two-sublattice Ising model.

The extensive literature on this material includes specific heat, magnetic resonance, magnetization, susceptibility, optical and neutron scattering experiments, and one may safely say that it is one of the most extensively investigated antiferromagnetic materials. Recently, Wolf, Landau, Keen and Schneider have started a series of papers which aim to give a complete picture of the magnetic and thermal properties (Landau *et al.* 1971, Wolf *et al.* 1972). Since there is no use in reproducing all the information, we refer the reader to these publications for the full details as well as references to the earlier papers.

What is of interest to us in the present context is how closely DAG resembles a particular 3-d Ising model. Although calculations for the garnet structure are absent, we may make a comparison with lattices of a similar coordination number. In DAG about 80% of the total interaction is between the nearest neighbours. The dipolar contribution to this nearest-neighbour interaction is about twice as large as the exchange part (see, e.g., Norvell *et al.* 1969 for a table of the various interaction energies). Since there are four nearest neighbours one would expect the critical parameters to agree with an Ising model with coordination number 4, such as the diamond lattice. This is indeed observed from table 11. That the sum of the interactions with further neighbours is only 20% of the total interaction energy arises because the individual interactions have a tendency to cancel. The estimate $\theta = 3{\cdot}7 \pm 0{\cdot}2$ K used in table 11 was obtained by Ball *et al.* (1963), who deduced the effective interaction constant from both the heat capacity in the low-temperature region and the total magnetic energy.

The observed value of the critical exponent for the magnetization $\beta = 0{\cdot}26 \pm 0{\cdot}02$ (Norvell *et al.* 1969) is lower than the accepted Ising prediction, $\beta = 0{\cdot}312$ (see § 4.4). An accurate value for the amplitude was not reported. The critical exponents associated with the staggered susceptibility and the correlation length, also determined by these authors, will likewise be discussed in § 4.4. Moreover, in that section the results of the measurement of the specific heat singularity are compared with Ising model predictions.

Lastly we note that DAG is a suitable material for a study of the field dependence of the thermodynamic properties, since the fields needed to bring the system into the paramagnetic phase (for $T < T_c$) are lower than 10 kOe. Elaborate measurements have been reported by Landau *et al.* (1971), from which some very interesting features have emerged (see § 4.5).

$DyPO_4$

This compound has the tetragonal zircon structure in which the magnetic ions form a diamond-like lattice. The extensive study of this compound by Wright *et al.* (1971) includes magnetic susceptibility, heat capacity and optical measurements. In addition the temperature dependence of the sublattice magnetization has been obtained by Rado (1969) from the magnetoelectric susceptibility. The nature of the magnetic structure (antiferromagnetic alignment along the *c* axis) was confirmed by the neutron diffraction studies of Scharenberg and Will (1971), and of Fuess *et al.* (1971).

The excellent agreement between the experimental data and the theoretical results of Essam and Sykes (1963) and Sykes *et al.* (1965) on the diamond $S=\frac{1}{2}$ Ising lattice (nearest-neighbour interactions only ; $z=4$) is illustrated in fig. 50 (*a*), (*b*), (*c*), where the plots of the heat capacity, the parallel susceptibility and the sublattice magnetization have been reproduced from the references cited. As mentioned above, the theoretical information is obtained from high and low-temperature series expansions. An attractive point of the diamond lattice, in this respect, is that the convergence of the low-temperature expansions appears to be much faster than for other 3-d lattices. Discussion of the critical behaviour of the specific heat will again be postponed to § 4.4.

It must be stipulated that in the case of the specific heat and the susceptibility the only adjustable parameter in the series expansions is the effective exchange J/k. In the present case J/k was determined by comparing the observed exponential decay of C_m and $\chi_{\parallel}$ below T_c with the theoretical expressions for the limiting low-temperature forms of these quantities $[\sim \exp(-z|J|/kT)]$. The values obtained in this way are wholly consistent and in agreement with that derived from the optical work. As a final check one may compare the experimentally obtained critical temperature $T_c/\theta = 0{\cdot}678 \pm 0{\cdot}01$ (using $T_c = 3{\cdot}390$ K and $J/k = -2{\cdot}50$ K) with the theoretical value 0·676. As concerns the sublattice magnetization, comparison with theory in fig. 50 (*c*) involves the low-temperature expansion for $T/T_c < 0{\cdot}8$, and in the critical region a fit to the power law taking for the exponent $\beta = 0{\cdot}314$ and for the amplitude $B = 1{\cdot}661$, values that are within the uncertainty of theory (see also § 4.4).

Before leaving $DyPO_4$ the question must be posed why in fact there is over the whole temperature range such a striking agreement with the

Fig. 50

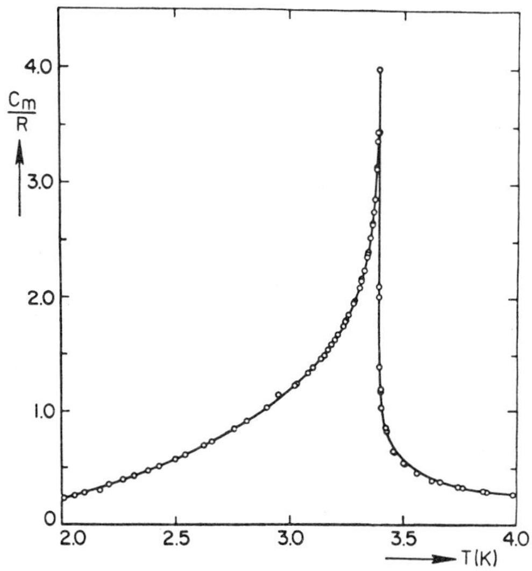
4.0
$\frac{C_m}{R}$
3.0
2.0
1.0
0
2.0
2.5
3.0
3.5
4.0
T(K)

(*a*)

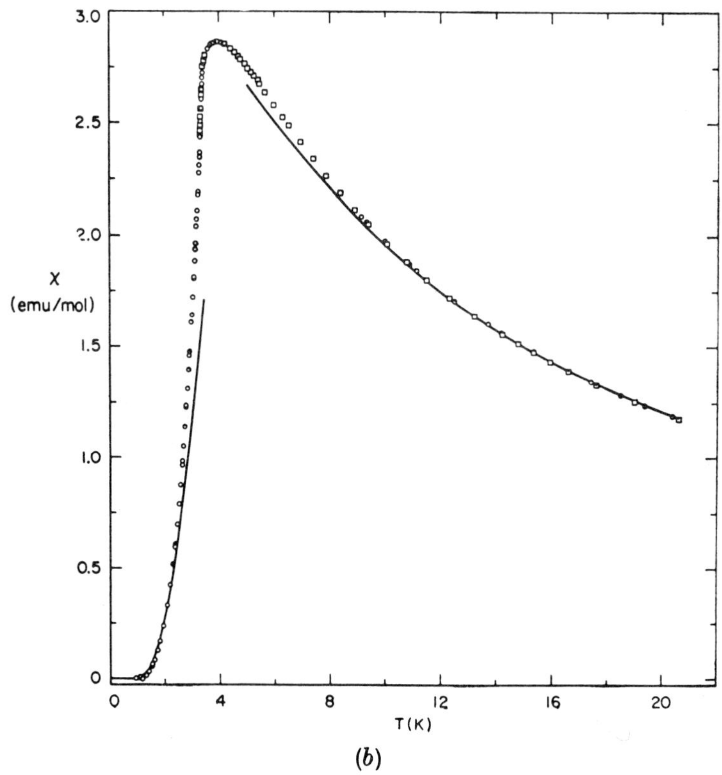
3.0
2.5
2.0
χ
(emu/mol)
1.5
1.0
0.5
0
0
4
8
12
16
20
T(K)

(*b*)

Fig 50 (*continued*)

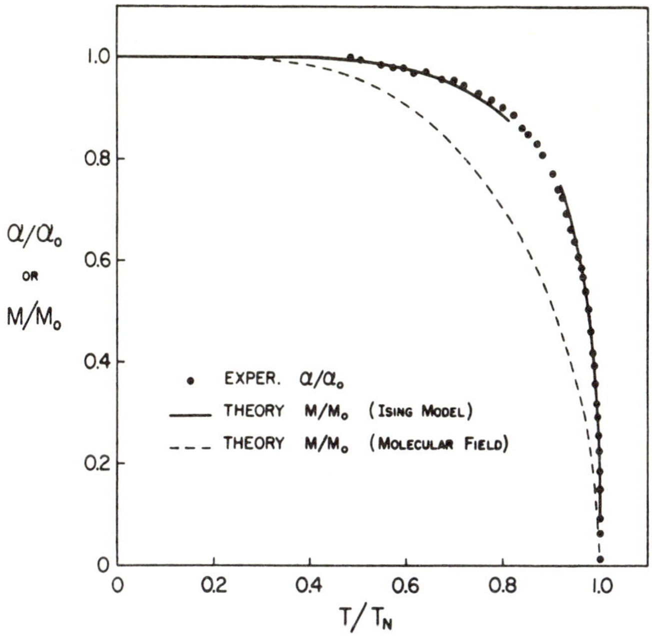

(*c*)

Magnetic and thermal data on $DyPO_4$ compared with predictions for the $S=\frac{1}{2}$ Ising model on the diamond lattice ($z=4$). For the references see the text. (*a*) Magnetic specific heat. The solid curve is not a line drawn through the experimental data but is the theoretical prediction obtained from high and low-temperature series expansions ! (*b*) Antiferromagnetic parallel susceptibility. The solid curves are the predictions of the high and low-temperature expansions for $\chi T/C$. (*c*) Temperature dependence of the spontaneous magnetization derived from the magnetoelectric susceptibility. The solid curves are the low-temperature expansion and the power-law behaviour with $\beta=0{\cdot}314$ and $B=1{\cdot}661$.

nearest-neighbour only Ising model, in view of the substantial contributions to the magnetic interaction from the long-ranged dipole interactions that will exist in this material. Indeed, about 50% of the nearest-neighbour interaction constant J/k cited above is of dipolar origin. The only way out of the dilemma seems to be contained in the conclusion drawn by Wright *et al.* (1971) from their experiments, namely, that the dipolar and exchange interaction of a given ion with neighbours other than first tend to cancel. They found that, although the individual interactions with further neighbours are considerable, their sum amounts to only a few per cent of the nearest-neighbour J/k. By this fortunate coincidence $DyPO_4$ distinguishes itself from the other two dysprosium compounds, in which large further neighbour interactions have to be reckoned with.

In addition to the above-mentioned examples there are some other more or less anisotropic materials that have been analysed in terms of the 3-d Ising model. For instance, the specific heat of CoF_2, which has a ground doublet (effective spin $\frac{1}{2}$) with fairly anisotropic g values, is in reasonable agreement with 3-d Ising predictions (Stout and Catalano 1955, Wielinga 1971). A complication here is the fairly high T_c value (37·70 K) and the presence of higher energy levels that have a relatively small separation from the lowest doublet. In any case, we expect that table 11 comprises most of the clear-cut examples of the 3-d Ising antiferromagnet with $S=\frac{1}{2}$ known at present.

Turning now to Ising-like compounds of higher (effective) spin value, there exist some Ni^{2+} compounds ($S=1$) and Fe^{2+} compounds ($S=2$), that receive consideration. However, we repeat here the warning given in § 2.2 that in these materials with higher spin values the anisotropy is not as complete as in the Dy and Co compounds. As examples we mention $Ni(CN)_2NH_3C_6H_6$, in which the single-ion anisotropy and the exchange were found to be about equal (Takayanagi and Watanabe 1970). Furthermore, a number of Fe^{2+} compounds with effective $S=2$ typically have an anisotropy that is about twice as large as the exchange. Such a value for the anisotropy is certainly large enough for the critical behaviour to be Ising-like. On the other hand, the spin-wave dispersion will still be of the anisotropic Heisenberg form, the energy not being independent of the wave-vector as in the Ising limit. Moreover at temperatures $T \gg T_c$ the Ising $S=2$ formalism will break down.

As our last example we will therefore consider FeF_2, which has also been the subject of much research during the past 20 years. FeF_2 has the rutile crystal structure and is isomorphous to MnF_2 and CoF_2. In the body-centred tetragonal magnetic lattice the moments are aligned along the c axis, with the moments at the cell corners antiparallel to the central spin. The a_0/c_0 ratio is about 1·4. In FeF_2 the splitting of the ^5D state by a cubic field results in an orbital triplet lying lowest. The orbital degeneracy of this level is removed by an orthorhombic distortion, leaving an orbital singlet as the lowest level, with five-fold spin degeneracy. Spin–orbit coupling (and spin–spin interactions) will further split this orbital state, but since the first excited level is about 1600 K away, whereas the spin–orbit coupling constant is only $\simeq 90$ K, one may describe the ground state by an $S=2$ spin Hamiltonian. The anisotropy is mainly uniaxial (DS_z^2) and the exchange term contains small contributions from next-nearest neighbours in addition to the (antiferromagnetic) nearest-neighbour interaction. Hutchings *et al.* (1970) have measured the spin-wave dispersion, finding the interactions other than between nearest neighbours to be only about 5% of the nearest-neighbour exchange. From their results one can calculate an effective exchange $J/k=-2·69$ K, leading to a Curie–Weiss $\theta=86$ K, which compares favourably with $\simeq 82$ K as measured by Foner (1964). For the anisotropy constant Hutchings *et al.* obtained $D/k=-4·65$ K.

The transition temperatures vary slightly in the various publications, but most agree to $T_c = 78{\cdot}3 \pm 0{\cdot}2$ K. Such a high value is a disadvantage in specific heat measurements, since it will be difficult to separate the magnetic specific heat from the lattice contribution. In spite of this, Stout and Catalano (1955) have succeeded in making a reliable subtraction of the lattice part, mainly because they were able to estimate the phonon contribution from the heat capacity of the isomorphous, non-magnetic zinc compound. According to their results, about 87% of the total magnetic entropy $R \ln (2S+1)$ is gained below T_c, which may be compared to the value of 92% predicted for the f.c.c. Ising lattice with $S=2$ (Domb and Miedema 1964). For the b.c.c. Ising model there are no theoretical calculations available, but one may obtain an estimate of about 88·5% by considering the difference between f.c.c. and b.c.c. for $S=\frac{1}{2}$.

Another critical parameter that can be compared with theory is the transition temperature T_c/θ. With $T_c = 78{\cdot}3$ K and $\theta = 86$ K we derive $T_c/\theta = 0{\cdot}91$. The value for Ising b.c.c. with $S=2$ may be obtained in a similar manner as above from the f.c.c. result (0·864), giving $T_c/\theta = 0{\cdot}84$. The rather large difference might be due to a temperature dependence of the exchange, since θ was calculated from interaction energies measured at 4·2 K. Analyses of high-temperature data (χ, C_m) by Lines (1967 a) (see also Domb and Miedema 1964), although subject to large errors, have indeed yielded a higher (6%) estimate for the exchange, bringing down T_c/θ to 0·86. In particular the magnetic energy parameters are not in agreement with Ising model predictions (in contrast to the entropy parameters), and point to a larger value for θ. This may be understood from the expected failure of the Ising $S=2$ formalism at high temperature, mentioned above.

On the whole, however, FeF_2 can be considered as being a fairly good example of a b.c.c. Ising model with $S=2$. To illustrate the effect of a change in spin value, we have therefore compared in fig. 51 its magnetic specific heat with that of $CoCs_3Cl_5$, which as we have seen above approximates the $S=\frac{1}{2}$ b.c.c. Ising model. One may observe that the experiment neatly confirms the expected increase in C_m with S. The effects of short-range order, reflected in the high-temperature tail, are predicted to become less important by increasing S and this is also apparent from fig. 51.

The specific heat in the critical region has been studied by Salamon and Ikushima (1971). The critical behaviour was found to be similar to that observed in compounds with lower S (see § 4.4). Other experiments on FeF_2 include Mössbauer (Wertheim and Buchanan 1967), ultrasonic (Shapira 1970), N.M.R. (Gottlieb and Heller 1971) and neutron diffraction studies (Hutchings *et al.* 1972 a). Some of these bear upon the critical behaviour and will also be mentioned in § 4.4.

In concluding the discussion of highly anisotropic 3-d magnets we mention the only Ising-like ferromagnets that to our knowledge have been found so far, namely the group of compounds $R(OH)_3$ with $R = Tb$, Dy

Fig. 51

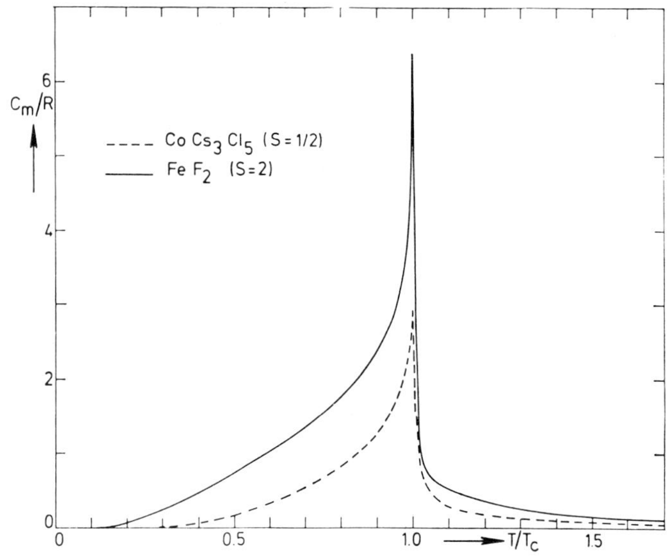

Comparison of the specific heats of two highly anisotropic materials with different S. The (effective) number of nearest neighbours is eight in both cases. This figure illustrates the qualitative differences arising from a change in S. For references to the data see the text.

and Ho (Wolf *et al.* 1968). Of these, $Tb(OH)_3$ seems to be the best approximation of the Ising model, the first excited level lying 170 K above the ground doublet (Scott *et al.* 1969). In the case of Dy and Ho, specific heat measurements indicated the existence of lower-lying excited levels (Meissner and Wolf 1969). The transition temperatures are all within the liquid helium region ($T_c = 3{\cdot}71$ K for $Tb(OH)_3$).

From the evaluation of the dipolar and exchange interactions in $Tb(OH)_3$ by Skjeltrop and Wolf (1971), it turns out that the range dependence of the interactions is quite complex, due to cancellations between dipolar and non-dipolar contributions within the various shells of neighbours on the one hand and between the total contributions of these shells on the other. For instance, the dipolar part of the first neighbour interaction ($z_1 = 2$) is nearly wholly cancelled by the exchange contribution. The second neighbour interactions ($z_2 = 6$) appear to be predominating, but further neighbour contributions are substantial, to say the least. Therefore, although the interactions are of the Ising type, one would not expect to find agreement with a nearest-neighbour only model, except perhaps very close to T_c. Indeed, the critical behaviour of the ferromagnetic susceptibility, which according to Wolf *et al.* (1968) obeys a power law with a critical index γ near to the 3-d Ising value of 1·25, seems to be confined to a rather narrow range above T_c. The value of the Curie–Weiss constant obtained by Skjeltrop and Wolf is $\theta = 4{\cdot}47$ K, in agreement with the results deduced from susceptibility, caloric and optical measurements (Scott and Wolf 1969). This leads to $T_c/\theta = 0{\cdot}83$, pointing to a fairly high effective coordination number.

Next we turn to the representatives of the 3-d Heisenberg model. Evidently, since a large number of materials exist that are fairly isotropic, one has to choose a certain maximum value for the anisotropy allowed for a compound in order to be considered a reasonable approximant of the Heisenberg model. Although there is a certain amount of arbitrariness in such a choice, we have confined ourselves to substances in which the ratio H_A/H_E of anisotropy to exchange field is not larger than about 1%. In some respects this is already too large a value, for instance, as far as the critical behaviour is concerned. In the preceding pages it has been mentioned that the presence of anisotropy will cause a changeover from Heisenberg to Ising character if T_c is approximated closely enough. The distance of the changeover point (or region) from T_c depends on H_A/H_E and theoretical work shows that in the case of an anisotropy of 1% the changeover already occurs outside the critical region, which for the magnetization and the susceptibility extends to $1 - T_c/T \simeq 10^{-2}$ and 10^{-1}, respectively. On the other hand, 1% will be low enough to justify comparisons of the general appearance of the specific heat curve with Heisenberg model predictions.

The examples that remain after applying this criterion have been compiled in table 12. It may be seen that (thanks to the existence of the cubic perovskite structure) we have been provided with two extremely isotropic materials, $KNiF_3$ and $RbMnF_3$, having $S = 1$ and $S = \frac{5}{2}$, respectively. For $S = \frac{1}{2}$ we are not quite as fortunate, which originates for most part from the fact that Cu^{2+} compounds are particularly notorious for their Jahn–Teller distortions (e.g. $KCuF_3$).

$CuCl_2 . 2H_2O$

This is the only example of a 3-d antiferromagnetic Cu compound that we could find (Cu compounds tend to be ferromagnetic). At the same time it is one of the earliest investigated antiferromagnets. Although by no means an ideal example, we shall argue that its behaviour agrees qualitatively with what is expected for a $S = \frac{1}{2}$, 3-d Heisenberg system. In this respect the present analysis is a bit different from earlier treatments (Marshall 1958, Nagai 1963, Oguchi 1955, and Hewson *et al.* 1965), in which the effects of short-range order as observed in the specific heat and susceptibility were ascribed mainly to a chain-like character. However, it appears that the relative importance of short-range order is certainly not larger than expected from the calculations for a 3-d $S = \frac{1}{2}$ Heisenberg antiferromagnet. That these effects are larger than commonly observed in 3-d systems, arises merely from the low spin value. There is thus no *a priori* reason to assume a chain model in order to explain the observed behaviour.

For instance, the specific heat, reproduced in fig. 52 from data of Clay and Staveley (1966), shows no sign of a ‘ shoulder ’ above the peak at T_c,

Table 12. Critical energy and entropy parameters of theoretical 3-d Heisenberg models. In addition to the nearest-neighbour $S=\frac{1}{2}$ and $S=\infty$ models (1) the equivalent second (1, 2) neighbour models with $S=\frac{1}{2}$ have been listed. The values refer to ferromagnets. In case of T_c/θ, the values for antiferromagnets ($S=\frac{1}{2}$) have been added (plus and minus signs). The additional numbers in parentheses give the number of equivalently interacting magnetic neighbours. The theoretical values may be found in Rushbrooke *et al.* (1973). For references to the experimental data see the text.

Model or compound	S	T_c (K)	J/k (K)	T_c/θ	S_c/R	$\frac{(S_\infty - S_c)}{R}$	$\frac{(S_\infty - S_c)}{S_c}$	$\frac{-E_0}{RT_c}$	$\frac{-E_c}{RT_c}$	$\frac{(E_c - E_0)}{RT_c}$	$\frac{-E_c}{(E_c - E_0)}$
Heisenberg, s.c. (1) ($z=6$)	$\frac{1}{2}$			0·56 (+) 0·64 (−)	0·43	0·26	0·60	0·89	0·60	0·30	2·03
Heisenberg, b.c.c. (1) ($z=8$)	$\frac{1}{2}$			0·63 (+) 0·70 (−)	0·45	0·24	0·53	0·79	0·46	0·33	1·39
Heisenberg, f.c.c. (1) ($z=12$)	$\frac{1}{2}$			0·67 (+) 0·72 (−)	0·46	0·23	0·50	0·75	0·43	0·31	1·38
Heisenberg, s.c. (1, 2) ($z=18$)	$\frac{1}{2}$			0·76 (+)	0·50	0·19	0·38	0·66	0·31	0·35	0·88
Heisenberg, b.c.c. (1, 2) ($z=14$)	$\frac{1}{2}$			0·71 (+)	0·48	0·21	0·44	0·72	0·36	0·36	1·00
Heisenberg, f.c.c. (1, 2) ($z=18$)	$\frac{1}{2}$			0·76 (+)	0·50	0·19	0·38	0·66	0·31	0·35	0·88

Heisenberg, s.c. (1) ($z=6$)	∞			0·72 (+)		0·42		2·07	0·69	1·38	0·50
Heisenberg, b.c.c. (1) ($z=8$)	∞			0·77 (+)		0·34		1·95	0·53	1·43	0·37
Heisenberg, f.c.c. (1) ($z=12$)	∞			0·79 (+)		0·31		1·89	0·47	1·42	0·33
Antiferromagnets											
$CuCl_2 . 2H_2O$	$\frac{1}{2}$	4·36	?	?	0·43	0·23	0·54				
NdGaG	$\frac{1}{2}$	0·516	$\simeq -0·34$	$\simeq 0·76$	0·46	0·26	0·58	0·91	0·55	0·36	1·58
SmGaG	$\frac{1}{2}$	0·967	$\simeq -0·60$	$\simeq 0·81$	0·42	0·27	0·65	0·87	0·53	0·34	1·53
$KNiF_3$	1	246	−44	0·72							
$RbMnF_3$	$\frac{5}{2}$	83·0	−3·40	0·70							
MnF_2	$\frac{5}{2}$	67·33	−1·76	0·79	1·53	0·26	0·17				
Ferromagnets											
$Cu(NH_4)_2Cl_4 . 2H_2O$	$\frac{1}{2}$	0·701	0·23	0·77							
$CuK_2Cl_4 . 2H_2O$	$\frac{1}{2}$	0·877	0·30	0·74	0·47	0·22	0·46	0·73	0·39	0·33	1·18
$CuRb_2Cl_4 . 2H_2O$	$\frac{1}{2}$	1·02									
$Cu(NH_4)_2Br_4 . 2H_2O$	$\frac{1}{2}$	1·83	0·60	0·74							
$CuRb_2Br_4 . 2H_2O$	$\frac{1}{2}$	1·87	0·63	0·74							
EuO	$\frac{7}{2}$	69		0·80				1·75	0·51	1·24	0·41
EuS	$\frac{7}{2}$	16·4		0·81							

Fig. 52

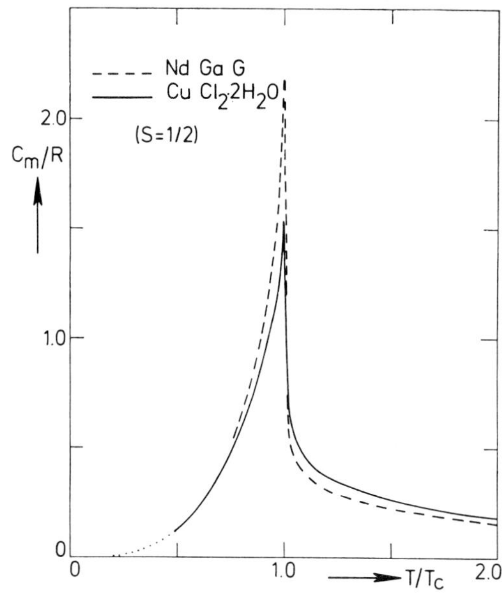

Specific heats of $CuCl_2 . 2H_2O$ and neodymium gallium garnet (NdGaG). Both materials have $S=\frac{1}{2}$ and are fairly isotropic, in particular the Cu compound. For references to the data see the text.

which would be indicative of a pronounced chain character. The entropy change above T_c is about 33% of $R \ln 2$ (Friedberg 1952), which is of the same order as (but even less than) expected for the Heisenberg s.c. ferromagnet (38%). The difference between T_c and the temperature T_{max} of the maximum in $\chi_{\parallel}$ was reported to be about 15% of T_{max} (Van der Marel *et al.* 1955). De Jongh (unpublished) has used the high-temperature series expansion for the $S=\frac{1}{2}$, s.c., Heisenberg antiferromagnet (Baker *et al.* 1967 b) to locate the susceptibility maximum. In combination with the prediction for the transition temperature given by Rushbrooke and Wood (1963), the difference is found to be 20% of T_{max} (see also below). Again the agreement is satisfactory.

Nevertheless the exchange will likely be a bit smaller in one crystallographic direction. Indication of a chain character may be deduced from the orthorhombic crystal structure, in which chains of octahedra linked by edges are found along the *c* axis. The copper ions are in the centres of the octahedra and each ion has its two nearest neighbours along the *c* axis, being connected with these via two Cu–Cl–Cu paths. The superexchange paths connecting it with the four next-nearest neighbours, which are in the *a*–*b* plane, involve at least two ligands (Cl or O). The nearest-neighbour interaction is antiferromagnetic, the (net) coupling within the *a–b* plane is ferromagnetic. This leads to the picture of antiferromagnetic chains coupled by a weaker (ferromagnetic) coupling. Alternatively one may conceive the structure as consisting of ferromagnetic layers, with an antiferromagnetic coupling J_{af} between them that is stronger than the intralayer interaction J_f. Evidence for a slight canting of the moments out of the *a–b* plane (which is favoured by the dipolar coupling) into the direction of the *c* axis has been given by Umebayashi *et al.* (1968). This effect has been predicted by Moriya and later by Joshua (1970).

The net antiferromagnetic interaction may best be estimated from the field needed to saturate the system at temperatures $T \ll T_c$. From the measurements of Van der Sluys *et al.* (1967) one deduces $H_c = (150 \pm 10)$ kOe. Considering that the experiment was performed on a powdered sample and that the orthorhombic anisotropy does not exceed 1%, one may safely put $H_c = 2H_E^{af}$, so that $H_E^{af} = (75 \pm 5)$ kOe. The g values have been measured by Gerritsen *et al.* (1955), who obtained $g_a = 2{\cdot}19$; $g_b = 2{\cdot}04$ and $g_c = 2{\cdot}25$ (see also Rao and Narasinhamurty 1963). With the g value for a powder, $g = 2{\cdot}16$, we then calculate the antiferromagnetic exchange to be $z_{af}J_{af}/k = -(11 \pm 1)$ K. Assuming two antiferromagnetic neighbours this leads to an antiferromagnetic exchange along the c axis of strength $J_{af}/k \simeq -5{\cdot}5$ K. The ferromagnetic interaction in the *a–b* plane cannot be deduced from the existing data. One may only estimate it to be about five times smaller than J_{af} in the following way. From the theoretical value $T_c/\theta \simeq 0{\cdot}6$ for a s.c., $S = \frac{1}{2}$, ferro or antiferromagnet and the measured $T_c = 4{\cdot}36$, one obtains a mean value for the interaction per magnetic neighbour of about 7/3 K. With six neighbours and $z_{af}|J_{af}|/k \simeq 11$ K, $z_{af} = 2$, this leaves a ferromagnetic interaction of about 1 K.

Evidence for zero-point spin deviations is found from the perpendicular susceptibility extrapolated to $T = 0$, in a similar way as discussed in preceding sections. Using the value $z_{af}J_{af}/k = -11$ K derived from H_c, we calculate the $\chi_\perp$ in the b direction at 0 K to be $\chi_\perp^b(0) = 3{\cdot}5 \times 10^{-2}$ cm³/mole from the MF formula : $\chi_\perp(0) = N_0 g^2 \mu_B^2/4z|J|$. This may be compared to the experimental value $\chi_\perp^b(0) = 2{\cdot}56 \times 10^{-2}$ cm³/mole found by Van der Marel *et al.* (1955) for this direction. Since the anisotropy fields are not larger than 1% of the exchange field, one is apt to explain this large reduction of (27 ± 5)% of $\chi_\perp(0)$ as being nearly wholly due to zero-point motions. Spin-wave theory (see, e.g., Keffer 1966) predicts a 25% reduction of $\chi_\perp(0)$ in the case of a s.c. isotropic antiferromagnet, so that once again the order of magnitude is correct.

The orthorhombic anisotropy may be estimated in the following way. From the spin-flop field $H_{SF} \simeq 6{\cdot}5$ kOe (Hardeman and Poulis 1955, Butterworth and Zidell 1969), the anisotropy within the *a–b* plane is calculated as $H_A^I \simeq 280$ Oe from the relations $H_{SF}^2 \simeq 2H_E^{af}H_A^I \simeq H_c H_A^I$. From AFMR results Joenk (1962) has found the out-of-plane anisotropy

H_A^{II} to be about three times as large as H_A^{I}. Thus we have $H_A^{I}/H_E \simeq 4\times 10^{-3}$ and $H_A^{II}/H_E \simeq 1{\cdot}2\times 10^{-2}$.

Summing up we may say that $CuCl_2 . 2H_2O$ has a fairly small anisotropy and behaves *grosso modo* as expected for a 3-d $S=\frac{1}{2}$ Heisenberg antiferromagnet. In spite of its imperfections it remains the best example of this particular model available at present.

NdGaG

In a recent paper, Onn *et al.* (1967) have reported specific heat measurements on a number of gallium garnets, some of which appear to approximate the antiferromagnetic, $S=\frac{1}{2}$, Heisenberg model. We have therefore included Nd and Sm gallium garnet in table 12. From the total entropy changes, which are near to $R \ln 2$, one concludes that only the lowest doublet is populated. The critical parameters derived from the specific heat are in reasonable agreement with the predictions for a Heisenberg model with low coordination number (which is likely to be $z=4$ in these garnets). Note that the theoretical values apply to ferromagnets. Estimates of the dipolar contributions to the exchange yield rather small values.

However, it is very likely that, although the g values are not quite as anisotropic as in DAG (Wolf *et al.* 1962), the anisotropy in these garnets will still be far larger than in $CuCl_2 . 2H_2O$ (quantitative values have not yet been obtained). This will be the explanation for the fact that the specific heat curve of NdGaG, also plotted in fig. 52, lies above the result for the Cu compound. The latter is therefore most probably a better approximation.

$KNiF_3$

This cubic perovskite is an extremely close approximation of a 3-d nearest-neighbour only Heisenberg system. Although not yet known quantitatively, the anisotropy will be very small. Dipolar contributions cancel because of the cubic symmetry, which is retained also at low temperatures (Okazaki and Suemune 1961 b, Scatturin *et al.* 1961). As concerns the crystal-field anisotropy, we may compare $KNiF_3$ with K_2NiF_4, since in both cases the Ni^{2+} ions are surrounded by an octahedron of F^- ions. In K_2NiF_4 this octahedron has a small tetragonal distortion of about 1·5%, which, as we have seen in the above pages, gives rise to a uniaxial anisotropy of the order of 10^{-3}. Since for $KNiF_3$ there is no indication of such an effect, one expects the anisotropy to be at least one or two orders of magnitude smaller. Furthermore, there exists evidence that next-nearest neighbour interactions are a mere 5×10^{-3} of the nearest-neighbour exchange (Yamaguchi and Sakamoto 1969). The only setback is the high position of the transition point, $T_c = 246$ K (Nouet *et al.* 1972) which greatly hampers an accurate determination of the magnetic specific heat.

Lines (1967 b) has given an extensive discussion of $KNiF_3$ and has been able to obtain a fairly accurate estimate of the exchange by analysing the measured powder susceptibility (Hirakawa *et al.* 1960) in terms of the high-temperature series expansion. From the fit of the paramagnetic susceptibility to this expansion he obtains $J/k = 43 \pm 2$ K. The series expansion also indicated the position of the maximum in the susceptibility. Relating this to the experimentally observed $T(\chi_{max}) = 275$ K, Lines finds $J/k \simeq 45$ K. Furthermore, with $J/k = 43$ K, the experimental $T_c/\theta = 0{\cdot}716$ is in good agreement with the prediction $T_c/\theta = 0{\cdot}721$ for an s.c. Heisenberg antiferromagnet with $S = 1$ (Rushbrooke and Wood 1963, Rushbrooke *et al.* 1973), also derived from analyses of series expansions. Although Chinn *et al.* (1971) obtained $J/k = -50{\cdot}8 \pm 0{\cdot}6$ K from the analysis of their two-magnon Raman scattering experiments, we will adhere to the value $J/k = -44$ K, since also in the case of K_2NiF_4 the J/k determined by these authors was more than 10% higher than the other results (see above).

Also in the case of $KNiF_3$, having $S = 1$, one should expect a substantial reduction of the $\chi_\perp(0)$ arising from zero-point motions. For $S = 1$, spin-wave theory predicts a 13% reduction. Unfortunately there is only a powder susceptibility measurement available, but one may put $\chi_p(0) = \frac{2}{3}\chi_\perp(0) + \chi_{v.v.}$ and calculate the experimental $\chi_\perp(0)$ from the measured $\chi_p(0) = 9{\cdot}0 \times 10^{-6}$ cm^3/g and the temperature independent Van Vleck term $\chi_{v.v.} = 2{\cdot}4 \times 10^{-6}$ cm^3/g, as determined by Lines (this will include the diamagnetic contribution). The result $\chi_\perp(0) = 1{\cdot}53 \times 10^{-3}$ cm^3/mole may be compared with the value $\chi_\perp(0) = N_0 g^2 \mu_B{}^2/4z|J| = 1{\cdot}80 \times 10^{-3}$ cm^3/mole calculated with $g = 2{\cdot}25$ and $J/k = 44$ K, yielding an apparent reduction of $\simeq 15\%$, in reasonable agreement with expectation, considering the uncertainties involved.

Another test of zero-point spin deviation in this material has been accomplished by Hutchings and Guggenheim (1970) who, from neutron diffraction measurements, deduced the effective moment to be $\langle S_z \rangle = 0{\cdot}851$ ($\pm 0{\cdot}050$). In this experiment the reduction of the observed moment arises from the combination of zero-point effects (prediction: $\langle S_z \rangle = S - 0{\cdot}078$ for zero anisotropy) and of covalency (prediction still rather uncertain). As regards the order of magnitude, the observed reduction ($\simeq 15\%$) of the magnetic moment is in accord with theoretical calculations that take into account both covalency and zero-point effects. The experimental accuracy, however, was too limited to enable a choice between the different theoretical estimates of both effects.

$RbMnF_3$

This isomorphous compound is at least as ideal as $KNiF_3$. Its anisotropy has been measured and is a minute $H_A/H_E \simeq 5 \times 10^{-6}$! From X-ray studies Teaney *et al.* (1966) concluded that departures from cubic symmetry larger than a few parts in 10^5 were not present. Hardly any thermal expansion effect is seen at T_c ($= 83{\cdot}0$ K). Second-neighbour interactions will be of the same order as in $KNiF_3$. In short: $KNiF_3$

and $RbMnF_3$ are most certainly the best approximations of the nearest-neighbour only Heisenberg model known at present.

The spin-wave dispersion relation in $RbMnF_3$ has been measured by Windsor and Stevenson (1966) with neutron diffraction. It has been reproduced in fig. 58 of § 4.2 as being the best example available of the isotropic antiferromagnetic dispersion relation. The anisotropy gap is so small (0·4 K) that it is not discernible, so that all the data essentially fall on a sine curve, representing the behaviour in the absence of anisotropy. The authors found no detectable next-nearest neighbour interactions, and determined the nearest-neighbour exchange as $J/k = -3{\cdot}4 \pm 0{\cdot}3$ K, in agreement with earlier determinations. From AFMR experiments, Teaney *et al.* (1962) and Freiser *et al.* (1963) calculated the anisotropy, which corresponds to a field of 4·5 Oe only (four-fold symmetry).

A more accurate value for J/k may be deduced from the susceptibility data (single crystal) of Breed (1969). Due to the small value of H_A, the critical field is only about 3 kOe. Accordingly, for fields exceeding this value measurements in any direction will yield the perpendicular susceptibility, since also with the field parallel to the easy axis the moments will have swung to the perpendicular orientation. This of course mimics the behaviour of an ideally isotropic Heisenberg antiferromagnet, which does not differentiate between parallel and perpendicular, since the $\chi_{\parallel}$ is only defined for $H_A \neq 0$. Evidently this is the reason why one expects the χ of the ideal isotropic model to be identical (at least at $T=0$) to the $\chi_{\perp}$ as given by the spin-wave theory in the limit $H_A \rightarrow 0$.

Breed's experimental results in fields $H > H_{SF}$ are shown in fig. 53 (*a*), where they have been fitted to the series expansion in the high-temperature region. The H.T.S. curves for $S=\frac{5}{2}$ and $S=\frac{1}{2}$ shown here have been calculated (De Jongh, unpublished) using the coefficients of the s.c. lattice (7 and 10, for $S=\frac{5}{2}$ and $\frac{1}{2}$, respectively) given by Rushbrooke and Wood (1958), Stephenson *et al.* (1968) and Baker *et al.* (1967 b). Most gratifyingly, the maximum in χ is indicated by the series ($T_{max}/\theta = 0{\cdot}80$ and 0·77 for $S=\frac{1}{2}$ and $\frac{5}{2}$, respectively). Below the maximum the inaccuracy of the predictions (obtained by extrapolation to an infinite number of terms) increases very rapidly, so that the apparent agreement between theory and experiment for $T_c < T < T_{max}$ for $RbMnF_3$ may be fortuitous. The T_c's indicated in the figure ($T_c/\theta = 0{\cdot}64$ and 0·72 for $S=\frac{1}{2}$ and $\frac{5}{2}$, respectively) are theoretical values for the s.c. lattice obtained by Rushbrooke and Wood (1963), that should be accurate within a few per cent. The experimental value for $RbMnF_3$ is $T_c/\theta \simeq 0{\cdot}70$, using $T_c = 83{\cdot}0$ K (Teaney *et al.* 1966, Golding 1971) and $J/k = 3{\cdot}40 \pm 0{\cdot}05$ K derived from the fit of the χ data for $T < 120$ K†. The horizontal arrows labelled *a* and *b* in fig. 53 indicate the values for $\chi_{\perp}(0)$ predicted by the spin-wave theory (see Keffer 1966) for the s.c. lattice with $S=\frac{5}{2}$ and $S=\frac{1}{2}$, respectively. One may see that good agreement is obtained

† For $T > 100$ K a systematic decrease of J with T was found, that can be correlated to the observed thermal expansion (Teaney *et al.* 1966).

Fig. 53

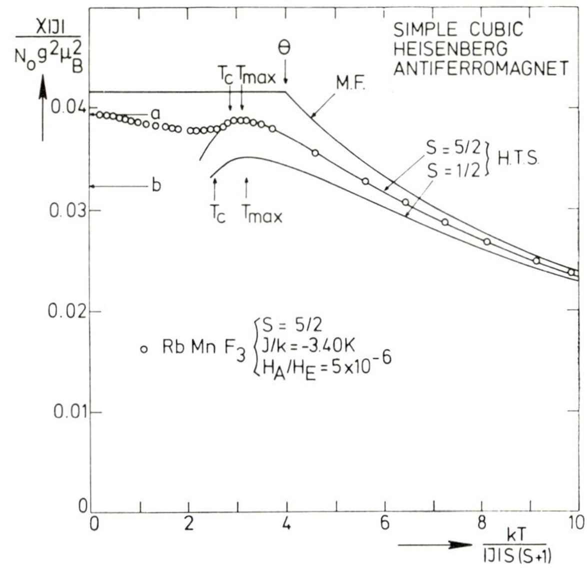

(*a*)

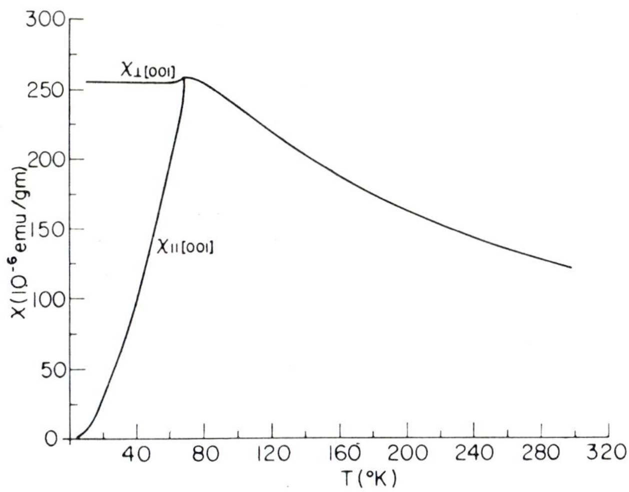

(*b*)

(*a*) Measurements of the (perpendicular) susceptibility of $RbMnF_3$ by Breed (1969), fitted to the high-temperature series expansion (H.T.S.) for the s.c., $S=\frac{5}{2}$, Heisenberg antiferromagnet. The molecular field prediction (MF) and the H.T.S. result for $S=\frac{1}{2}$ are also shown. The horizontal arrows *a* and *b* indicate the spin-wave predictions for $\chi_{\perp}(0)$. The vertical arrows indicate the predicted positions of the maxima in the susceptibility (T_{max}) and of the transition temperatures (T_c). Both follow from H.T.S. expansions, the latter having been reported by Rushbrooke and Wood (1963). (*b*) The perpendicular and parallel susceptibilities of MnF_2, which is a typical example of a fairly isotropic 3-d antiferromagnet.

for $RbMnF_3$, and that the predicted reduction of $\chi_{\perp}(0)$ with respect to the MF theory (about 5% for $S=\frac{5}{2}$) is verified, the experimental error in χ being of the order of 1%. As outlined in the preceding pages, this reduction is the consequence of zero-point spin deviations. Another test of the spin-wave theory is provided by the work of Montgomery (1966) who found clear evidence for a T^3 dependence of the magnetic specific heat at low temperatures, in accordance with simple spin-wave theory (see fig. 66, § 4.2).

Critical indices for the staggered susceptibility, sublattice magnetization and inverse correlation range have been deduced by Lau *et al.* (1969) from inelastic neutron scattering near T_c. Together with the indices for the specific heat reported by Teaney (1966) and Golding (1971) they form a more or less complete set of critical parameters that is of great value in testing Heisenberg model and scaling law predictions (§ 4.4).

MnF_2

Similarly extensive information regarding the critical indices is available for MnF_2 (Schulhof *et al.* 1970, 1972, Heller 1966, Teaney 1965). By comparing the results for the two manganese compounds one may study the influence of anisotropy, since this is much larger in MnF_2 ($H_A/H_E \simeq 1{\cdot}6 \times 10^{-2}$). This matter will be taken up in § 4.4.

The (rutile) structure of MnF_2 is body centred tetragonal, the direction of the moments being along the c axis. ($a_0 = 4{\cdot}87$ Å ; $c_0 = 3{\cdot}31$ Å). The main interactions are a weak ferromagnetic interaction along [001] and an antiferromagnetic interaction along [111] of magnitude $J/k = -1{\cdot}76$ K. Apparently the exchange paths are such as to make the coupling between nearest neighbours (along the c axis) much weaker than that between the central spin and those at the cell corners (next-nearest neighbours). The accurate value for J/k has been calculated by Trapp and Stout (1963) from their measurements of the perpendicular susceptibility (taking into account zero-point spin deviation). The estimate $0{\cdot}3 \pm 0{\cdot}1$ K for the exchange along the c axis (and $\simeq 0$ K for the exchange along [100] and [010]) follows from the work of Brown *et al.* (1961) and Okazaki *et al.* (1964), who used paramagnetic resonance and neutron diffraction techniques, respectively.

The (uniaxial) anisotropy may be calculated from the spin-wave gap as measured by zero-field, zero-temperature AFMR by Johnson and Nethercot (1959). They found $\omega/\gamma = (9{\cdot}33 \pm 0{\cdot}05) \times 10^4$ Oe, in excellent agreement with the value $9{\cdot}3 \pm 0{\cdot}2 \times 10^4$ Oe obtained by Jacobs (1961) for the spin-flop field. With the aid of the formula $(\omega/\gamma)^2 = 2H_E H_A - H_A{}^2$ and the above-mentioned value for the antiferromagnetic exchange, one deduces $H_A = 8220$ Oe ($H_A/H_E \simeq 1{\cdot}6 \times 10^{-2}$), which should be accurate within 2%. It turns out that the anisotropy is for the most part of dipolar origin. Keffer (1952) calculated the dipolar contribution to be 8300 Oe. Correction for zero-point reduction (2·4%) reduces this to 8100 Oe. The remaining part of about 100 Oe, due to crystal-field effects, is considerably smaller than Keffer's estimate of 500 Oe, yet it is about twice as large as that reported for K_2MnF_4 (Folen 1972).

The critical temperature is $T_c = 67{\cdot}34$ K (Heller 1966), although values differing slightly from sample to sample have been reported (Teaney 1965). With the exchange constants given above one calculates $T_c/\theta \simeq 0{\cdot}79$, which may be compared with the prediction 0·75 for the b.c.c. Heisenberg antiferromagnet with $S=\frac{5}{2}$ (Rushbrooke and Wood 1963). The entropy gained at T_c is 85·4% of $R \ln 6$ (Stout and Catalano 1955), whereas for the b.c.c. *ferro*magnetic $S=\frac{5}{2}$ Heisenberg model the value is 81% (Rushbrooke *et al.* 1973). It is not likely that these discrepancies may be attributed to the anisotropy. Dalton and Wood (1967) have studied the influence of anisotropy on the critical parameters of $S=\frac{1}{2}$ Heisenberg ferromagnets. From their results one can conclude that an anisotropy of 1–2% is too small to produce shifts of this magnitude from the pure Heisenberg values. More probably the origin will be a somewhat higher effective coordination number than $z=8$, due to the various interactions present.

The susceptibility of MnF_2 is shown in fig. 53 (*b*). The perpendicular susceptibility is quite similar to that of $RbMnF_3$. Since the spin-flop field is about 10^5 Oe, the parallel susceptibility can be easily measured in fairly high fields. The behaviour shown in fig. 53 (*b*) is typical for fairly isotropic 3-d antiferromagnets. We also mention the spin-wave analysis of the susceptibility by Kanamori and Itoh (1968) (experimental data of Trapp and Stout 1963).

Next we turn our attention to the known examples of the isotropic 3-d ferromagnet. These are also very few in number, in fact there are only two magnetic systems that receive consideration ; the series of compounds $M_2CuX_4 \cdot 2H_2O$, where M = K, Rb, Cs or NH_4 and X = Cl or Br, and EuO and EuS. As we shall see, these materials are still far from ideal, at least when one wants to make a comparison with nearest-neighbour only Heisenberg models. To the above list one may add $CrBr_3$, but with the proviso that only the critical behaviour is considered, since farther away from T_c the layered character of this compound will have its influence on the thermodynamic behaviour. For this reason we have not included $CrBr_3$ in this section, but we will mention the results obtained in the critical region in § 4.4.

$M_2CuX_4 \cdot 2H_2O$(M = K, Rb, Cs *or* NH_4 ; X = Cl *or* Br)

The salts of this general formula have a body-centred tetragonal unit cell. Each Cu^{2+} ion is surrounded by an approximate octahedron of four chlorine or bromine ions and two water molecules, the latter lying along the *c* axis, whereas the halogen ions are within the *a*–*a* plane. Since the c_0/a_0 ratios are of the order of 1·05, the magnetic structure may be considered as being approximately b.c.c.

The earlier measurements were caloric and magnetic experiments on $K_2CuCl_4 \cdot 2H_2O$ and $(NH_4)_2CuCl_4 \cdot 2H_2O$ by Miedema *et al.* (1963), subsequently extended to $Rb_2CuCl_4 . 2H_2O$ and $(NH_4)_2CuBr_4 . 2H_2O$ (Miedema *et al.* 1965). The main results of these investigations are the heat capacity data shown in fig. 54 and the spin-wave specific heat, to be discussed in § 4.2. Later publications of various authors have been concerned mainly with $(NH_4)_2CuBr_4 . 2H_2O$, in particular with the critical behaviour. This is another magnetic substance of which there is a more or less a complete set of critical indices available (see § 4.4). Lastly, the compound $Rb_2CuBr_4 . 2H_2O$ may be added to the above list of isomorphous salts.

Fig. 54

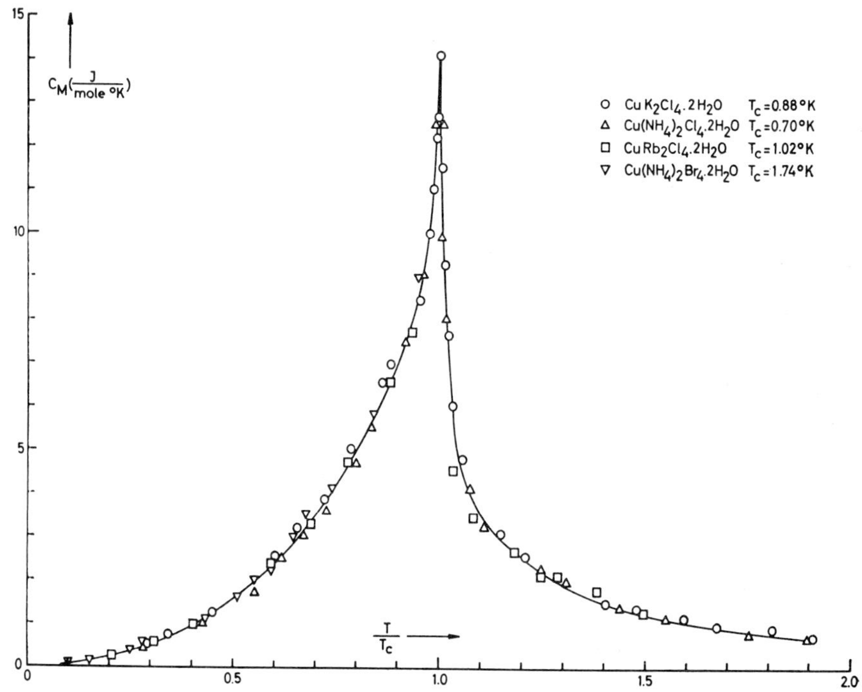

Heat capacities of four isomorphous ferromagnetic copper salts ($S=\frac{1}{2}$), the magnetic structure of which is approximately b.c.c., although with substantial next-nearest neighbour interactions. (After Miedema *et al.* 1965.)

One may observe that the specific heat data of the four isomorphous salts in fig. 54 seem to fall on a single curve, in spite of the expected differences in anisotropy and other deviations that may exist. This led Miedema *et al.* to conclude that the common curve in fig. 54 will be a good approximation of the b.c.c. Heisenberg ferromagnet with mainly nearest-neighbour interactions. However, although qualitatively the curve is indeed representative for the $S=\frac{1}{2}$ Heisenberg ferromagnet, a closer comparison of the critical parameters of the specific heat with

Heisenberg model predictions points to the presence of substantial further neighbour interactions (cf. table 12). This was already recognized by Wood and Dalton in 1966, who analysed the experimental data on $(NH_4)_2CuCl_4 \cdot 2H_2O$ and $K_2CuCl_4 \cdot 2H_2O$ and concluded the relative strength of the next-nearest neighbour interaction to be about 25% of the nearest-neighbour exchange. They did not mention the anisotropy as a possible mechanism to explain the observed shifts in the critical parameters with respect to the nearest-neighbour model. It turns out indeed that the anisotropy is too small to have an appreciable effect. For the bromine compounds the anisotropy is $H_A/H_E \simeq 7 \times 10^{-3}$ (Velu *et al.* 1972, Suzuki and Watanabe 1967, 1971), for the chlorine compounds it will be probably even less. In fact Ford and Jeffries (1966) have produced the estimate $H_a < 10$ Oe for $K_2CuCl_4 \cdot 2H_2O$, leading to $H_A/H_E < 6 \times 10^{-4}$. One may estimate the effect of an anisotropy of 1% on T_c/θ and on $-E_c/RT_c$ with the aid of the calculations of Dalton and Wood (1967) on the anisotropic Heisenberg model. Interpolating between their values one obtains an upward shift of 1% in T_c/θ and a downward shift of 3% in $-E_c/RT_c$, both with respect to the values of the fully isotropic b.c.c. ferromagnet (0·63 and 0·46 respectively). Clearly the effect is much to small too account for the experimentally found deviations from the nearest-neighbour model (table 12).

Recently, Van Amstel *et al.* (1974) have measured the (effective) exchange constants of $Cu(NH_4)_2Br_4 \cdot 2H_2O$ and $CuRb_2Br_4 \cdot 2H_2O$ and also re-examined the earlier data on J/k of the other compounds. Their results for J/k and T_c/θ are listed in table 12. Comparing these T_c/θ values with those of the equivalent neighbour Heisenberg models also given in table 12, one is apt to conclude that the further neighbour interactions are even more substantial than as estimated by Wood and Dalton. These results, combined with the critical parameters of the specific heat, point to an effective number of equivalent neighbours of at least 17. Van Amstel *et al.* point out that such a high number is quite possible in view of the crystal structure, since an examination of the superexchange paths connecting the various neighbours shows that the interactions between first, second and third neighbours may very well be comparable in strength. Summing up one may say that these materials are fairly isotropic but certainly not good examples of the nearest-neighbour only, b.c.c., Heisenberg ferromagnet. Instead the properties resemble those of the equivalent neighbour model with first and second neighbours.

The other experimental work that is not mentioned here is mostly concerned with measurements of the susceptibility and magnetization. The discussion of these papers is postponed to § 4.4.

EuO *and* EuS

These magnetic semiconductors have the rock-salt structure, in which Eu^{2+} ions form a f.c.c. lattice. The large spin $S = \frac{7}{2}$ and the very small anisotropy, which amounts to $H_A/H_E \simeq 2\times10^{-4}$ for EuS (Franzblau *et al.* 1967) and $H_A/H_E \simeq 7\times10^{-4}$ for EuO (Miyata and Argyle 1967), make a comparison with the classical Heisenberg model appropriate. A disadvantage of EuO is that the magnetic properties are rather strongly dependent on stoichiometry. McGuire *et al.* (1972) have reported evidence of considerable contributions to the exchange from excess Eu. With increasing Eu concentration the Curie–Weiss θ was found to vary from 76 K to 84 K (the θ's were determined from data above 200 K). Moreover, both EuO and EuS have the same complication as the Cu compounds discussed above, namely, the existence of substantial next-nearest neighbour interactions.

The reported values for the exchange constants vary somewhat according to the method of measurement. For EuO, Boyd (1966) has obtained $J_1/k=(0{\cdot}75\pm0{\cdot}025)$ K and $J_2/k=(-0{\cdot}098\pm0{\cdot}004)$ K from a spin-wave theoretical analysis of the magnetization. Henderson *et al.* (1970) applied a similar analysis to the low-temperature specific heat, finding $J_1/k=(0{\cdot}75\pm0{\cdot}02)$ K and $J_2/k=(-0{\cdot}084\pm0{\cdot}02)$ K. Contrastingly, Passell *et al.* (1971) deduced $J_1/k=(0{\cdot}602\pm0{\cdot}008)$ K and $J_2/k=(0{\cdot}155\pm0{\cdot}014)$ K from the spin-wave dispersion curve, as measured with neutron diffraction. Since Passell *et al.* seem to be quite sure about the positive sign of J_2, there is a clear disagreement between their result and the specific heat and N.M.R. analyses. An even larger positive value for J_2 was reported by Menyuk *et al.* (1971), who obtained $J_1/k=(0{\cdot}53\pm0{\cdot}005)$ K and $J_2/k=(0{\cdot}26\pm0{\cdot}1)$ K. The discrepancies between these various J/k values could be caused by the effect of an excess of europium, described above (for EuS there is a reasonable agreement between the J/k values of Passell *et al.* and those derived from N.M.R. and specific heat measurements, so that the methods of analysis do not seem to be at fault). Since the authors give no information about the chemical analyses of their samples it is not possible to decide which determination is best. Also the resulting T_c/θ values give no clear indication. The nearest-neighbour only f.c.c. ferromagnet with $S=\frac{7}{2}$ is predicted to have $T_c/\theta=0{\cdot}78$ (Rushbrooke and Wood 1958), for the equivalent next-nearest neighbour model the value is 0·85 (Rushbrooke *et al.* 1973). The combinations of Boyd and of Henderson *et al.* both yield $T_c/\theta=0{\cdot}78$, whereas Passell's gives 0·81 and that of Menyuk *et al.* 0·83 (in all these calculations $T_c=69$ K has been adopted). Although one expects the value to lie in between 0·78 and 0·85, probably near 0·80 in view of the ratio $|J_2/J_1|$, there is certainly no clear choice possible. We would conclude therefore that $J_1/k\simeq0{\cdot}6$ K and $|J_2/J_1|\simeq0{\cdot}2$ with evidence in favour of J_2 to be positive. The critical energy parameters listed in table 12 are taken from the work of Argyle *et al.* (1967).

Fortunately the situation is better in EuS. Callaway and McCollum (1963) obtained $J_1/k=(0{\cdot}17\pm0{\cdot}02)$ K and $J_2/k=(-0{\cdot}013\pm0{\cdot}032)$ K from the specific heat; Charap and Boyd (1964), $J_1/k=(0{\cdot}20\pm0{\cdot}01)$ K and $J_2/k=(-0{\cdot}08\mp0{\cdot}02)$ K from the magnetization. The values of Passell *et al.* (1971)

are $J_1/k = (0{\cdot}234 \pm 0{\cdot}016)$ K and $J_2/k = (-0{\cdot}098 \pm 0{\cdot}014)$ K. We conclude therefore to $J_1/k = (0{\cdot}21 \pm 0{\cdot}03)$ K and $J_2/k \simeq -0{\cdot}09$ K, yielding with $T_c = 16{\cdot}4$ K a value $T_c/\theta = 0{\cdot}81$.

Of the many interesting experiments we mention further the work of Matthias *et al.* (1961), who discovered the ferromagnetism in EuO, the N.M.R. measurements of Heller and Benedek (1965) on EuS, the thermal expansion experiment of Argyle *et al.* (1967) on EuO (critical energy parameters), the specific heat measurements of Teaney *et al.* (1968) on EuS, and the neutron diffraction work of Als-Nielsen *et al.* (1971) that produced values for the critical indices β, γ and ν. We will discuss the results of some of the above papers in the next section, in the sections on spin-wave theory (4.2) and on critical behaviour (4.4).

3.3.3. *Concluding remarks*

The above discussion concludes the list of interesting examples of simple magnetic model systems given in this section and in the preceding ones. Within the outline of the present review, 3-d systems play their role in that they are the closest approximation of the molecular field model, in accordance with the general picture given in the introduction that explains the qualitative behaviour in terms of correlation functions. It is gratifying to observe how close the agreement between theory and experiment is on this point. Thus we have seen how in 3-d antiferromagnets the difference between T_c and the temperature at which the maximum in χ occurs is typically of the order of 10% of $T(\chi_{max})$, whereas it is a huge 50% in the 2-d cases. Likewise, the changes in entropy and energy above T_c are very much smaller in three than in two dimensions, as expected. Moreover, also in quantitative respect the experiments satisfactorily confirm the theoretical results.

Lastly we have observed the importance of quantum-mechanical effects from the influence of the spin value on the specific heat or the quantity T_c/θ. The experimental results do corroborate the theoretical predictions in that within a given magnetic lattice the differences between the observed behaviour and MF theory are enhanced by decreasing S, becoming especially apparent for $S = \frac{1}{2}$.

Since the critical behaviour found in 3-d systems will be treated in a § 4.4, we will now proceed to the last section.

§ 4. Special topics. Further comparison of theory and experiment

In the preceding sections we have had occasion many times to fit experimental results to existing theories. In most cases specific heat or susceptibility measurements were concerned.

The present section will be devoted to a closer examination of to which extent various existing theoretical approaches have been or can be checked by the experimental work. Evidently we shall rely heavily on the examples already presented above.

The main topics that we shall consider are spin-wave theory, series expansions, critical behaviour, and the field-dependent properties of magnetic substances. In the first section, moreover, some of the new developments in the application of neutron scattering to the study of magnetic substances will be discussed. This clearly is appropriate, since this technique has played a very important role recently in the research on 1-d and 2-d systems, as well as in testing spin-wave theory and in the determination of critical exponents.

4.1. *Neutron diffraction*

In this section we will briefly mention some of the principles of this technique, in particular in connection with the study of spin waves and its important contribution to the field of 1-d and 2-d magnetism. An extensive review of the theory of magnetic neutron scattering has been given by Marshall and Lowde (1968). The interested reader may also consult the recent book of Marshall and Lovesey (1971).

Besides the nuclear scattering common to all solids, arising from the diffraction of the incident neutron beam by the nuclei of the atoms in magnetic crystals, there is an additional magnetic scattering, due to the interaction between the magnetic moment of the neutron and that of the electrons. One may distinguish between elastic scattering, that may be used for the determination of the magnetic structure, and inelastic scattering which supplies a means of studying the magnetic excitations by measuring the changes in energy on collision, caused by the creation or annihilation of a magnon. In this way the spin-wave dispersion at low temperature or the evolution of a spin wave of a particular wave-vector with temperature may be studied.

Since the intensity of the magnetic Bragg scattering at zero field is proportional to the square of the spontaneous magnetization, the temperature dependence of the Bragg peaks yields the behaviour of the magnetization or sublattice magnetization and thus the critical exponent β. In addition to the coherent Bragg scattering, which evidently vanishes at T_c, an important contribution to the elastic scattering at small values of wave-vector k arises from the critical or diffuse scattering. This critical scattering becomes predominant as T_c is approached and since it is due to fluctuations of regions of short-range order it is also present above T_c. Although it is therefore of dynamic origin and thus partly inelastic, the critical scattering near to T_c and for small angles of scattering is for most part elastic, the conditions being that the transit time of a neutron through a region of correlated spins is short compared with the fluctuations in the magnetization, and that the inelasticity of the scattering is small as compared to kT. It may be shown that in this so-called quasi-elastic approximation the critical scattering is simply proportional to the wave-vector dependent susceptibility so that the peak in the critical scattering for $k=0$ observed at T_c reflects the susceptibility divergence.

Neutron scattering therefore provides us with a method of not only measuring the ferromagnetic susceptibility but also the *staggered* susceptibility of an antiferromagnet, which at first sight sounds like a rather hypothetical quantity in view of the apparent difficulty in realizing a staggered field in the laboratory. Since an antiferromagnet in a staggered field is equivalent to a ferromagnet in a normal field, one may indeed directly compare the critical exponent γ obtained from the staggered susceptibility of an antiferromagnet with that of its ferromagnetic counterpart, or with theoretical predictions for the latter (§4.4). The sublattice magnetization of an antiferromagnet is evidently also a staggered quantity, but, as we have seen, in this case other measuring techniques besides neutron diffraction are available (N.M.R., Mössbauer effect, magnetoelectric effect).

In addition to the static susceptibility $\chi(0)$ the behaviour as a function of wave-vector is of interest. At $T=T_c$ theory predicts the wave-vector dependent susceptibility to diverge as a function of k according to (Ritchie and Fisher 1972, Fisher and Burford 1967)

$$\hat{\chi}_c(\mathbf{k})/\chi_0(\mathbf{k}) \sim 1/k^{2-\eta} \quad (k\to 0\,;\ T=T_c). \tag{4.1}$$

In this expression the wave-vector dependent susceptibility is normalized by the susceptibility $\chi_0(\mathbf{k})$ of a paramagnetic system, and is defined by

$$\hat{\chi}(\mathbf{k})/\chi_0(\mathbf{k}) = \sum_{\mathbf{r}} \exp(i\mathbf{k}\,.\,\mathbf{r})\chi(\mathbf{r}). \tag{4.2}$$

Moreover $\chi(\mathbf{k})$ for $T \geqslant T_c$ is predicted to behave as (Fisher and Burford 1967)

$$\hat{\chi}(\mathbf{k},\,T)/\chi_0(\mathbf{k}) \sim (\kappa^2+\phi^2k^2)^{1/2\eta}/(\kappa^2+\psi k^2) \tag{4.3}$$

where $\phi(T)$ is a slowly varying function of order 0·05, $\psi \simeq 1$, and κ has the dimension of a length and is called the effective inverse range of correlation. Instead of (4.3) the approximate formula

$$\hat{\chi}(\mathbf{k},\,T)/\chi_0(\mathbf{k}) \sim 1/(\kappa^2+k^2)^{1-\eta/2} \tag{4.4}$$

is often used. The inverse range of correlation is expected to vanish at T_c according to

$$\kappa(T) \sim (1-T_c/T)^{\nu} \quad \text{for} \quad T\to T_c^{+}. \tag{4.5}$$

A similar formula applies for $T<T_c$. Thus the wave-vector dependent scattering intensity involves the two critical exponents η and ν in addition to the exponent γ describing the (static) susceptibility divergence

$$\hat{\chi}(0)/\chi_0 \sim (1-T_c/T)^{-\gamma}. \tag{4.6}$$

Theory predicts these three exponents to be related by $(2-\eta)\nu=\gamma$. In the classical Ornstein–Zernike theory $\eta=0$ and $\nu=0{\cdot}5$ so that $\gamma=2\nu=1$, the mean-field value. According to the Ornstein–Zernike theory (4.4) thus reduces to

$$\hat{\chi}(k)/\chi_0 \sim 1/(\kappa^2+k^2). \qquad (4.7)$$

For the 2-d Ising model $\nu=1$ and $\eta=\frac{1}{4}$, whereas for the 3-d Ising model $\nu \simeq 0{\cdot}64$ and $\eta \simeq 0{\cdot}06$. The Lorentzian form (4.7) of the scattering will therefore give a reasonable description of the scattering in 3-d crystals, but will be wholly inadequate in the 2-d case.

The parameter $\kappa(T)$, which is defined more formally as the reciprocal of the k^2 term in the expansion of $1/\hat{\chi}(k)$ in powers of k (Fisher and Burford 1967), should be distinguished from the true inverse range of correlation κ_t that appears in the exponential factor $\exp(-\kappa_t r)$, through which according to general theoretical expectations the decay of the correlations $\Gamma(\mathbf{r})$ is thought to be predominated. In the critical region κ and κ_t are expected to be approximately equal, at least proportional, but away from T_c they will differ considerably (Fisher and Burford 1967).

Apart from the difference in the values for the critical exponents, there are other marked differences between the elastic scattering in systems of magnetic dimensionality one, two or three. This was pointed out by Birgeneau *et al.* (1969, 1970 b) in the work on 2-d K_2NiF_4 and by Skalyo *et al.* (1970) in their study of the linear chain $CsMnCl_3 . 2H_2O$. The argument is exemplified in fig. 55, taken from the latter paper. In the usual 3-d situation, Bragg scattering will be observed at the reciprocal lattice points. However, if the long-range order extends only in two or in one dimensions, the Bragg condition will occur with respect to lines and planes in reciprocal space, respectively. Thus in 2-d systems, for instance, instead of Bragg peaks one will observe Bragg ridges. A similar argument also applies for the diffusive part of the scattering which should thus take the form of a ridge or a plane in 2 d and 1-d systems, respectively. Accordingly, the $\mathbf{k}$ dependence of the susceptibility will be such that $\mathbf{k}$ is measured from the reciprocal lattice line or plane, respectively.

Reversing the argument, the observation of such 1-d or 2-d magnetic scattering provides convincing proof for the apparent lower dimensionality of the magnetic system. In the preceding sections we have already mentioned how the measurements of the spin-wave dispersion curves by inelastic scattering give evidence for the nature of the magnetic ordering by showing the lack of dispersion in the directions perpendicular to the magnetic reciprocal lattice planes or lines in 1-d or 2-d systems, respectively (figs. 24 and 37). Additional proof is supplied by the elastic measurements, and as an example we have reproduced in fig. 56 (*a*), (*b*), (*c*), the pioneering work of Birgeneau *et al.* (1969, 1970 b) on K_2NiF_4. The existence of a ridge may be readily established from scans of the type A and B shown in fig. 56 (*a*). In fig. 56 (*b*) it is seen from the data taken at 99 K and 95 K (the transition point is at 97·23 K) that such 2-d behaviour is indeed present as far as the critical scattering is concerned. Scan B

Fig. 55

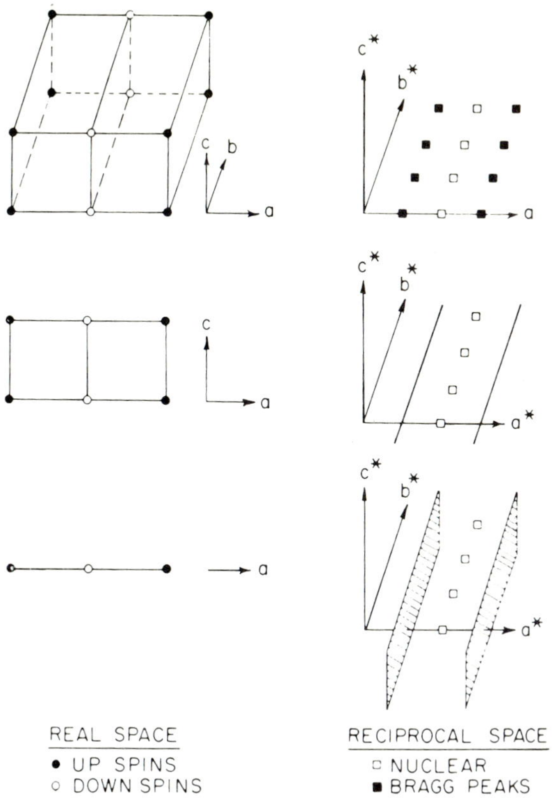

Illustration of the type of magnetic Bragg scattering that may be expected from various types of magnetic ordering in a 3-d crystal. From top to bottom we have 3, 2, and 1-d magnetic ordering, giving rise to magnetic Bragg peaks, lines and planes, respectively. (After Skalyo *et al.* 1970.)

along $(1, 0, l)$ (along the top of the ridge) gives a fairly constant value far above the background, the decrease in intensity at large l being caused by geometrical factors. On the other hand, scan A along $(h, 0, 0{\cdot}25)$, perpendicular to the ridge, shows a sharp peak with a linewidth determined by the instrumental resolution. In cooling through the transition, sharp Bragg peaks appear on top of the ridge at the magnetic reciprocal lattice points, as is observed from the data taken at 95 K. Approximate integration indicated that the intensity in the Bragg peak is just that lost by the ridge. As shown in fig. 56 (*c*) the Bragg peak intensity increases very rapidly as the temperature is lowered, at the same time the ridge intensity decreases. It may be seen that the ridge reaches its limiting intensity and linewidth at T_c, and that the behaviour in the paramagnetic region is analogous to the critical scattering in a 3-d system, the difference being that the temperature scale is greatly expanded. From the measured linewidth one may obtain the length over which the

Fig. 56

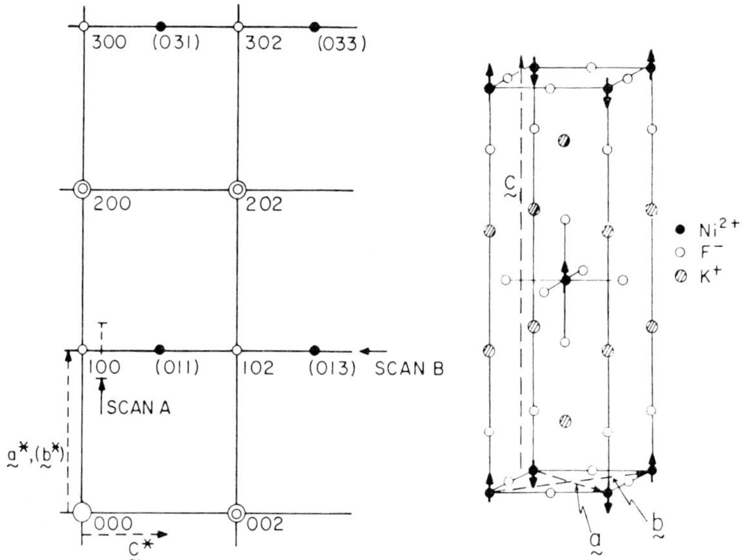
300
(031)
302
(033)
200
202
100
(011)
102
(013)
SCAN B
SCAN A
a*,(b*)
000
002
c*
c
a
b
Ni2+
F−
K+

(*a*)

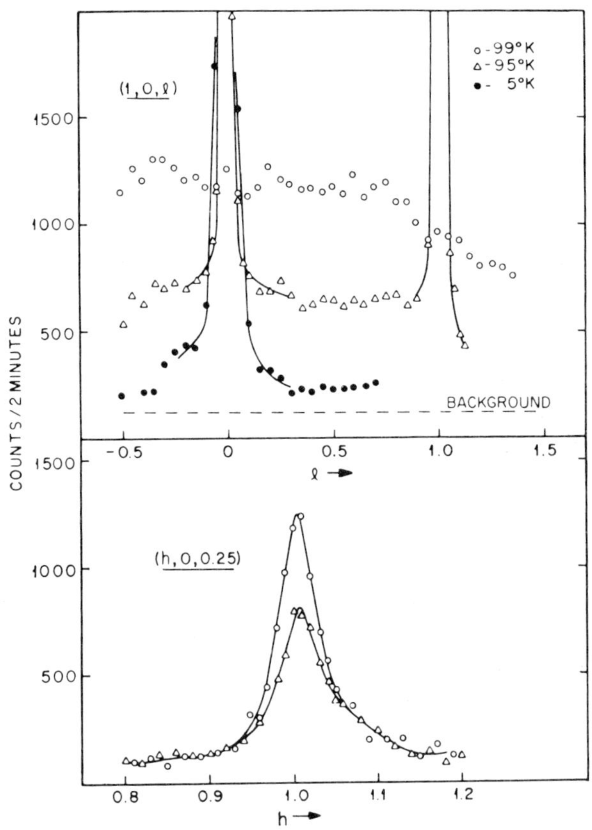
(1,0,ℓ)
o-99°K
Δ-95°K
•- 5°K
1500
1000
500
BACKGROUND
-0.5
0
0.5
1.0
1.5
ℓ
COUNTS/2 MINUTES
(h,0,0.25)
1500
1000
500
0.8
0.9
1.0
1.1
1.2
h

(*b*)

Fig. 56 (*continued*)

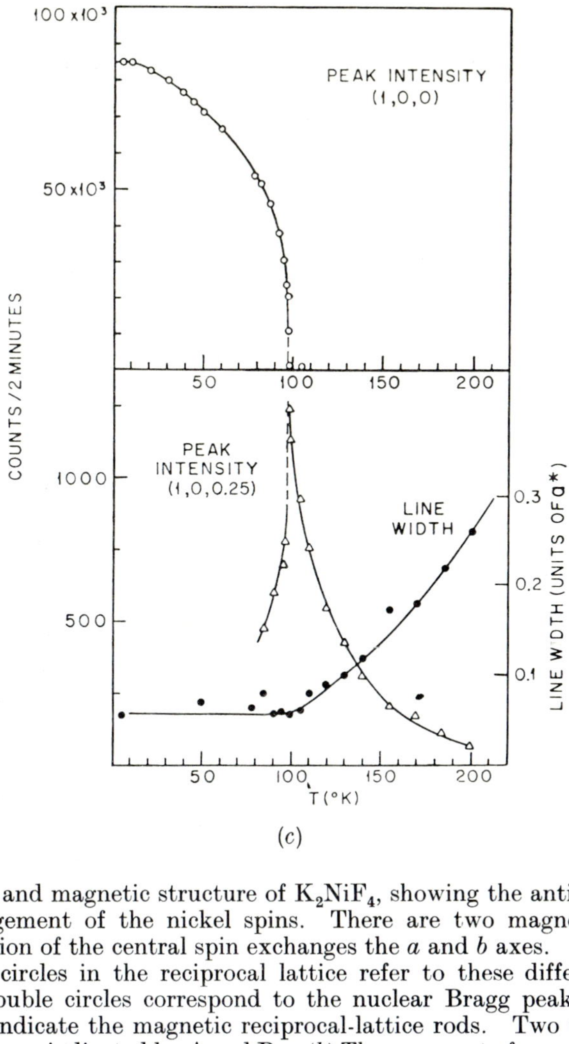

(*c*)

(*a*) Chemical and magnetic structure of K_2NiF_4, showing the antiferromagnetic arrangement of the nickel spins. There are two magnetic domains; inversion of the central spin exchanges the *a* and *b* axes. The open and filled circles in the reciprocal lattice refer to these different domains, the double circles correspond to the nuclear Bragg peaks. The thick lines indicate the magnetic reciprocal-lattice rods. Two types of scans have been indicated by A and B. (*b*) The upper set of curves corresponds to scans of the type B along the top of the ridge, the lower set to scans A across the ridge. (*c*) Top: scattering intensity at the (1, 0, 0) peak as a function of temperature. Bottom: scattering intensity of the (1, 0, 0·25) peak and linewidth for scan A as a function of temperature. (After Birgeneau *et al.* 1970 b.)

spins are correlated. As Birgeneau *et al.* point out, in K_2NiF_4, even at $T = 2T_c$, the estimated length is still 23 Å (within the planes), whereas Cooper and Nathans (1966) have reported that in $KMnF_3$ at $T = 1{\cdot}1\ T_c$ the correlation length has already decreased to 12 Å.

Since no evidence of 3-d critical scattering is observed both above and below T_c, the conclusion is that the diffuse scattering is indeed wholly 2-d in form. This supports the picture, obtained by various pieces of evidence, that in this system the occurrence of a phase transition is really caused by the 2-d properties and that the 3-d correlations play a very minor role. The most probable explanation for the occurrence of long-range order (cf. § 3.2.1) is that as the temperature is lowered to the value at which the staggered susceptibility is predicted to diverge, the correlations become of such a long range that the anisotropy may trigger the occurrence of long-range order within the layers ; and once long-range order is established in two dimensions it will of necessity occur also in the third, since even the smallest interplanar interaction is then amplified by N, the number of spins within a plane. That the long-range order is most probably established by the anisotropy rather than by the (much smaller) interplanar coupling has already been discussed in § 3.2. It is also in accord with the results found by Birgeneau *et al.* (1971 b) for the behaviour of the parallel and the perpendicular susceptibility.

In contrast to the diffuse scattering, the Bragg scattering is 3-d in form (fig. 56 (*b*)) and no evidence of 1-d or 2-d Bragg scattering has to our knowledge been found as yet. However, the temperature dependence of the sublattice magnetization, as deduced from the intensity of the Bragg peak, displays a 2-d character as close to T_c as one part in 10^4 (fig. 47). Evidently the critical point must be even more closely approached before it becomes apparent also from this aspect that the long-range order is in reality not confined to the layers but extends to the third dimension too.

In concluding this section we draw attention to the recent work of Birgeneau *et al.* (1971 a) and Hutchings *et al.* (1972 b) on the linear chain antiferromagnet $[(CD_3)_4N][MnCl_3]$. As in the case of $CsMnCl_3 . 2H_2O$, planes of critical scattering were observed from 40 K down to the lowest temperature reached (1·1 K, $T_c = 0{\cdot}84$ K). The high spin value $S = \frac{5}{2}$ justifies a comparison of the data with the exact theory of Fisher (1964) for the classical linear Heisenberg chain, and in fact it was found that the dependence of the scattering on both wave-vector and temperature could be accounted for by this theory qualitatively as well as quantitatively. The interesting point is of course that this is one of the few cases in which an exact theoretical result is available (one does not expect much difference in behaviour between $S = \frac{5}{2}$ and $S = \infty$). As an illustration we show in fig. 57 the behaviour of the zero-angle cross section $d\sigma(q = 0)/d\omega$ and the inverse correlation length κ as a function of temperature. The fit to theory only involves the exchange constant, and the value obtained agrees favourably with those derived from other measurements. Since for the ideal chain the transition to long-range order occurs at $T = 0$, both theoretical curves vanish at the origin. Due to the extremely small value of the interchain coupling in the experimental system, its actual transition temperature is so low that no deviations from the ideal behaviour can be observed in these plots.

Fig. 57

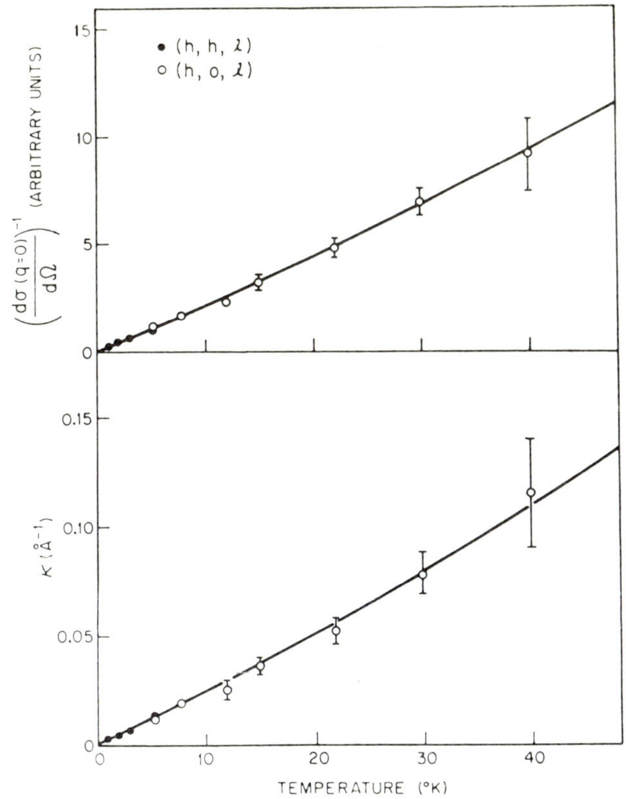

The zero wave-vector scattering cross section (top) and the inverse correlation length (bottom) of the linear chain antiferromagnet $[(CD_3)_4N][MnCl_3]$ as a function of temperature. Solid curves are fits to Fisher's exact theory for the classical linear Heisenberg chain. (After Birgeneau *et al.* 1971 a.)

4.2. *Spin wave theory*

Spin-wave theory has proven to be a most valuable tool in describing the low-temperature properties of magnetic substances, even in its most simple form, in which no account is taken of the interactions between the individually excited spin waves (unrenormalized spin-wave theory). We may cite from Dyson's (1956) well-known paper on the theory of spin-wave interactions : " The practical conclusion is simply this, that the linear Bloch theory with non-interacting spin waves is good enough for all practical purposes ", meaning that in the temperature range $0 < T < \frac{1}{2}T_c$, in which spin-wave theory is expected to be applicable quantitatively as a low-temperature approximation, the effects of renormalization are not likely to be observable. We shall not describe here the way in which spin-wave theory and its experimental verification have evolved historically, since there exists an extensive review paper by

Keffer (1966), covering both the theoretical and the experimental results obtained before 1965. Instead we shall confine ourselves to some prominent results that have been obtained after the appearance of Keffer's work.

Perhaps the most fundamental probe available to the experimentalist to study spin waves is neutron scattering. Quantities like the heat capacity and the magnetization, the observed behaviour of which may be compared with spin-wave predictions, involve an integral over the entire spin-wave spectrum. However, in a neutron scattering experiment one can select a spin wave of a particular wave-vector $\mathbf{k}$ and study its dependence on temperature. As mentioned above, the observation of these spin-wave modes is possible by measuring the energy changes involved in the inelastic scattering events in which magnons are created or annihilated. By combining measurements of different k at a particular temperature one obtains the quantity that is of prime importance, the dispersion relation. Knowledge of this functional relationship between energy and wave-vector is a priority to all calculations of thermodynamic properties. In the absence of anisotropy the dispersion is of the form $\hbar\omega = 4Js(1-\cos ka)$ for ferromagnets, or, for small k, $\hbar\omega \simeq (2JSa^2)k^2$, a being the lattice constant. In the presence of anisotropy a term $g\mu_B H_A$ is added to the $(1-\cos ka)$ term. In the case of antiferromagnets, the theory suffers from the fact that the ground state is unknown, because the fully aligned Néel state is not an eigen-state of the Hamiltonian. However, one may take the Néel state as a starting point and afterwards apply corrections for the fact that it is only an approximation of the true ground state. We shall come back to the effects of zero-point spin deviation at the end of this section.

The dispersion for antiferromagnets thus obtained is of the form $\hbar\omega = 2|J|S \sin ka$, which, for small k, reduces to the linear relationship $\hbar\omega = 2|J|Ska$, in contrast to the quadratic dispersion in ferromagnets. The effect of anisotropy is much more pronounced in antiferromagnets than in ferromagnets because in the former it is enhanced by interplay with the exchange field. This can be readily seen from the expression for the $k=0$ mode, which becomes $\hbar\omega = g\mu_B(2H_AH_E + H_A{}^2)^{1/2}$ when an anisotropy field H_A is present. Consequently the ' anisotropy gap ' is usually quite large in antiferromagnets, since H_E is of the order of 10^5–10^{-6} Oe. But in the compound $RbMnF_3$ the anisotropy is so small that the presence of the gap could not be detected within the experimental limits. The dispersion relation, as reported by Windsor and Stevenson (1966), is shown in fig. 58. Since the gap is only about 0·4 K, all measuring points lie on the sine curve appropriate to the fully isotropic case.

Other antiferromagnetic dispersion curve have already been shown in figs. 24 and 37. In those cases the lack of dispersion in one or two directions in reciprocal space was used to demonstrate that the magnetic character was 2-d or 1-d in form, respectively. An example of a ferromagnetic dispersion relation is given in fig. 59, showing data of Passell *et al.* (1971) on EuO.

Fig. 58

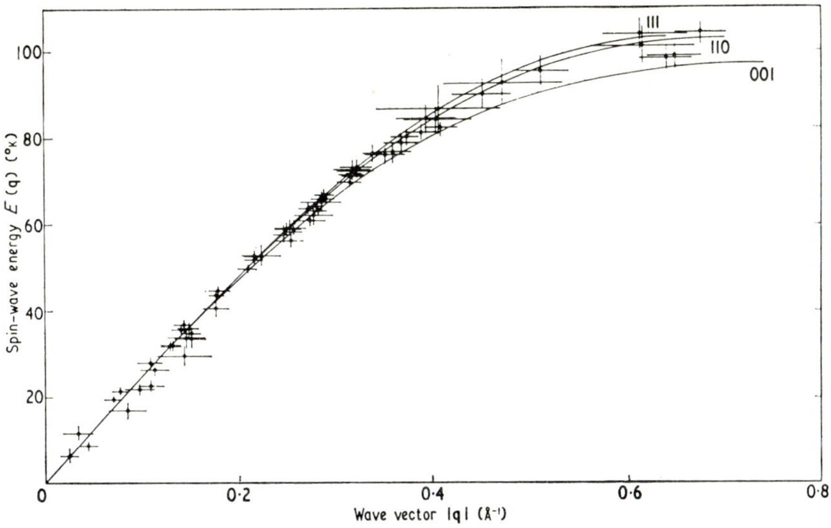

Spin-wave dispersion relation in a single crystal of $RbMnF_3$ at 4·2 K, with **k** vectors distributed over a [110] plane. Solid curves are dispersions along the indicated directions calculated with a nearest-neighbour exchange $J/k = -3{\cdot}4$ K ($\pm 0{\cdot}3$ K). The anisotropy gap is so small (0·4 K) that the measuring points near to the origin do not deviate from the linear behaviour expected for a fully isotropic antiferromagnet. (After Windsor and Stevenson 1966.)

Fig. 59

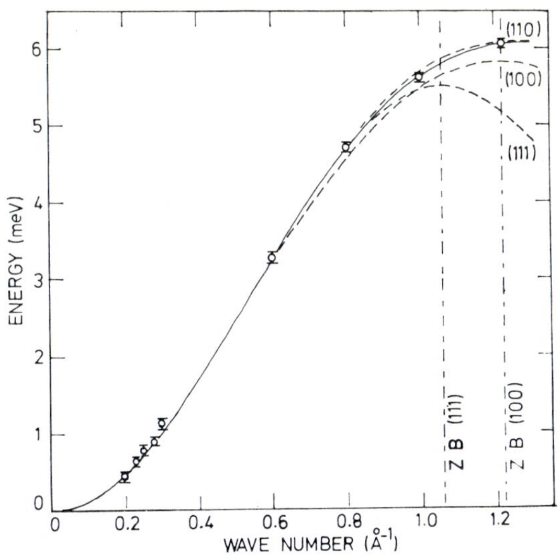

Spin-wave dispersion in a polycrystalline sample of EuO. The solid curves have been calculated for a number of directions using best fits for the first and second-neighbour interactions: $J_1/k = 0{\cdot}602 \pm 0{\cdot}008$ K and $J_2/k = 0{\cdot}155 \pm 0{\cdot}014$ K. The zone boundaries in the (100) and (111) directions have been indicated. (After Passell *et al.* 1972.)

The effects of interactions between magnons has been mentioned above. Dyson (1956) distinguishes two reasons why the picture of a linear superposition of spin waves is incorrect. Since each spin wave reduces the magnetization by one unit of $\hbar$, so that it will be completely reversed if $2N_0S$ spin waves are excited, it is clear that there must be some *repulsive* interaction between them in order to prevent the unlimited excitement of spin waves (kinematical interaction). Secondly there is an *attractive* interaction arising from the fact that the energy required to excite a spin wave will be lower if the spins are already partially reversed (dynamical interaction). From Dyson's calculations it emerges that the kinematical interaction will be negligible except when magnons of very short wavelength are excited, which is the case when the critical point is closely approached. For practical purposes therefore the dynamical interaction will be predominating, so that one expects the spin-wave energies to decrease when the temperature is raised. As cited above, in the temperature region $T < \frac{1}{2}T_c$, which is the range in which experimental results on the magnetization may be adequately fitted to the truncated series expansions in $kT/|J|S$ given by the spin-wave theory, the effect of the dynamical interaction term will be too small to be detected. A possible way of observing the effect, however, is to follow the evolution of the spin-wave modes with temperature, with the aid of neutron scattering. Such a study has been accomplished by Turberfield *et al.* (1965) in MnF_2. Figure 60 (*a*) shows the spectra of neutrons of initial wavelength 3·0 Å as a function of energy gain. At $T \simeq 0$ ($T/T_c \simeq 0{\cdot}06$) a sharp magnon line is observed, whose shape is determined by the experimental resolution. As the temperature is raised the line broadens and the spin-wave energy decreases. As T_c ($= 67{\cdot}33$ K) is approached the full width at half maximum becomes comparable with the energy itself. Finally the spin-wave peak is lost in critical scattering.

These spin-wave peaks were used to obtain the dispersion at different temperatures, as shown in fig. 60 (*b*). The solid lines were calculated by Low (1965) taking into account the dynamical interactions. Since the observed differences with the experimental data are less than 10%, whereas the observed linewidths are of the same order or larger, these measurements constitute a clear verification of the renormalization of the spin-wave energies.

In a similar experiment on $RbMnF_3$ ($T_c = 83$ K), Nathans *et al.* (1968) were able to show that short wavelength magnon-like excitations may even persist to temperatures above T_c. The temperature dependence of a magnon with wave-vector $k = 0{\cdot}2$ Å^{-1} near T_c is depicted in fig. 61. As the temperature is raised through T_c one observes a broadening and an energy renormalization of the two spin-wave peaks on both sides of the narrow central peak (zero energy transfer). The latter is due to Bragg scattering and accordingly vanishes at T_c. Above T_c the magnon peaks are still clearly in evidence and in addition a central diffusive peak is observed.

Fig. 60

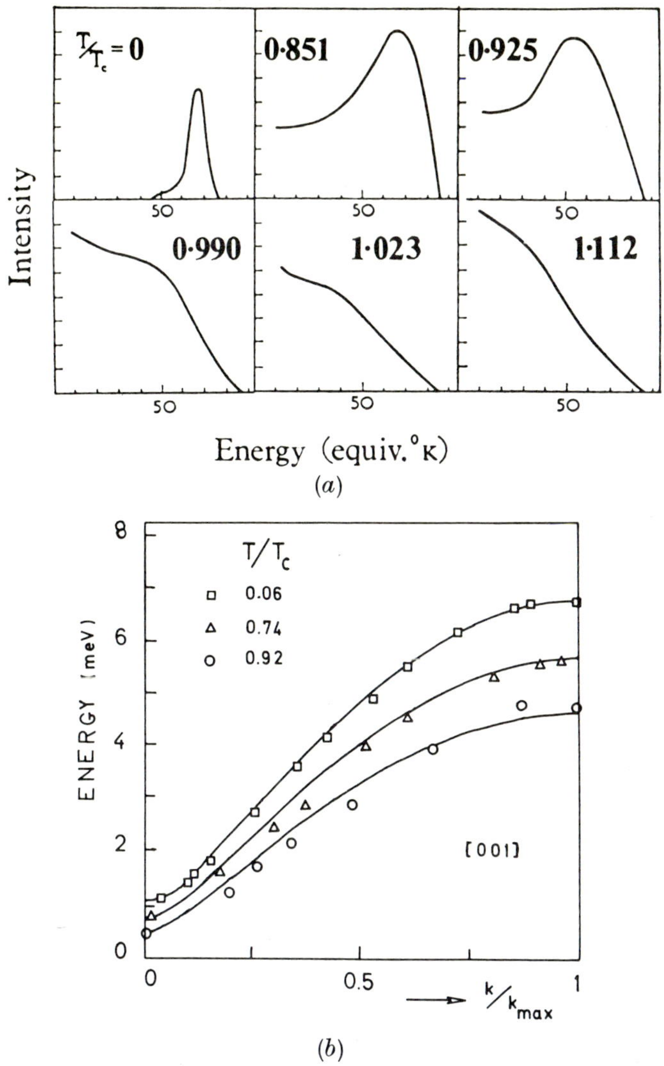

(*a*) Spin-wave energy spectra of neutrons of initial wavelength 3·0 Å, as measured at different temperatures in MnF_2 ($T_c = 67{\cdot}33$ K). (*b*) Spin-wave dispersion curves for MnF_2 at different temperatures. The experimental points have been obtained from spectra such as shown in fig. 60 (*a*). The solid curves have been calculated from spin-wave theory taking into account dynamical interactions between pairs of magnons. (After Turberfield *et al.* 1965.)

The presence of magnons or magnon-type excitations in the paramagnetic regime needs some further consideration. A simple physical model explaining the possibility of such paramagnetic spin waves has been put forward by Marshall (1965). Even in the absence of long-range order, above T_c, there will exist regions of correlated spins as a consequence of the short-range order. The extent of such a region is determined by the correlation length. One may imagine that damped excitations, or quasi spin waves of wavelengths shorter than the correlation length may exist, propagating within these slowly varying correlated regions with lifetimes limited by the characteristic times associated with changes in the local order. As the temperature is raised the correlation length decreases so that one expects the paramagnetic magnons to broaden and finally disappear.

Fig. 61

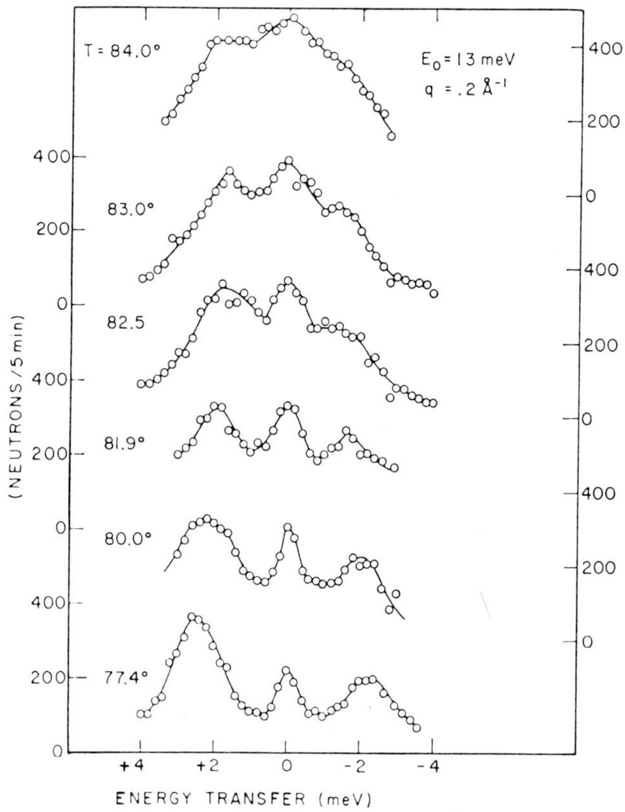

Temperature dependence of the inelastic scattering at $k = 0{\cdot}2$ Å^{-1} in $RbMnF_3$ near the critical point $T_c \simeq 83$ K. Two magnon peaks are observed, corresponding to energy gain and loss according to whether a magnon is annihilated or destroyed in the scattering process. In addition there is a central Bragg peak for $T < T_c$. The peaks for energy gain are higher because of a weighing term in the resolution function of the instrument. (After Nathans *et al.* 1968.)

The usefulness of this concept is clearly demonstrated by the recent experiments on the 1-d and 2-d Heisenberg systems, in which the paramagnetic region extends to much lower temperatures relative to J/k, as compared with 3-d systems. It has been discussed in the preceding sections that these 1-d and 2-d systems would ideally have no long-range order except at $T=0$, the fact that experimentally long-range order is found below finite transition temperatures T_c being the consequence of the existing deviations from the ideal model. We have also seen that for the 2-d Heisenberg antiferromagnets 2-d spin-wave theory is an excellent low-temperature approximation of the behaviour observed in the ordered region. From figs. 29, 37 and 38 and it may be seen how the magnon dispersion, the antiferromagnetic parallel and perpendicular susceptibilities and the sublattice magnetization may all be fitted to the predictions of 2-d spin-wave theory.

But we are now particularly interested in the behaviour above T_c, that is in the temperature range where there is indeed no long-range order, also experimentally. As a consequence of the low dimensionality the length over which the spins are correlated will be considerably larger than in 3-d systems at the same temperature relative to T_c. Thus one expects

Fig. 62

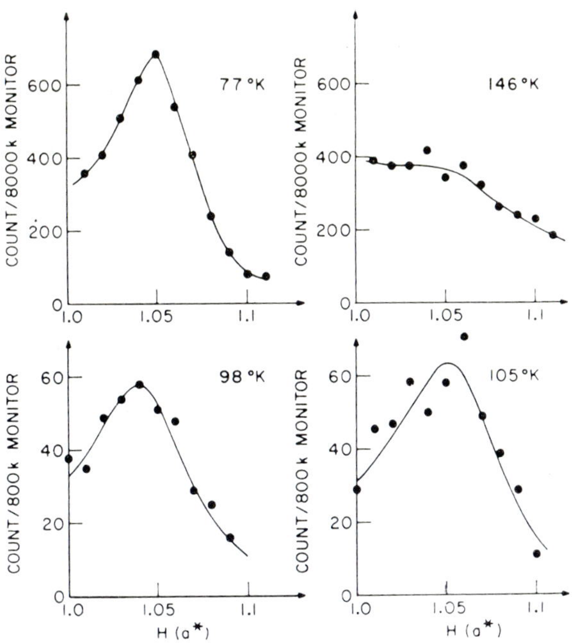

Temperature dependence of the (0·05, 0, 0) magnon in K_2NiF_4 ($\lambda \simeq 110$ Å). The transition temperature is $T_c=97{\cdot}2$ K. Note that these scans were taken at constant energy, those of fig. 61 at constant k. (After Skalyo *et al.* 1969 a.)

more pronounced magnon modes above T_c. In fig. 62 the temperature dependence is shown of a magnon of wavelength $\simeq 110$ Å in the 2-d antiferromagnet K_2NiF_4, for which $T_c = 97{\cdot}2$ K. The surprising feature emerging from these measurements of Skalyo *et al.* (1969 a) is that there is no appreciable difference between the magnon spectrum observed at $T = 1{\cdot}08\ T_c$ with that at $T = 0{\cdot}8\ T_c$ (in linewidth or in k value), even for such a relatively long wavelength. It is only at $T = 1{\cdot}5\ T_c$ that the 110 Å magnon mode is lost in the diffusive mode. In the above we have already mentioned that in K_2NiF_4 the estimated correlation length at $T = 2\ T_c$ is 23 Å, whereas in $KMnF_3$ at $T = 1{\cdot}1\ T_c$ the correlation length is down to 11 Å already, explaining the difference in behaviour observed in comparing figs. 61 and 62.

The impression left by the experiments shown in these figures is that as far as magnons with finite (not too small) k are concerned, nothing special happens at T_c itself, the prime quantity determining their gradual disappearance as the temperature is raised being the correlation length. Also for the renormalization, T_c is not the important parameter. In fig. 60 (*b*) one observes that at $T/T_c = 0{\cdot}92$ the magnon energies in MnF_2 have decreased by about 30–40%, whereas in K_2NiF_4, even at $T/T_c = 1{\cdot}1$, there is still no measurable renormalization effect (fig. 62). The difference may be brought back to the fact that in 2-d systems T_c is much lowered with respect to J/k. The predictions of spin-wave theory for the thermodynamic quantities, on the other hand, are in the form of series expansions in powers of $kT/|J|S$ and the renormalization is taken into account by additional terms in the expansion. Therefore the appropriate parameter is not T/T_c but $kT/|J|S$. At the transition temperature $kT/|J|S$ equals about 15 in MnF_2 and 2 in K_2NiF_4, so that in MnF_2 the ratio of thermal to exchange energy at T_c is about eight times as large as in K_2NiF_4, explaining the lack of renormalization in the latter as compared to the former.

In the 1-d Heisenberg antiferromagnet $[(CD_3)_4N][MnCl_3]$ the picture just sketched is even more clearly confirmed. In this substance the interchain interaction is so weak that the transition to long-range order is as low as 0·84 K, whereas the Curie–Weiss temperature is about 75 K ! In their recent study Hutchings *et al.* (1972 b) report that the magnon spectra, measured at $T = 1{\cdot}9$ K, are fully accounted for by a simple two-sublattice spin-wave theory for the linear Heisenberg chain. The dispersion relation of these 1-d spin waves at 4·4 K is a perfect sine curve and the value for the intra-chain exchange calculated from the fit to the theoretical dispersion is in agreement with that obtained from the susceptibility. Thus in spite of the fact that there is no long-range order, spin wave-like excitations are observed over most of the Brillouin zone. This at first sight surprising result is corroborated by the recent theoretical work of McLean and Blume (1973) on the spin-wave excitations in linear Heisenberg chains, who were able to account for the observed spectra at least qualitatively, in particular for the temperature dependence of the

Fig. 63

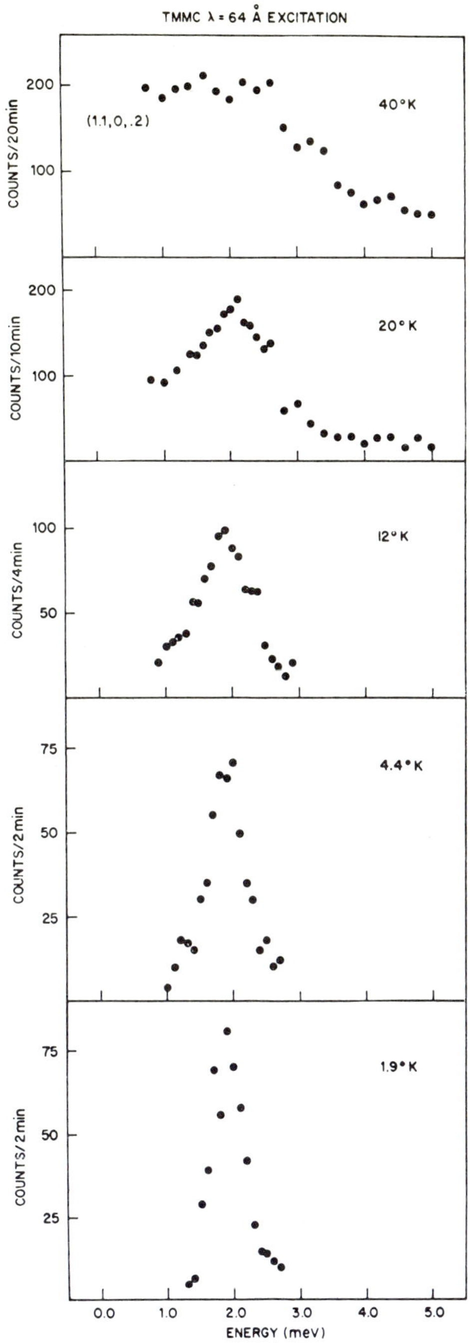

Temperature dependence of the $\lambda = 64$ Å excitation in the 1-d Heisenberg antiferromagnet $[(CD_3)_4N][MnCl_3]$. The transition to long-range order in this salt occurs at 0·84 K. (After Hutchings *et al.* 1972 b.)

magnon lines. As our last example we have reproduced in fig. 63 the evolution of the $\lambda = 64$ Å excitation with temperature. Clearly the fact that T_c is at 0·84 K is of no interest here, rather one should note that at 20 K the correlation length l is about 22 Å, even smaller than the magnon wavelength. This suggests that the condition for the existence of these excitations is $k = 2\pi/\lambda > \kappa$, where κ is the inverse correlation length, rather than $\lambda < l$. The results for other k values confirm this. The most recent measurements of Birgeneau *et al.* (1972 a) for values of wave-vector $k^* = k - \pi/a < 0{\cdot}015\ \pi/a$ likewise indicate that the excitations are changed from magnon-like to overdamped behaviour as k is varied from greater to less than κ. Clearly, these exciting new developments have opened up new areas of theoretical and experimental research in the field of spin waves.

We now turn to a number of experimental verifications of spin-wave theoretical predictions for the behaviour of thermodynamic quantities, like the heat capacity and the magnetization. Let us first consider the specific heat. The predicted limiting low-temperature behaviour for systems of different dimensionality d may be conveniently memorized by the mnemonic formula $C \sim T^{d/n}$, where d is the dimensionality and n is defined as the exponent in the dispersion relation $\omega \sim k^n$. For phonons and antiferromagnetic magnons $n = 1$, for ferromagnetic magnons $n = 2$. Thus the lattice heat capacity of a 3-d system goes with T^3, which is the well-known low-temperature Debye approximation, the spin-wave specific heat of a 3-d ferromagnet as $T^{3/2}$, of a 2-d ferromagnet as T, etc. . . . One should not forget that these terms are only the first terms in series expansions in $kT/|J|S$. For instance the specific heat of a 3-d ferromagnet is approximated to higher order by (Dyson 1956)

$$C_m/R = a_0(kT/JS)^{3/2} + a_1(kT/JS)^{5/2} + a_2(kT/JS)^{7/2} + a_3(kT/JS)^4. \qquad (4.8)$$

In fig. 64 (*a*) the low-temperature heat capacity of $Cu(NH_4)_2Br_4 \cdot 2H_2O$ as measured by Miedema *et al.* is plotted as $C_m(T/T_c)^{3/2}$ versus T/T_c. In this way the first coefficient a_0 is obtained from the intercept at $T = 0$ and the second coefficient a_1 from the derivative at $T = 0$. The value for J/kT_c following from the comparison of the experimental and theoretical values for a_0 is in reasonable agreement with the experimentally determined $J/kT_c = 0{\cdot}35$. Also a_0/a_1 agrees with theory. This may also be inferred from fig. 64 (*b*) where the data for $T > 0{\cdot}2\ T_c$ are compared with the Dyson series (4.8) using $J/kT_c = 0{\cdot}35$. The agreement between experiment and theory may be extended to slightly higher temperatures (up to $kT/JS \simeq 3$) if instead of the truncated series (4.8) the full expression for C_m involving an integral over the Brillouin zone is used. In the calculation renormalization was accounted for, the correction being about 13% at $T = 0{\cdot}6\ T_c$. The fit is quite good, although it should be borne in mind that the result is rather sensitive to the J/kT_c value, and recent experiments suggest that $J/kT_c = 0{\cdot}34$ would be more accurate for this salt (Van Amstel *et al.* 1974).

Fig. 64

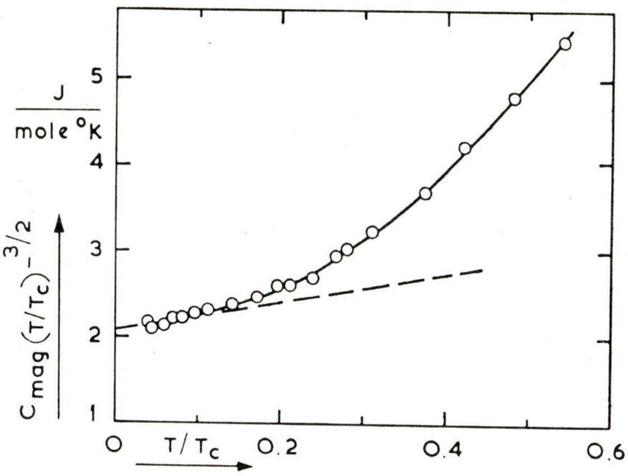

(*a*)

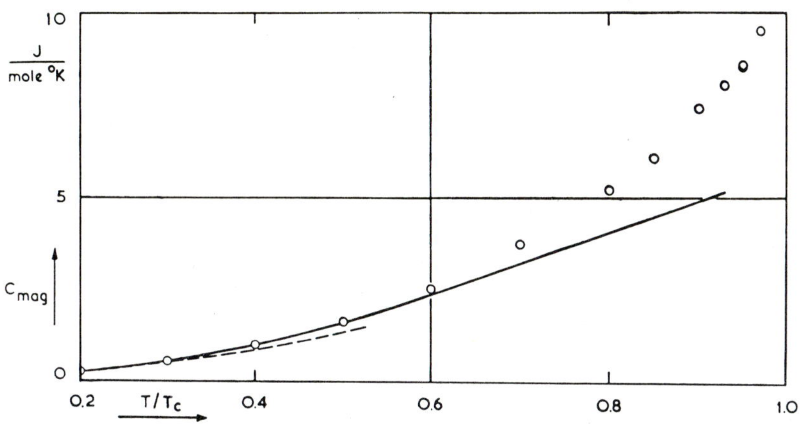

(*b*)

(*a*) The magnetic heat capacity of $Cu(NH_4)_2Br_4 \cdot 2H_2O$ in the spin-wave region. The data are plotted as $C_m(T/T_c)^{-3/2}$ versus T/T_c in order to demonstrate the existence of the $T^{3/2}$ and $T^{5/2}$ terms in the Dyson series for C_m. (*b*) The same data for $T > 0{\cdot}2T_c$. The broken line is the prediction from the Dyson series (eqn. (4.8)), the solid curve was calculated using the full expression for the heat capacity, integrated over the Brillouin zone. In both cases $J/kT_c = 0{\cdot}35$ has been assumed, in accordance with other experimental determinations of this quantity. (After Miedema *et al.* 1965.)

According to the spin-wave theory the heat capacity of a ferromagnet will be reduced on applying a magnetic field. This has been verified by Passenheim *et al.* (1966), whose data on EuS are shown in fig. 65. Also in this case the spin-wave predictions were calculated by integrating over the Brillouin zone, but a correction for renormalization effects was unfortunately not applied. The values for the first and second-neighbour exchange giving best fits are in agreement with those obtained by Passell *et al.* (1971) from the spin-wave dispersion curve. Note that at $T=4$ K, $kT/JS \simeq 5{\cdot}5$.

Fig. 65

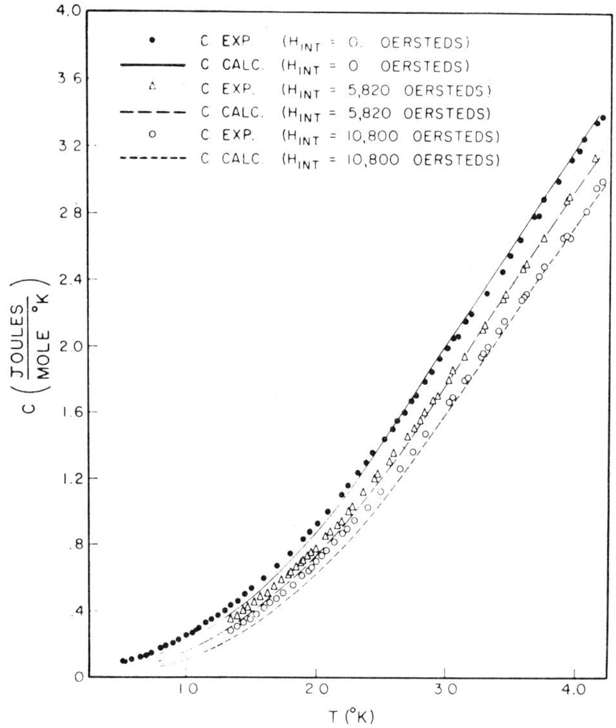

Magnetic heat capacity of EuS measured at two different field values and $H=0$. The solid curves have been calculated from spin-wave theory. (After Passenheim *et al.* 1966.)

The T^3 spin-wave term in the magnetic specific heat of a 3-d antiferromagnet is more difficult to demonstrate experimentally. This arises because of the above-mentioned large anisotropy gaps occurring in antiferromagnets even when the anisotropy is fairly small, as a consequence of the inter-play between H_E and H_A. Nevertheless, Montgomery (1966) has found evidence for the T^3 term in the extremely isotropic compounds $RbMnF_3$ and $KMnF_3$. His data on $RbMnF_3$ have been reproduced in fig. 66. In zero magnetic field the data could be accurately fitted to the equation $C=27{\cdot}7\ T^{-2}+0{\cdot}334\ T^3$ (C in mJ/mole K). The first term is the hyperfine specific heat, the second is the sum of the lattice and spin-wave contributions that both have the same temperature dependence. The slight increase observed on application of a field of 14·1 kOe is attributed to impurities.

Fig. 66

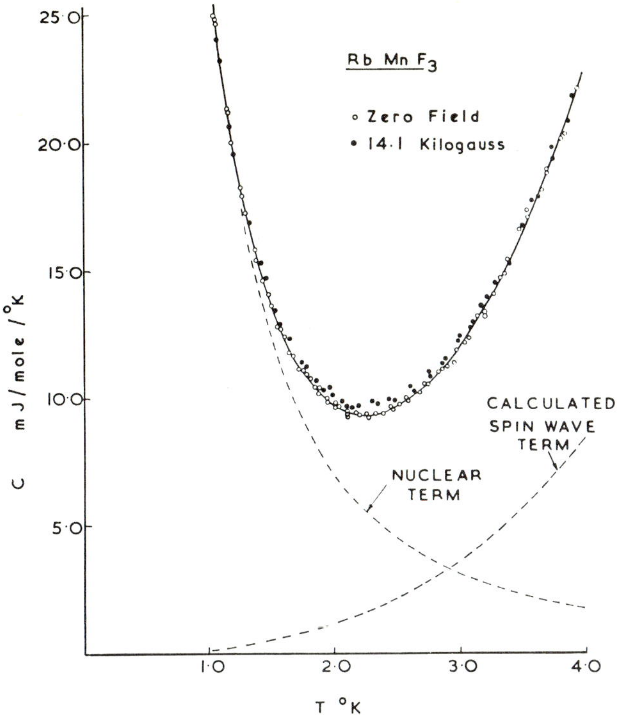

The low-temperature specific heat of the 3-d Heisenberg ferromagnet $RbMnF_3$ as measured by Montgomery (1966).

Applications of spin-wave theory in describing the temperature dependence of the spontaneous magnetization may be found in Keffer's review paper (e.g. MnF_2). Of the recent results we mention the analysis of Charap and Boyd (1964) of their N.M.R. data on EuS and Wielinga and Huiskamp's results (1969) on $(NH_4)_2CuBr_4 . 2H_2O$, that nicely fit the calculations of Loly (1968) (see also Velu *et al.* 1972). Although appreciable next-nearest neighbour interactions are present in these ferromagnetic compounds, this discrepancy is not too serious in the case of the spontaneous magnetization, since the results turn out to be rather insensitive to the presence of next-nearest neighbour interactions. Also one may incorporate a second-neighbour interaction within the theory. Nevertheless it is a pity that a clear-cut, nearest-neighbour only ferromagnetic insulator is yet to be discovered.

Turning next to systems of lower dimensionality we show in fig. 67 the measurements of Colpa (1972 b) of the low-temperature specific heat of one of the 2-d ferromagnetic copper compounds. As expected, the heat capacity depends linearly on temperature. However, we have mentioned in § 3.2.2 that the values for the exchange constants of these salts,

Fig. 67

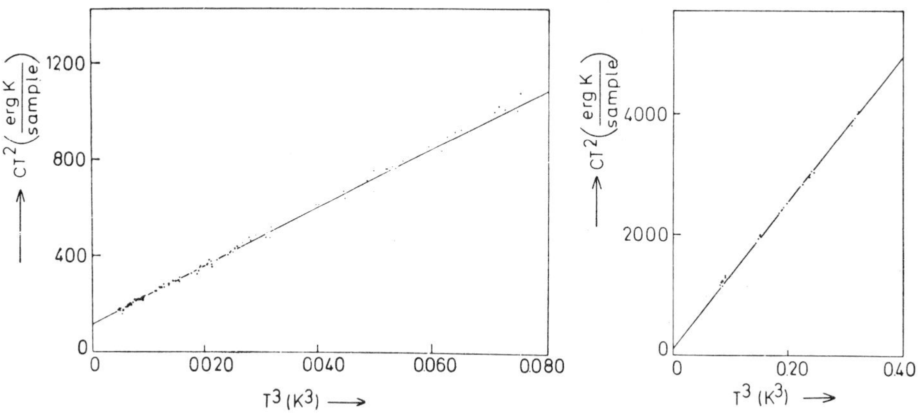

The low-temperature specific heat of the 2-d Heisenberg ferromagnet $(CH_3NH_3)_2CuCl_4$, as measured by Colpa (1972 b).

calculated from the coefficients of the linear term, are systematically lower (10%) than those obtained from the high-temperature susceptibility. A possible temperature effect may not be excluded, although the exchange is in that case mostly increasing with decreasing temperature. As explained in § 3.2.2 we suspect the discrepancy to be due to a failure of spin-wave theory to account quantitatively for the observed behaviour if both the dimensionality and the spin value are low. Quite generally, one may expect spin-wave theory to become a better approximation as the dimensionality and the spin value are increased. In particular, for magnetic chains, the inadequacy of the spin-wave approach is exposed. For an antiferromagnetic Heisenberg chain spin-wave theory predicts a linear temperature dependence of the specific heat. The calculations of Bonner and Fisher (1964) on finite $S=\frac{1}{2}$ Heisenberg chains corroborate this qualitatively, but yield an amplitude of the linear term that is a factor 3 times smaller than the spin-wave result. Weng (1969) has found this factor to be only 1·4 for the case $S=1$†. For ferromagnetic chains the heat capacity at low temperatures varies as the square root of the temperature. In this case the amplitude estimates from the calculations on closed finite rings differ a factor 1·3 from the spin-wave result for both $S=\frac{1}{2}$ and $S=1$. In § 3.1.2 we have seen (fig. 20 (*a*), (*b*)) that the experimental specific heat data for $S=\frac{1}{2}$ are in remarkably good agreement with the curve obtained by Bonner and Fisher, also near to the region where the dependence on temperature is approximately linear.

† It has been suggested that the $S=\frac{1}{2}$ antiferromagnetic linear chain is actually not a boson but a fermion problem. For recent experimental indications to this end see Ehrenfreund *et al.* (1973).

Concerning the antiferromagnetic susceptibility of Heisenberg chains, spin-wave theory erroneously predicts a diverging $\chi_\perp$ as $T \rightarrow 0$, in contrast with the finite, spin-dependent values obtained in better treatments (Griffiths 1964 a, Bonner and Fisher 1964, Fisher 1964, Weng 1969). For this particular example the perturbational approach of Davis (1960) yields better results than spin-wave theory in that it does give finite values for $\chi_\perp(T=0)$, although these are in poor agreement with the other calculations. For $S=\frac{1}{2}$ Davis obtains $\chi_\perp(0)/\chi_\perp{}^0=0{\cdot}11$, where $\chi_\perp{}^0$ is the molecular field prediction for the perpendicular susceptibility. This may be compared to Griffiths' result $\chi_\perp(0)/\chi_\perp{}^0=0{\cdot}405$. For $S=\frac{5}{2}$ Davis's value is $\chi_\perp(0)/\chi_\perp{}^0=0{\cdot}83$, whereas Weng finds 0·59.

For the 2-d antiferromagnets with $S>\frac{1}{2}$ the experimental values of $\chi_\perp(0)/\chi_\perp{}^0$ are in close agreement with spin-wave theory, in any case much closer than with the predictions obtained by Davis (1960) and Walker (unpublished) from a perturbation theoretical approach (see De Jongh 1972 b, c and also below). For the only example with $S=\frac{1}{2}$ presently available ($CuF_2 \,.\, 2H_2O$), the reverse is the case. This could indicate a similar discrepancy with spin-wave theory as found above for the specific heat of the 2-d Heisenberg ferromagnets with $S=\frac{1}{2}$. On the other hand, the value of $|J'/J|$ of this compound is fairly high ($\simeq 10^{-2}$), which could also be responsible for the fact that $\chi_\perp(0)/\chi_\perp{}^0$ is higher than predicted by 2-d spin-wave theory.

For the 2-d antiferromagnets with $S>\frac{1}{2}$, on the other hand, an extremely good agreement with spin-wave theory (including renormalization) has been found for a large variety of measurements, the only adjustable parameters being the exchange and the (small) anisotropy. Since the values for these parameters may be independently determined by methods not involving the magnon approach, a wholly consistent picture is obtained. As mentioned in § 3.2, in compounds such as K_2NiF_4 and K_2MnF_4 the spin-wave dispersion (Birgeneau *et al.* 1969, 1973), the sublattice magnetization (De Wijn *et al.* 1971, 1973 b) and the parallel and perpendicular susceptibility (Breed 1967, 1969) can all be fitted to the spin-wave theory for a 2-d Heisenberg antiferromagnet including a small anisotropy. The exchange constants obtained in these various experiments are in excellent accord, and are moreover equal within 1% to those obtained from fitting the high-temperature susceptibility to the exact series expansions (De Jongh 1972 c). From the good agreement in two dimensions one would expect an even better accordance in three dimensions. Indeed, we have already discussed many examples that support this conclusion.

We will end this section by reviewing the experimental evidence found for the existence of zero-point spin deviations in antiferromagnetic substances, as predicted by Anderson (1952). As mentioned earlier, the fully aligned Néel state is not an eigen-state of the Hamiltonian, so that even at zero temperature the spins will be subject to deviations from this orientation. However, as a starting point spin-wave theory takes the

Néel state as the approximate ground state, correcting afterwards for the zero-point motions. These corrections include a shift of the ground-state energy. Since the presence of anisotropy stabilizes the two-sublattice ground state, in the limit $\alpha = H_A/H_E \to \infty$ the energy becomes equal to the Ising approximation $E_I = -2z|J|S^2N_0$. For finite α we have

$$E(\alpha) = E_I\{1 + e(\alpha)/zS\},$$

where $e(\alpha)$ varies from $e(0) \simeq 1/4z$ to zero for $\alpha \to \infty$. As a second effect, the magnetic moment per site is no longer given by $g\mu_B S$ but by $g\mu_B(S - \Delta S)$, the anisotropy dependent spin reduction $\Delta S(\alpha)$ likewise reducing to zero as $\alpha \to \infty$. Values for $e(\alpha)$ and $\Delta S(\alpha)$ may be calculated from spin-wave theory or other methods (e.g. Davis 1960, Keffer 1966, Breed 1969, Lines 1970, Colpa *et al.* 1971). The value of $\Delta S(0)$ is roughly given by $1/2z$ for the various lattices.

The effects of zero-point spin deviation constitute yet another example of a phenomenon that is considerably more easily studied in lower dimensional magnets, since they become the more pronounced the lower the dimensionality (and the lower the spin value). In addition to the approximate expressions for $e(0)$ and $\Delta S(0)$ given above, we may quote the explicit values of $\Delta S(0)$ for the s.c. and the quadratic lattice, which are 0·078 and 0·197, respectively. During the past ten years substantial efforts have been made in finding experimental proof for the existence of the zero-point deviations, mainly by detecting their influence on the expectation value of the magnetic moment. Initially the experiments were performed on 3-d antiferromagnets, the magnetic moment being either measured directly with neutron diffraction, or indirectly via the hyperfine field, i.e. the field exerted by the magnetic moment on the nucleus of the magnetic atom. The results were rather disappointing in that no definite proof for the existence of spin reduction could be obtained. This was partly due to the smallness of the effect in three dimensions, in particular for the manganese compounds on which most of the investigations were performed. These have $S = \frac{5}{2}$, so that ΔS is only about 3% of S (that nevertheless manganese was chosen arises because the Mn^{2+} ion is in an $\mathscr{S}$ state, so that one is not troubled by orbital contributions to the magnetic moment). Such a small effect is difficult to detect convincingly, more so since in both methods of measurement one is troubled by additional mechanisms that complicate the analysis. For instance, in determinations of $\langle S \rangle$ by neutron diffraction, where $\langle S \rangle$ denotes the expectation value of the magnetic moment in the preferential direction, one is hampered by covalency effects. Covalency likewise reduces the magnetic moment since it delocalizes small parts of the moments of neighbouring magnetic ions onto the ligand in between them. A cancellation of these small parts will reduce the effective moment. Furthermore, covalency affects the form factor of the magnetic ion for neutron diffraction, accurate knowledge of which is indispensable in order to obtain $\langle S \rangle$.

In deducing $\langle S\rangle$ from the hyperfine structure interaction (h.f.s.) $A\mathbf{S}\,.\,\mathbf{I}$, one measures the hyperfine field, which is proportional to $A\langle S\rangle$, from specific heat or N.M.R. experiments. Consequently the value of the h.f.s. constant A has to be known to a sufficiently high accuracy to enable a reliable determination of $\langle S\rangle$. Earlier estimates of A from E.S.R. experiments in magnetically diluted diamagnetic isomorphs led to contradictory results, even to negative values of ΔS in 3-d antiferromagnets. As an explanation for this unphysical result, Owen and Taylor (1966, 1968) and Huang *et al.* (1966, 1967) suggested that the h.f.s. constant will be larger in the antiferromagnetic salts than in the diamagnetic isomorphs. This would be the consequence of a transfer of unpaired electron spin from one magnetic ion to its nearest magnetic neighbours via the intervening ligand (super-transferred hyperfine interaction). Such a process will affect the splitting of the nuclear energy levels and thus the value of A. In 3-d, $S=\frac{5}{2}$, antiferromagnets the two effects are expected to be of comparable magnitude, accounting for the observed behaviour.

In spite of this possible explanation one could still doubt the experimental evidence for zero-point effects. Recently, however, various groups of workers have together provided convincing experimental proof by exploiting the newly discovered 2-d antiferromagnetic materials, in particular K_2MnF_4 and Rb_2MnF_4. Values for $\langle S\rangle$ of the same order of magnitude as predicted by 2-d spin-wave theory were obtained in K_2MnF_4 by Loopstra *et al.* (1968) and by Rubinstein and Folen (1968), using neutron diffraction and magnetic resonance, respectively. In 1970 Walstedt *et al.* reported on N.M.R. measurements in the two Mn compounds, finding values for ΔS in seemingly complete agreement with theory. At about the same time Colpa *et al.* (1971) evidenced the much larger spin reduction in two as in three dimensions, by comparing the h.f.s. contributions to the heat capacity of Rb_2MnF_4 and $RbMnF_3$. The various results for ΔS in these manganese compounds have been gathered and reviewed by Schrama (1973 a, b). From his experiments on X_2MF_4 (X = K, Rb; M = Mg, Zn, Cd) and XMF_3 diluted with manganese, he concludes to an A value slightly different from that adopted by Walstedt *et al.* (1970) on the basis of fewer experimental results for A. Nevertheless, in calculating the spin reductions from the h.f.s. fields measured by Walstedt *et al.* using the new A values, Schrama ultimately obtained quite similar results for ΔS as Walstedt *et al.* because he also had to introduce a value for the predicted effect $\Delta A_{\mathrm{s.h.i.}}$ upon the hyperfine constant due to the super-transferred hyperfine interaction, which differs from that calculated by Owen and Taylor and Huang *et al.* This point is clarified in table 13. In the first and the second row values for the spin reduction in the listed compounds are given as predicted by spin-wave and perturbation-theory, respectively. The predictions have been corrected for the anisotropy, the correction being only substantial ($\pm 15\%$) for the spin-wave predictions for the 2-d salts. In the third row the experimental values are given, uncorrected for the super-transferred hyperfine interaction, in the fourth

Table 13. Experimental values for the zero-point spin reduction in 2-d and 3-d manganese compounds, compared with theoretical predictions from spin-wave and perturbation theory. The experimental results have been calculated by Schrama (1973 a, b) from the hyperfine fields measured by Walstedt *et al.* (1970), using values for the hyperfine constant A newly derived by him from ENDOR experiments in Mn^{2+}-doped diamagnetic isomorphs. The experimental reductions are given for three different values for the correction $\Delta A_{\text{s.h.i.}}$ per magnetic neighbour to the hyperfine constant due to the supertransferred hyperfine interaction. Third row: zero correction; fourth row: theoretical correction calculated by Owen and Taylor spin-wave theory in two as well as in three dimensions ($z=4$ and 6, respectively). The numbers in brackets indicate the estimated uncertainties in the last digit(s).

	$\Delta A_{\text{s.h.i.}}/z$ (10^{-4} cm^{-1})	$KMnF_3$ ($\alpha=7\times10^{-4}$)	$RbMnF_3$ ($\alpha=5\times10^{-6}$)	K_2MnF_4 ($\alpha=3{\cdot}9\times10^{-3}$)	Rb_2MnF_4 ($\alpha=4{\cdot}7\times10^{-3}$)
Spin-wave theory		0·077	0·078	0·169	0·167
Perturbation theory		0·0621	0·0623	0·108	0·108
Experiment	0	−0·021 (13)	−0·021 (13)	0·103 (9)	0·107 (9)
Experiment	0·26	0·023 (13)	0·023 (13)	0·131 (9)	0·135 (9)
Experiment	2·25 × 0·26	0·078 (26)	0·078 (26)	0·166 (17)	0·170 (17)

Table 14. Values of $\alpha = H_A/H_E$ and $\chi_\perp(0)$ of quasi 2-d Heisenberg compounds. Using these, the experimental ratios $\chi_\perp(0)/\chi_\perp{}^0(\alpha)$ are obtained, where $\chi_\perp{}^0(\alpha)$ is the anisotropy dependent MF prediction for the perpendicular susceptibility $\chi_\perp{}^0(\alpha) = N_0 g^2 \mu_B{}^2/4z|J|(1+\alpha/2)$. The J/k values used are listed in table 6. The experimental ratios are compared with those predicted by spin-wave theory for the 2-d quadratic antiferromagnet.

Compound	S	$\alpha = H_A/H_E$	$\chi_\perp(0)_{exp.}$ (cm³/mole)	$\chi_\perp(0)/\chi_\perp{}^0(\alpha)$	
				(Exp.)	(Theor.)
$CsFeF_4$	$\frac{5}{2}$	7×10^{-3}	$6{\cdot}06 \times 10^{-3}$	0·92	0·90
$RbFeF_4$	$\frac{5}{2}$	$6{\cdot}5 \times 10^{-3}$	$7{\cdot}31 \times 10^{-3}$	0·94	0·90
Rb_2MnF_4	$\frac{5}{2}$	$4{\cdot}7 \times 10^{-3}$	$2{\cdot}32 \times 10^{-2}$	0·92	0·90
K_2MnF_4	$\frac{5}{2}$	$3{\cdot}9 \times 10^{-3}$	$2{\cdot}01 \times 10^{-2}$	0·90	0·90
$BaMnF_4$	$\frac{5}{2}$	$3{\cdot}1 \times 10^{-4}$	$3{\cdot}05 \times 10^{-2}$	0·89	0·89
$BaNiF_4$	1	2×10^{-2}	$2{\cdot}98 \times 10^{-3}$	0·80	0·78
Rb_2NiF_4	1	1×10^{-2}	$2{\cdot}2 \times 10^{-3}$	0·77	0·77
K_2NiF_4	1	2×10^{-3}	$1{\cdot}83 \times 10^{-3}$	0·76	0·74
$CuF_2 . 2H_2O$	$\frac{1}{2}$	$3{\cdot}7 \times 10^{-3}$	$5{\cdot}26 \times 10^{-3}$	0·63	0·50

row the same but now corrected for the $\Delta A_{s.h.i.}$ per magnetic neighbour as calculated by Owen and Taylor and Huang *et al.* Finally in the last row a correction $\Delta A_{s.h.i.}$ has been applied that is 2·25 times as large as the theoretical value. Note that for $\Delta A_{s.h.i.} = 0$ negative values for ΔS result in three dimensions, as mentioned above. Note also that by using the theoretical $\Delta A_{s.h.i.}$ the agreement with spin-wave theory is still unsatisfactory, especially in three dimensions where the experiment is moreover far outside both spin-wave and perturbation theoretical predictions. With the 'corrected' $\Delta A_{s.h.i.}$ per magnetic neighbour, on the other hand, the agreement with spin-wave theory is perfect in two as well as in three dimensions ($z = 4$ and 6, respectively). Taking also into account the success of spin-wave theory in describing the properties of 2-d ($S > 1$) and 3-d antiferromagnets, as witnessed above, one would conclude that the spin-wave predictions for ΔS are most probably correct and that the experimental results indicate the calculated effect on A due to the super-transferred hyperfine interaction to be quantitatively in error.

Other experimental values for ΔS, encountered in the preceding pages, are $0{\cdot}20 \pm 0{\cdot}03$ for K_2NiF_4 ($S = 1$) (De Wijn *et al.* 1971) and 0·24 for $Cu(HCOO)_2 . 4H_2O$ ($S = \frac{1}{2}$) (Dupas and Renard 1970 b). In both cases the theoretical value is 0·18, taking into account the anisotropy. Also for $KNiF_3$ a considerable reduction has been observed (Hutchings and Guggenheim 1970) but in this case the zero-point effect could not be deduced unambiguously, due to the covalency effects. In addition, in some of the chain compounds considerable reductions of ΔS have been observed. For these systems, however, spin-wave theory is apparently in error since it predicts a reduction that diverges in the limit of zero anisotropy (Kubo 1952), in disagreement with the calculations of Griffiths (1964 a, b, c) and Bonner and Fisher (1964).

In a different approach De Jongh (1972 b, c) has shown how the existence of zero-point deviations can also be deduced experimentally from their effects upon the perpendicular susceptibility. This has already been mentioned in § 3.2.1 (cf. fig. 28). In this case both the spin reduction and the shift in the ground-state energy come into play, since both have the effect of lowering $\chi_\perp(0)$ from the MF value $\chi_\perp^0$ (cf. eqn. (3.6)). Using the known values of J/k and α, theoretical and experimental results for the quantity $\chi_\perp(0)/\chi_\perp^0$ have been calculated for a number of 2-d antiferromagnets with different spin value (De Jongh 1972 c), as listed in table 14. Note the good agreement for $S=1$ and $S=\frac{5}{2}$. In the case of $S=\frac{1}{2}$ the experimental reduction is too small, as discussed above. A similar agreement for the perpendicular susceptibility is also found for the 3-d salts $RbMnF_3$ and $KNiF_3$ (see § 3.3.2). It is remarked that these findings for $\chi_\perp(0)$ show that covalency does not affect the perpendicular susceptibility.

Summing up we may state that in the past five years quite convincing experimental evidence for the existence of zero-point spin deviations has been obtained, thanks mainly to the discovery of the quasi 2-d Heisenberg antiferromagnets. As concerns the XY antiferromagnet, some evidence for the presence of zero-point effects in this model has been encountered

Fig. 68

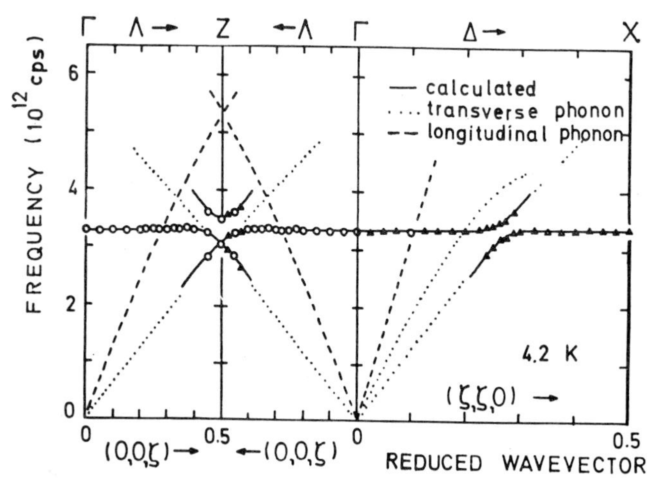

Magnon and phonon dispersion curves as measured in $FeCO_3$ by Wrege *et al.* (1971). The interesting features are firstly the independence of the wave vector of the magnon dispersion (solid curve : calculated), as expected for an Ising system, and secondly the apparent magnon–phonon coupling (note that the coupling only occurs with selected phonon modes).

in the 2-d compounds $CoCl_2 \cdot 6H_2O$ and $CoBr_2 \cdot 6H_2O$. Lastly we mention that in some types of antiferromagnets zero-point effects will be entirely negligible, as for instance in the case of ferromagnetic layers with a very weak antiferromagnetic interlayer coupling. This is due to the fact that the antiferromagnetic intersublattice (=interlayer) interaction is so much smaller than the ferromagnetic intrasublattice (=intralayer) interaction (De Jongh 1972 a).

We would like to end this section with the remark that in many analyses of experiments the magnon and phonon effects are taken to be simply independent. That this will not always be a valid assumption is evidenced by the calculations of Kittel (1958) on the magnon–phonon interaction (see also Rives *et al.* 1969 for more recent references). This interaction is best illustrated by the fact that the magnon and phonon dispersion curves will intersect at some point in the ω–k frame. As an experimental verification we show in fig. 68 the dispersion relations of $FeCO_3$, as reported by Wrege *et al.* (1971). Note that the magnetic excitations interact selectively with a particular type of phonon mode. A second reason for reproducing the measurements on this Ising-like substance (with effective spin $S=\frac{1}{2}$) is that the magnon dispersion curve forms a clear example of the wave-vector independent magnon dispersion appropriate for an Ising system. Another good example of magnon–phonon coupling is provided by the recent neutron diffraction study of Rainford *et al.* (1972) on FeF_2. A theoretical treatment has been given by Lovesey (1972). In the two examples cited the coupling arises predominantly from the phonon modulation of the crystal field, so that the interaction between spins and lattice is via the orbital moment by means of the spin–orbit coupling. As another mechanism for the magnon–phonon interaction we mention the phonon modulation of the exchange integral.

4.3. *Series expansions*

In this section we want to stipulate the important role of series expansions in the field of phase transitions. We have already witnessed in the preceding pages how this approximate method of obtaining information about the thermodynamic behaviour has been quite successful, in particular in the many cases where exact closed-form solutions are lacking.

The series for the various thermodynamic quantities are mostly expansions in a high or low-temperature parameter. They are exact in the sense that the calculated coefficients are exact results, but at the same time they are approximations valid only in a certain temperature range, since the number of terms that can be obtained is limited. At first sight this might limit their applicability rather severely. However, the technique of analysing the truncated series has been highly developed and one has managed to derive predictions from them in ranges of temperatures widely outside their apparent validity, for instance even in the critical region.

This being no place to review the existing methods of analysis, we only mention that at the root of many lies the assumption that the critical behaviour of thermodynamic functions is in the form of simple power laws, as we have already encountered above for the magnetization. In the next section the reader will find a list of the most frequently used critical exponents associated with these power laws. In many cases the predicted values for the exponents have been obtained from the analyses of series expansions. The assumption of a power-law behaviour in the critical region is not an arbitrary choice, but is based upon the fact that in the exactly soluble models like the 2-d Ising model the critical behaviour is of this form. The soluble models also present a testing ground for the methods of analysis of the series, since one may compare the exactly known behaviour of a thermodynamic function with that derived from a series expansion analysis.

We will now briefly mention some of the successful results obtained with the series expansion technique. Besides values for the critical exponents they yield fairly accurate estimates of the critical temperature, as has for instance been done by Rushbrooke and Wood (1958, 1963) for the 3-d Heisenberg model. This is mostly accomplished by locating the temperature at which the (staggered) susceptibility diverges, since the series for the susceptibility are usually rapidly converging. As we have seen, such an analysis for the 2-d Heisenberg models has led to the postulate of a new magnetic phase with an infinite susceptibility but without a spontaneous magnetization.

Apart from the critical behaviour, predictions for the temperature dependence of thermodynamic functions over quite a wide temperature range have been obtained. Thus the analysis of Baker *et al.* (1964) of the series for the specific heat of the ferro and antiferromagnetic Heisenberg chain nicely confirms the calculations of Bonner and Fisher (1964) down to $kT/|J| \simeq 0{\cdot}4$. Sykes and Fisher (1962) deduced the behaviour of the antiferromagnetic susceptibility for the 2-d and 3-d Ising model for all temperatures. For the 3-d Ising and Heisenberg models and for the 2-d Heisenberg model all the quantitative knowledge about the thermodynamic behaviour is based upon series expansion analyses (in the case of the isotropic models spin-wave theory, of course, provides an additional source of information).

In summing up, series expansions have yielded important contributions to our knowledge of phase transitions. The many examples of excellent fits of experimental results to the predictions obtained from their analyses, as given in the preceding pages (cf. figs. 28, 34 (*b*), 40, 49, 50, 53), may serve as convincing proof of the usefulness and applicability of this approach.

4.4. *Critical behaviour*

If the transition point is approached closely enough one finally enters the critical region, i.e. the temperature range around T_c in which the behaviour of the thermodynamic functions is governed by the asymptotic power-law expressions (although more complex expressions have been proposed and may not be excluded *a priori*). For a discussion of these matters and definitions of critical exponents see, e.g., Fisher (1967) or Stanley (1971). As to the extent of the critical region around T_c one may say that—depending on the particular thermodynamic function—it is confined to at most $|T-T_c|/T_c < 10^{-2}$–10^{-1}. By this we mean that in this temperature range the behaviour of a function will not differ perceivably from the asymptotic power-law form, so that the usual log–log plot will result in a straight line with a slope equal to the value of the critical exponent. It is very important to realize the finite extent of the critical region and to know beforehand how close one has to come near T_c in order to expect the power-law behaviour. For instance, in the case of the specific heat of 3-d Ising models, which have only a weak singularity at T_c, one has to come closer than 10^{-10} of T_c in order to have a difference smaller than about 1% between the power-law behaviour and the full expression for the specific heat (Sykes *et al.* 1967, Wielinga 1971). Quite generally, in comparing experimental heat capacities with theory, one should adhere to the full theoretical expression (Domb and Bowers 1969). It is evident that the method of plotting experimental data on a double logarithmic scale and drawing a ' straight ' line through them without any further consideration will yield incorrect values for the critical exponents, also in cases where the singularities are stronger.

We now give the asymptotic expressions for the thermodynamic functions that we will consider. The theoretical predictions for the critical exponents are listed in table 15.

Specific heat :

$$C_m/R \sim A\,(1-T_c/T)^{-\alpha} \quad (T \to T_c^+;\ H=0), \tag{4.9}$$

$$C_m/R \sim A'(1-T/T_c)^{-\alpha'} \quad (T \to T_c^-;\ H=0). \tag{4.10}$$

Spontaneous magnetization :

$$M_s(T)/M_s(0) \sim B(1-T/T_c)^{\beta} \quad (T \to T_c^-;\ H=0). \tag{4.11}$$

Initial susceptibility :

$$\chi T/C \sim C_0\,(1-T_c/T)^{-\gamma} \quad (T \to T_c^+;\ H=0), \tag{4.12}$$

$$\chi T/C \sim C_0'\,(1-T/T_c)^{-\gamma'} \quad (T \to T_c^-;\ H=0). \tag{4.13}$$

Critical isotherm :

$$H \sim D|M(H)|^{\delta} \quad (H \to 0;\ T=T_c). \tag{4.14}$$

Inverse correlation length:

$$\kappa \sim N\,(1-T_c/T)^{\nu} \quad (T \to T_c^+;\ H=0). \tag{4.15}$$

$$\kappa \sim N'(1-T/T_c)^{\nu'} \quad (T \to T_c^-;\ H=0). \tag{4.16}$$

Wave-vector dependent susceptibility :

$$\hat{\chi}(k) \sim k^{\eta-2} \quad (k \to 0;\ T=T_c). \tag{4.17}$$

Table 15. Predictions for the critical indices for a number of theoretical models. Exact results have been obtained for the MF model, the spherical model and the $d=2$ Ising model. The other values have been derived from series expansion analyses. In compiling the published data we have assumed the indices to be independent of spin value and, within a given dimensionality, of the precise lattice structure. In some cases therefore the listed result is a mean of the values obtained for different S. For references see, for instance, Fisher (1967), Rushbrooke *et al.* (1973), Betts (1973), Sykes *et al.* (1972), Baker and Gaunt (1967), Ferer *et al.* (1971), Ritchie and Fisher (1972), Fisher and Burford (1967), Baker *et al.* (1967 b, 1970). In these papers other references, in particular to earlier work, may be found.

Model system	α	α'	β	γ	γ'	δ	η	ν	ν'
MF	0 (discont.)	0 (discont.)	$\frac{1}{2}$	1	1	3	0	$\frac{1}{2}$	$\frac{1}{2}$
Spherical ($d=3$)	-1	—	$\frac{1}{2}$	2	—	5	0	1	—
Ising ($d=3$)	$\simeq\frac{1}{8}$	$\frac{1}{8}-\frac{1}{16}$	0·312	1·25	1·25–1·31	$\simeq 5$	0·03–0·05	$\simeq 0·63$	—
XY ($d=3$)	$\simeq 0$ (logar.)	—	—	$\simeq 1·33$	—	—	—	—	—
Heisenberg ($d=3$)	$\simeq -0·1$	—	$\simeq 0·36$	$\simeq 1·40$	—	$\simeq 5$	0·03–0·04	$\simeq 0·71$	—
Ising ($d=2$)	0 (logar.)	0 (logar.)	$\frac{1}{8}$	1·75	1·75	15	0·25	1	1

As usual the primed exponents refer to the region below T_c. It is further noted that the power law for the susceptibility is different from the often used form $\chi \sim |T-T_c|^{-\gamma}$. We believe that the expression (4.12) is more correct in comparing experiment with theory, since in theoretical calculations one does not consider χ itself, but the ' energetic ' susceptibility $\chi T/C$. As pointed out by Wielinga (1971), in practical cases the γ value derived from a χ/C versus $(T-T_c)/T_c$ plot will tend to be a few per cent larger than that obtained from a $\chi T/C$ versus $(T-T_c)/T$ plot.

Equations (4.9) and (4.10) also contain the possibility of a logarithmically diverging specific heat, since this is included in the case $\alpha = 0$, which can be shown by considering that

$$\lim_{\alpha \to 0} \frac{x^\alpha - 1}{\alpha} = \ln x. \tag{4.18}$$

The case $\alpha = 0$ further comprises a finite discontinuity at T_c, as is predicted for the specific heat by the MF theory.

Of considerable interest, though probably not open to direct experimental investigation, are the so-called gap exponents Δ_n (Essam and Fisher 1963). Below T_c we may consider the successive field derivatives of the free energy F (in the limit $H \to 0$) :

$$\left.\begin{aligned} (\partial F/\partial H)_T &\equiv F^{(1)} \sim (1-T/T_c)^{-\Delta_1'} F^{(0)}, \\ &\vdots \\ (\partial^n F/\partial H^n)_T &\equiv F^{(n)} \sim (1-T/T_c)^{-\Delta_n'} F^{(n-1)}, \end{aligned}\right\} \quad (T < T_c).$$

Above T_c all odd field derivatives of $F(H, T)$ are zero for $H = 0$, so that one may define similarly

$$(\partial^{2n} F/\partial H^{2n})_T \equiv F^{(2n)} \sim (1-T_c/T)^{-2\Delta_{2n}} F^{(2n-2)} \quad (T > T_c).$$

Now since the specific heat $C(H=0) \sim (1-T/T_c)^{-\alpha'}$ is the second temperature derivative of F, it follows that $F^{(0)} \sim (1-T/T_c)^{2-\alpha'}$, and since $F^{(1)} \sim M_s(T) \sim (1-T/T_c)^{\beta}$, one obtains the relationship

$$\Delta_1' = 2 - \alpha' - \beta. \tag{4.19}$$

Furthermore, $F^{(2)} \sim \chi T/C \sim (1-T/T_c)^{-\gamma'}$ and hence

$$\Delta_2' = \beta + \gamma'. \tag{4.20}$$

Above T_c we have likewise

$$2\Delta_2 = 2 - \alpha + \gamma, \tag{4.21}$$

What is of importance is that in the case where the critical exponents are known exactly, as for the 2-d Ising model and the MF model, it is found that all gap exponents are equal. Thus in the MF theory $\alpha = \alpha' = 0$, $\beta = \frac{1}{2}$ and $\gamma = \gamma' = 1$, so that $\Delta_1' = \Delta_2' = \Delta_2 = \frac{3}{2}$. For the 2-d, $S = \frac{1}{2}$, Ising model $\alpha' = \alpha = 0$, $\beta = \frac{1}{8}$ and, most likely, $\gamma = \gamma' = \frac{7}{4}$, yielding

$$\Delta_1' = \Delta_2' = \Delta_2 = \tfrac{15}{8}.$$

Numerical studies for the 2-d Ising model (Essam and Hunter 1968) are indeed consistent with $\Delta_n' = \Delta_{2n} = \frac{15}{8}$ for all n. For the 3-d Ising model ($S=\frac{1}{2}$) the same authors obtained $\Delta' = \Delta \simeq \frac{25}{16} = 1{\cdot}56$, although the possibility $\Delta' = \Delta \simeq \frac{13}{8} = 1{\cdot}625$ could not be excluded. For instance eqns. (4.19) to (4.21) are satisfied with $\alpha = \alpha' = \frac{1}{8}$, $\beta = \frac{5}{16}$; $\gamma = \gamma' = \frac{5}{4}$ and $\Delta = \Delta' = \frac{25}{16}$. These values for α, α', β, γ and γ' are consistent with the best numerical estimates (cf. table 15). For the 3-d, $S=\frac{1}{2}$, Heisenberg model, Baker *et al.* (1967 b) found $2\Delta = 3{\cdot}63 \pm 0{\cdot}03$, whereas Stephenson and Wood (1968) concluded to $2\Delta \simeq 3{\cdot}45$ for the $S=\infty$ case. There is therefore strong evidence in favour of the general assumption that the gap exponents are equal for all n.

This assumption is also related to the scaling hypothesis. For instance, the relationship $\alpha' + 2\beta + \gamma' = 2$, obtained through equating Δ_1' and Δ_2', is one of the scaling laws, and in fact scaling theory predicts $\Delta_n' = \Delta_{2n}$ for all n. Originally put forward by Widom (1965), Domb and Hunter (1965) and Kadanoff (1966), the scaling hypothesis has proven to be a successful approach, to which many authors have contributed. For a recent review see, e.g., Hankey and Stanley (1972). Though not giving numerical values for the critical indices, scaling theory predicts relationships between them, the scaling laws. Accordingly, the number of independent critical exponents is restricted, so that from a knowledge of two all others can be derived. In addition, the hypothesis yields predictions concerning the form of the equation of state.

It can be argued (see, e.g., Stanley 1971) that the scaling hypothesis comes down to the assumption that the free energy $F(T, H)$ is a generalized homogeneous function, meaning that there exist two parameters a_T and a_H, such that

$$F(\lambda^{a_T} T, \lambda^{a_H} H) = \lambda F(T, H) \tag{4.22}$$

for any value of the number λ. It then transpires that the critical exponents can all be expressed in terms of the two parameters a_H and a_T. It is intriguing that such a unification of many various items can be brought about by the introduction of a fairly simple mathematical concept. We mention some of the scaling relations that are useful in the present context (primed and unprimed exponents are equal) :

$$\alpha + 2\beta + \gamma = 2, \tag{4.23}$$

$$\alpha + \beta(\delta + 1) = 2, \tag{4.24}$$

$$\gamma = \beta(\delta - 1), \tag{4.25}$$

$$\Delta' = \Delta = \beta + \gamma = \beta\delta = 1 + \tfrac{1}{2}(\gamma - \alpha) = 2 - \alpha - \beta, \tag{4.26}$$

$$\gamma = \nu(2 - \eta), \tag{4.27}$$

$$\nu d = 2 - \alpha, \tag{4.28}$$

$$2 - \eta = d(\delta - 1)/(\delta + 1). \tag{4.29}$$

Here d, as before, denotes the lattice dimensionality.

In view of the remark made in connection with the equalities between the gap exponents, it will come as no surprise that most of the above scaling relations are obeyed in the case of model systems for which the exponents are known exactly, or to high accuracy (2-d Ising, MF, and spherical model). In the case of the 3-d Ising and Heisenberg models the possible errors in the exponent values are larger, but within the uncertainties a set of critical exponents can be chosen that satisfy the scaling relations. Experimental results also seem to be in agreement; in any case no set of critical indices has yet been obtained on a magnetic substance that invalidates them, as we shall see below. It is also remarked that many of the scaling relations can be derived as inequalities on the basis of thermodynamic arguments (see, e.g., Fisher 1967, Stanley 1971) under certain assumptions. For instance, $\alpha+2\beta+\gamma\geqslant 2$ is the well-known Rushbrooke inequality, whereas the inequality $\alpha+\beta(\delta+1)\geqslant 2$ has been derived by Griffiths (1965). Evidently, the scaling relations are useful in the case of a model for which only two exponents have been determined from series expansions. In that way Betts (1973) has deduced a set of exponents for the 3-d XY model from the estimated α and γ.

Another interesting recent development is the bilinear form hypothesis (Betts *et al.* 1971, Stanley and Betts 1972). As a starting point this theory takes the universality hypothesis (Griffiths 1970 b, Kadanoff 1970), already quoted in the preceding sections, assuming the critical exponents to depend only on the lattice dimensionality d and the spin dimensionality D. The latter gives the dimensionality of the interacting spin vectors, thus $D=1$, 2, and 3 corresponds to the Ising, the XY and the Heisenberg model, respectively, the order parameter being a 1, 2 and 3-d vector. As shown by Stanley (1968 c) the limit $D\to\infty$ (also $S=\infty$) yields the so-called spherical model. The critical indices should be independent of spin value S, and, at least close enough to T_c, also of lattice anisotropy and spin-space anisotropy, by which are meant the dependence of the exchange on the direction in the crystal and on the spin components, respectively. Current theoretical evidence indeed supports the spin independence and seems to indicate that the exponents change discontinuously in going from a 3-d to a 2-d lattice or from an Ising to a Heisenberg Hamiltonian (retaining their 3-d or Ising values until the interaction in the third dimension or the anisotropy have indeed become zero). Lastly, in the case of short-range forces, the exponents are generally accepted to be independent of the range of the interaction (as long as this remains finite).

Taking then for a given dimension d a variation with D to be the only possibility, Stanley and Betts propose a dependence of each exponent λ on D of the form $\lambda(D)=\lambda(\infty)R_\lambda(D)$, where the factor $R_\lambda(D)$ is the bilinear form

$$R_\lambda(D)=\frac{b_\lambda+D}{c_\lambda+D}. \tag{4.30}$$

For instance, the γ values for the Ising, the XY and the Heisenberg model (1·25, 1·33 and 1·38) are closely reproduced by the expression $\gamma = 2(4+D)/(7+D)$, or likewise by $\gamma = 2(21+4D)/(36+4D)$. As pointed out by Rushbrooke *et al.* (1973), for the Heisenberg model one may take either of the so-obtained γ's and, if in addition $\delta = 5$ is assumed, calculate all other exponents from the scaling laws with γ and δ. For the spherical model $\delta = 5$, whereas numerical work on the Ising and Heisenberg 3-d models strongly suggests that δ is close to 5. On the other hand, the assumption $\delta = 5$ invariably leads to the result $\eta = 0$ in three dimensions, in view of the scaling relation $2 - \eta = d(\delta - 1)/(\delta + 1)$. Numerical studies seem to exclude the possibility that $\eta = 0$. Indeed, from $\eta = 0{\cdot}056 \pm 0{\cdot}008$ for the Ising model (Fisher and Burford 1967) and $\eta = 0{\cdot}043 \pm 0{\cdot}014$ for the Heisenberg model (Ritchie and Fisher 1972), one calculates $\delta = 4{\cdot}68 \mp 0{\cdot}08$ and $\delta = 4{\cdot}75 \mp 0{\cdot}08$ for Ising and Heisenberg models respectively. Contrastingly, the numerical estimates favour a δ value slightly higher than 5. Similar discrepancies have been encountered by Essam and Hunter (1968) in testing the relation $2\Delta = \gamma + d\nu$ and by Fisher and Burford (1967) and Ritchie and Fisher (1972) in the case of the equation $d\nu = 2 - \alpha$. In some cases the discrepancies are clearly outside the quoted errors in the numerical estimates. It is not quite clear whether the deficiency lies in the numerical calculations or whether the (dimension-dependent) scaling laws involving the correlation exponents ν and η are only approximately correct. Indications of the latter possibility have been found by Stell (1968) and Domb (1968).

In turning now to the experimental results, we shall first confine our attention to the 3-d compounds, and start with the specific heats. We shall thereto rely heavily on the recent review of Wielinga (1971). It is first remarked that in particular for specific heat measurements it is useless to compare the experimental data with theoretical predictions of the power-law form, since the asymptotic behaviour (eqn. (4.9)) is only followed in a region so close to T_c that it is widely outside experimental reach. Even in high quality magnetic single crystals a considerable rounding of the specific heat singularity (over a range of 10^{-3}–10^{-2} of T_c) is the rule rather than the exception and this limits meaningful analyses to the region $|T - T_c|/T_c > 10^{-2}$–10^{-3}, since the position of T_c becomes an additional unknown parameter.

The only way out is therefore to derive a closed-form theoretical prediction, valid over the whole temperature range. In the case of series expansions with only positive terms, a useful method has been introduced by Sykes *et al.* (1967) that was subsequently applied by Domb and Bowers (1969) and by Wielinga (1968, 1971). Starting from the series expansion of a thermodynamic function $F(T)$

$$F(T) = \sum_{n=0}^{N} a_n (J/kT)^n, \tag{4.31}$$

of which only the first N coefficients a_n are known, one assumes the asymptotic behaviour to be given by

$$F(T) \simeq R(1-T_c/T)^{-\rho}, \tag{4.32}$$

where ρ is some appropriate critical index. With the aid of the usual extrapolation methods (ratio method or Padé approximant values) values for ρ and T_c are derived from the N known coefficients a_n. In the next step one forms the binomial expansion

$$R(1-T_c/T)^{-\rho} = \sum_{n=0}^{\infty} Rb_n(J/kT)^n, \tag{4.33}$$

using the values obtained for ρ and T_c in calculating the coefficients b_n (for any n). The constant R is determined as the limiting value of the ratio's $R_n \equiv a_n/b_n$ for large n. The behaviour of $F(T)$ over the whole temperature range $0<1-T_c/T<1$ can now be calculated from the expression

$$F(T) = R(1-T_c/T)^{-\rho} + \sum_{n=0}^{N} (a_n - Rb_n)(J/kT)^n. \tag{4.34}$$

On substituting eqn. (4.33) into eqn. (4.34), one observes that the first N terms of the right-hand side of (4.34) are identical with the N known terms of the expansion (4.31). Thus expression (4.34) reduces to the truncated series (4.31) in the high-temperature region. In the neighbourhood of T_c the function $F(T)$ is very well approximated by the asymptotic term $R(1-T_c/T)^{-\rho}$ plus a constant $P(T_c)$, which is the value of the correction polynomial in (4.34) :

$$P(T) = \sum_{n=0}^{N} (a_n - Rb_n)(J/kT)^n, \tag{4.35}$$

evaluated at T_c. It is also clear that the relative magnitude of the constants R and $P(T_c)$ will determine the extent of the critical region, that is the temperature range around T_c in which the asymptotic power-law term is expected to account within 1% for the behaviour of $F(T)$. The amplitudes R are of the order of unity for the specific heat, the susceptibility and the magnetization. For the susceptibility the value of $P(T_c)$ is typically an order of magnitude smaller, but for the specific heat $P(T_c)$ is about equal to R, explaining the very narrow extent of the critical region in the latter case. For further details and numerical results for these parameters one is referred to the review of Wielinga (1971) and the papers cited therein.

In figs. 69 and 70 are plotted the specific heat data on the 3-d Ising compounds, treated above, for $T > T_c$ and $T < T_c$, respectively, together with theoretical predictions for 3-d Ising lattices obtained in the just described manner. Again we stipulate that only for relative temperatures exceeding 10^{-3} (in some cases even 10^{-2}) do the experimental results have a real meaning, in view of the uncertainties in T_c. For comparison the

Fig. 69

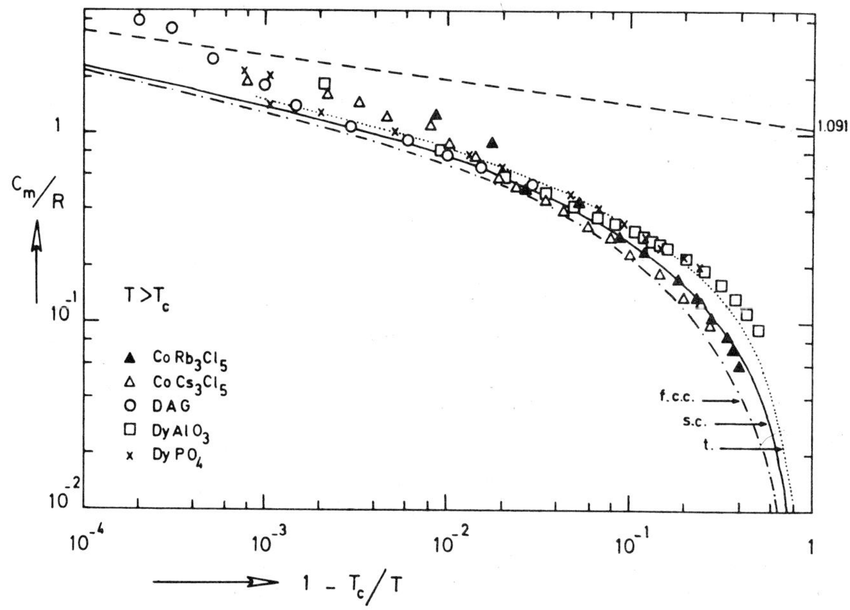

The magnetic specific heat of a number of 3-d Ising compounds for $T > T_c$. The dashed curve gives the asymptotic power law behaviour of the s.c. Ising model, assuming $\alpha = \frac{1}{8}$. The other curves are predictions for the s.c., b.c.c. and tetrahedral Ising lattices, calculated from series expansions by Wielinga (1971) and Blöte (private communication) in the manner described in the text.

dashed curve in fig. 69 displays the asymptotic power-law behaviour for the s.c. lattice, calculated with $A = 1{\cdot}091$ and $\alpha = \frac{1}{8}$ (Wielinga 1971). The experimental data are from Blöte and Huiskamp (1969 ; $CoRb_3Cl_5$), Wielinga *et al.* (1967 ; $CoCs_3Cl_5$), Keen *et al.* (1967 ; DAG), Cashion *et al.* (1968 ; $DyAlO_3$) and Wright *et al.* (1971 ; $DyPO_4$). In agreement with the discussion given in § 3.3.2, the data for $CoRb_3Cl_5$ and $CoCs_3Cl_5$ fit the s.c. and the f.c.c. curve, respectively, whereas the dysprosium compounds DAG and $DyPO_4$ are well described by the diamond Ising model. $DyAlO_3$ also shows a tendency expected for a low coordination number.

The theoretical curves in fig. 70 need some explanation. In the limit $T \to 0$ the s.c. and diamond curve coincide, which is explained by the exponential behaviour ($\exp(-zJ/kT)$) at the lowest temperatures. Since the energy content below T_c increases with the coordination number (cf. table 11), one expects the s.c. curve to lie above the diamond curve, as is indeed the case for $1 - T/T_c > 10^{-1}$. At about 10^{-1} there is a crossing point, but since obviously most of the energy content comes from the region $1 - T/T_c > 10^{-1}$, the net result for $(E_c - E_0)/RT_c$ is still a bit higher for the s.c. than for the diamond lattice.

Fig. 70

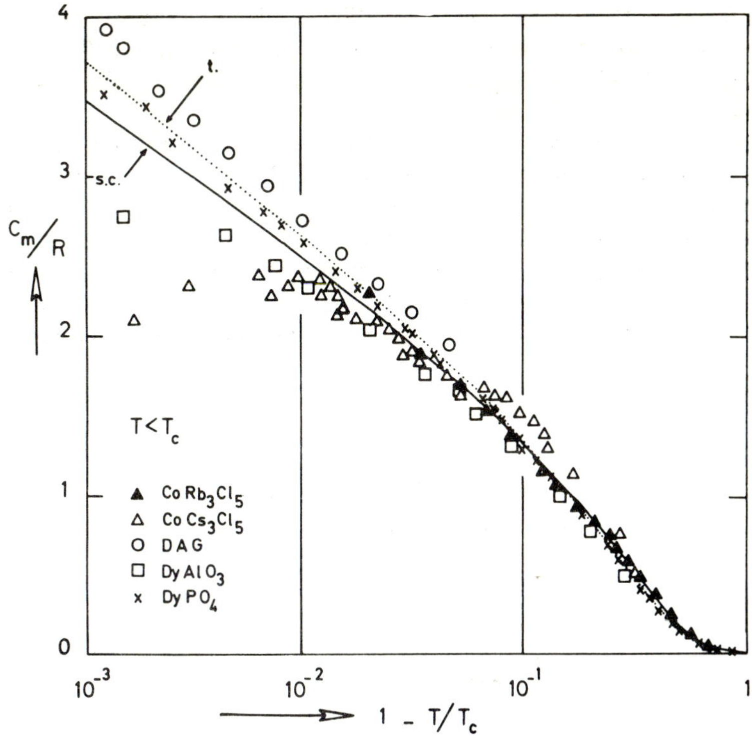

The magnetic specific heat of a number of 3-d Ising compounds for $T < T_c$. The drawn curves are predictions for the tetragonal and the s.c. lattice, calculated from series expansions by Wielinga (1971) and Blöte (private communication).

Theoretical results on the Heisenberg model have only been obtained for $T > T_c$ and spin values $S = \frac{1}{2}$ and $S = \infty$. As mentioned in the preceding pages, current expectations are that the Heisenberg specific heat does not diverge but displays a cusp at T_c, so that the heat capacity has a finite maximum, although the temperature derivatives still diverge on both sides of T_c. Instead of

$$C_m/R = A(1 - T_c/T)^{-\alpha} + P(T) \tag{4.36}$$

the high-temperature expression for the specific heat then becomes

$$C_m/R = A_0 - A_1(1 - T_c/T)^{-\alpha} + P_0(T), \tag{4.37}$$

where A_0 and A_1 are constants, $\alpha < 0$, and $P_0(T)$ is again a correction polynomial. The expression (4.37) yields a finite maximum equal to $A_0 + P_0(T)$ at the critical temperature. For the f.c.c., $S = \infty$, ferromagnet, Domb and Bowers (1969) obtained $\alpha = -\frac{1}{16}$ and $A_0 + P_0(T_c) = 10{\cdot}00$, taking $T_c/\theta = 0{\cdot}794$. However, with the same T_c/θ value, the choice

$\alpha = -\frac{1}{8}$ and $A_0 + P_0(T_c) = 5{\cdot}7$ leads to essentially the same specific heat values (within 1%) in the range $(1 - T_c/T) > 10^{-2}$. Moreover, Wielinga (1971) has shown that the same applies to the combination $T_c/\theta = 0{\cdot}792$ (Stanley 1967), with α varying from $-0{\cdot}02$ to $-0{\cdot}06$. With $\alpha = -0{\cdot}04$ the value of $A_0 + P_0(T_c)$ becomes 15·59.

Evidently it is difficult to differentiate experimentally between these various possible combinations. Domb and Bowers (1969) and Wielinga (1971) both compared their calculations with data of Van der Hoeven *et al.* (1968) on EuS. With its high spin ($S = \frac{7}{2}$) and its low anisotropy ($H_A/H_E \simeq 2 \times 10^{-4}$) this salt is the nearest approximation of the classical f.c.c. Heisenberg ferromagnet available. EuO would also qualify but unfortunately in this case the sample on which specific heat measurements

Fig. 71

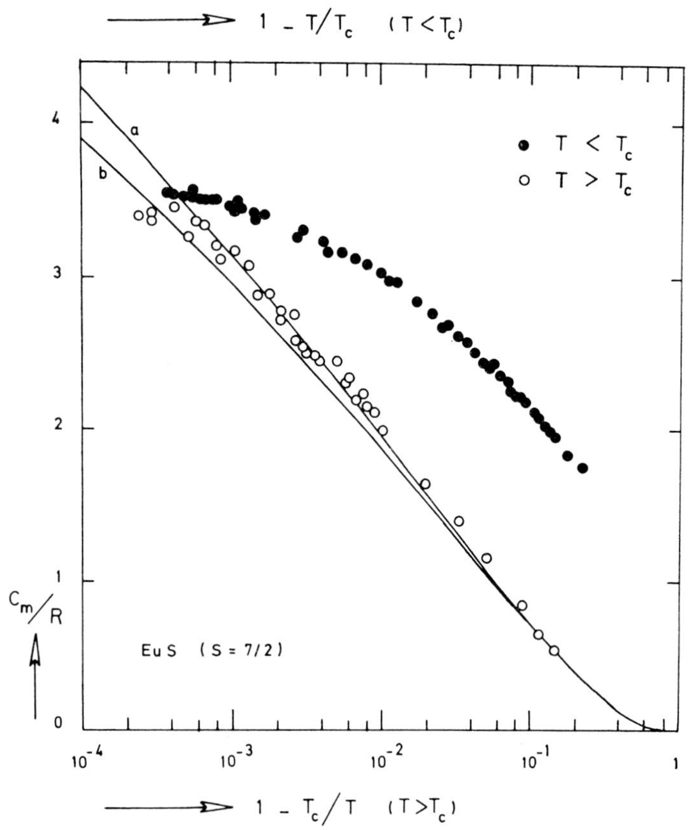

The magnetic specific heat of EuS (Van der Hoeven *et al.*, 1968), compared with the calculations of Wielinga (1971 ; curve *a*) and Domb and Bowers (1969 ; curve *b*) from the series expansion for the specific heat of the classical, f.c.c., Heisenberg ferromagnet ($T > T_c$).

were performed (Teaney *et al.* 1966) contained a substantial amount of impurity, that apparently influences the heat capacity considerably (Wielinga 1971). In fig. 71 the EuS data have been reproduced, together with the predictions of Wielinga (curve *a* : $\alpha = -0{\cdot}04$, $T_c/\theta = 0{\cdot}792$) and of Domb and Bowers (curve *b* : $\alpha = -\frac{1}{16}$, $T_c/\theta = 0{\cdot}794$). Since the data seem to favour the former calculation, one would think that Domb and Bowers' estimate of 10 R for the finite Heisenberg limit is too low. In this respect the following considerations are of importance. Firstly, the presence of substantial further neighbour interactions in EuS will tend to increase the asymmetry of the specific heat curve (cf. table 12), so that the experimental data on the high-temperature side will be lower than for the ideal

Fig. 72

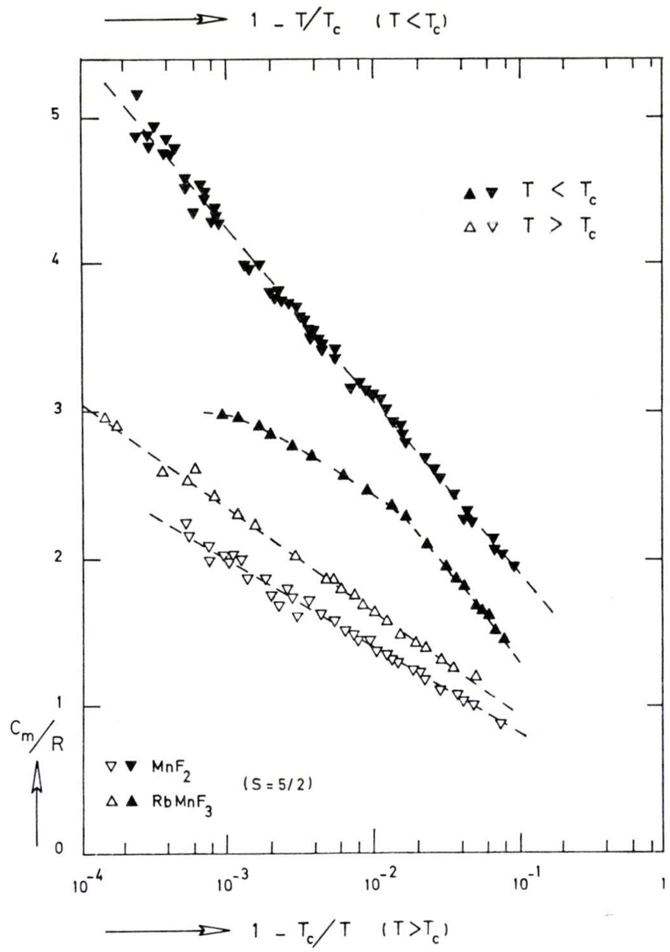

The magnetic heat capacity of $RbMnF_3$ (Teaney *et al.*, 1966) and MnF_2 (Teaney 1965) in the neighbourhood of T_c.

model. Secondly, the experimental specific heats are seen to decrease with decreasing spin value (cf. figs. 72 and 73 below), also implying that data taken on an $S=\frac{7}{2}$ compound should be somewhat below the $S=\infty$ prediction. Thus one could conclude to an α of the order of $-0{\cdot}04$ or even to a logarithmic divergence ($\alpha=0$, see Van der Hoeven *et al.* 1968) on the basis of the data above T_c. However, the experimental uncertainties and the sensitivity of the theoretical predictions to the exact value of T_c/θ do not warrant a firm conclusion, although a value $\alpha=-\frac{1}{8}$ does seem to be outside the limits.

In any case it is clear that the Heisenberg limit is, by far, not reached experimentally, since the rounded maximum reaches a height of about 3·6 R only. The mechanisms responsible for the experimentally observed rounding are not yet clear. Finite size effects are not expected to play a role for $|T-T_c|/T_c>10^{-6}$ (see, e.g., Fisher 1967). Various authors have suggested a distribution of transition temperatures throughout the sample as a possible explanation. This seems to be a plausible assumption and, indeed, calculations taking into account such a spread in T_c have produced specific heat curves that mimic quite well the observed behaviour (see, e.g., Wielinga *et al.* 1967). The fact that the experimental lattices are not rigid but compressible also has an effect, but it is generally accepted that this will sharpen the transition. For instance Domb and Wyles (1969), in comparing specific heat data on $GdVO_4$ and $MnCl_2 . 4H_2O$ with model calculations, have discussed this possibility to

Fig. 73

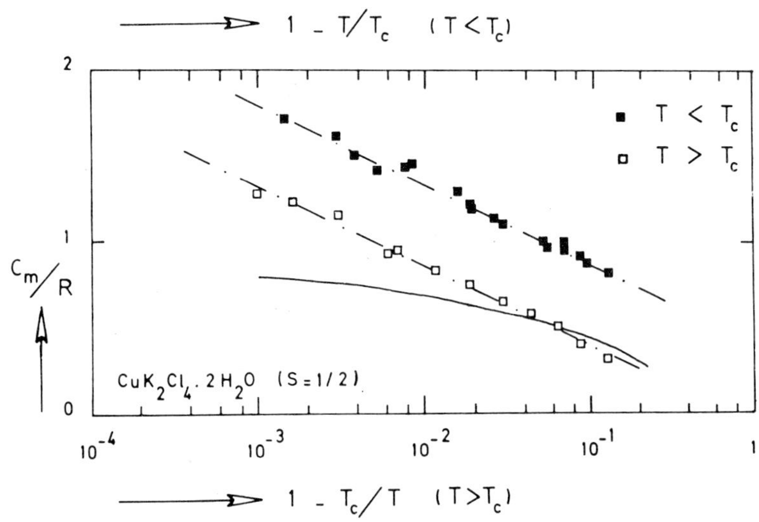

The magnetic specific heat of $CuK_2Cl_4 . 2H_2O$ in the neighbourhood of T_c (Miedema *et al.* 1965). The solid curve is the prediction obtained by Baker *et al.* (1967 b) from the series expansion of the $S=\frac{1}{2}$, b.c.c. Heisenberg ferromagnet ($T>T_c$).

explain the sharp uprise in the high-temperature specific heat of these salts, observed between $10^{-4} < 1 - T_c/T < 10^{-3}$. It is remarked that such an increase will lead to a high apparent value for the exponent α (in expression (4.9)), in accordance with Fisher's scheme for a renormalization of the critical exponents by hidden variables (Fisher 1968), such as impurity concentrations or extra degrees of freedom.

Returning to the EuS data of fig. 71, it is observed that the low-temperature results strongly indicate a negative value for the exponent α'. Thus Van der Hoeven *et al.* (1968) deduced $\alpha' = -0{\cdot}25 \pm 0{\cdot}03$ from their analysis, which also corroborates the theoretical expectation of a finite cusp in C_m at T_c.

A quite similar pattern is followed by the data on $RbMnF_3$ (Teaney *et al.* 1966) displayed in fig. 72. Notice the decrease in specific heat in going from $S = \frac{7}{2}$ to $S = \frac{5}{2}$ and the fact that the anomaly is becoming more symmetric around T_c. Also in this case the heat capacity is apparently diverging logarithmically above T_c in the region accessible to experiment, whereas for $T < T_c$ the downward curvature indicates a negative α'. In summing up the experimental and theoretical evidence just presented, we would therefore conclude that the indications for a negative α', with α' of the order of $\frac{1}{8}$, are rather strong, while the possibility of a negative α, with $0 < |\alpha| < \frac{1}{10}$ is also consistent with the results obtained.

For comparison we have included in fig. 72 the heat capacity data of MnF_2 (Teaney 1965). In view of the fairly large anisotropy ($H_A/H_E \simeq 1{\cdot}6 \times 10^{-2}$ compared to 5×10^{-6} in $RbMnF_3$) one may expect the critical behaviour to be Ising-like. Accordingly the specific heat should have a higher asymmetry around T_c. Moreover since the 3-d Ising specific heat most probably diverges logarithmically or with a small positive value for the exponent, the anomaly should be much sharper than in $RbMnF_3$ and in the experimental region the heat capacity should behave as if it were diverging logarithmically (note that in the region $10^{-3} < 1 - T_c/T < 10^{-1}$ the theoretical curves in fig. 70 may very well be approximated by straight lines). All these features are indeed confirmed by the MnF_2 data. Besides the anisotropy other effects will play a role, for instance, the higher coordination number ($z = 8$ compared to $z - 6$) and the considerable next-nearest neighbour interaction in MnF_2 will also increase the asymmetry of the specific heat. But we expect that the differences between the two manganese compounds are mainly due to the anisotropy (cf. the small effect of z in figs. 69 and 70), although the substantial thermal expansion observed in MnF_2 near T_c (Gibbons 1959) may also have an effect (Domb and Wyles 1969).

As our last example we show in fig. 73 heat capacity data on $CuK_2Cl_4 \cdot 2H_2O$ (Miedema *et al.* 1965). Although of less quality than the previous experiments (e.g. the value of T_c is known with less accuracy) they fit into the same pattern. Also in this case the anisotropy is considerable ($\simeq 1\%$) whereas the next-nearest neighbour interactions are such as to make an equivalent neighbour model with first and second neighbours applicable. This may explain the large deviation of the high-temperature specific heat from the prediction of Baker *et al.* (1967 b) for the $S=\frac{1}{2}$, b.c.c., Heisenberg lattice, which can be represented by the expression

$$C_m/R=(T_c/T)^2[0{\cdot}971-0{\cdot}668(1-T_c/T)^{0{\cdot}20}] \qquad (4.38)$$

for temperatures $0{\cdot}70<T_c/T<0{\cdot}95$, indicating a value 0·971 for the finite Heisenberg maximum for the case $S=\frac{1}{2}$. This value is most probably an underestimate in view of the large value adopted for $|\alpha|$. The decrease in specific heat with decreasing S is, however, confirmed by the experiment and again the behaviour is apparently logarithmic in the relative temperature range 10^{-3}–10^{-1}.

Turning next to the spontaneous magnetization we firstly remark that for this quantity the critical region extends appreciably farther away from T_c, due to the fact that the singularity is stronger than for the heat capacity. This is exemplified by fig. 74 where the predictions of various models are displayed (Wielinga 1971). It is seen that for most models the region in which the power-law behaviour is expected to hold with a high accuracy, say better than 1%, starts at about $1-T/T_c=4\times10^{-2}$. One can therefore safely conclude that log–log plots of experimental magnetization curves for $1-T/T_c<4\times10^{-2}$ will yield meaningful β and B values to compare with theory. Apart from the (exact) calculated results for the MF model and the quadratic Ising lattice, the curves for two cubic Ising models are shown, derived from series expansions in the manner described above. For the Heisenberg model series predictions are less conclusive (Baker *et al.* 1970), although for $1-T/T_c>10^{-1}$ the series analysis was in good accord with the curve obtained by Cooke and Gersch (1967), using second-order Green function theory. The results are represented by the solid curve in fig. 74. As a continuation in the critical region, we have drawn the broken curve, which has a slope 0·36 in agreement with most of the available predictions for the exponent β. In drawing this curve we have further assumed that the transition to the power-law behaviour occurs in a similar fashion as in the case of the 3-d Ising models.

In table 16 we have collected the values for β and B that have been found for the 3-d compounds considered in this paper. As mentioned before, the exponents are expected to be independent of spin value or, for a given dimensionality, of the precise lattice structure. The amplitude B, on the other hand, decreases with increasing S and coordination number (Fisher 1967) and, furthermore, both β and B will be affected by the anisotropy. This general pattern is followed by the results for B in table 16, moreover the β values for the anisotropic and the isotropic compounds tend to be closer to the Ising and to the Heisenberg prediction, respectively. In this respect it is surprising that just for $RbMnF_3$ the observed β value is within the uncertainty equal to the Ising result,

Fig. 74

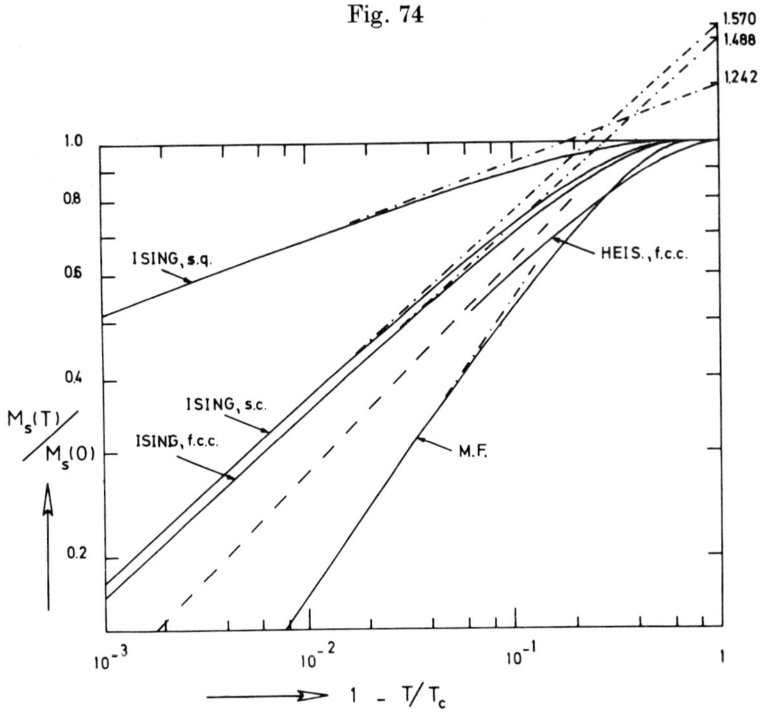

Theoretical predictions for the critical behaviour of the spontaneous magnetization of various model systems. Solid curves : closed-form expression, valid over the whole temperature range. Dot-dashed curves : asymptotic power-law behaviour. (After Wielinga 1971.)

being even lower than that obtained for MnF_2, which is considerably more anisotropic. Unfortunately, the $RbMnF_3$ value is only available in the literature as a result quoted in the abstract of a conference paper (Corliss *et al.* 1969), so that a check on the magnetization curve itself is not possible. All we can say is that according to the authors the result was obtained over one decade in relative temperature only, so that it might be changed by more extended measurements, such as the very careful N.M.R. study of MnF_2 by Heller (1966). In this work he was able to apply a correction for the thermal expansion of the lattice, whereby the exponent value changed from $0{\cdot}333 \pm 0{\cdot}003$ to the result $0{\cdot}335 \pm 0{\cdot}005$, listed in table 16.

Other apparent discrepancies are the β's of DAG and $FeCl_2$, which are too low as compared with the Ising prediction. In the case of $FeCl_2$ this may be due to the range in $1 - T/T_c$ in which the data were analysed. Below 10^{-3} there is only one measuring point near $1{\cdot}2 \times 10^{-4}$. If this is disregarded, the remaining data between 10^{-3} and 10^{-2} may be equally well fitted with a β of $0{\cdot}31$. In the case of DAG such an argument cannot be applied but there may be other explanations, as for instance the non-rigidity of the magnetic lattice. In fact, considering all the deviations

Table 16. Experimental values of the critical magnetization exponents β and δ and the amplitude B.

	S	β	B	Range of $1-T/T_c$	δ	Technique	Reference
Anisotropic compounds							
$DyPO_4$	$\frac{1}{2}$	0·314	1·661	1×10^{-4}–$3{\cdot}4\times10^{-2}$		Magnetoelectric effect	Rado (1969)
$Dy_3Al_5O_{12}$	$\frac{1}{2}$	0·26 ±0·02		1×10^{-3}–$5{\cdot}6\times10^{-2}$		Neutron scattering	Norvell *et al.* (1969)
$DyAlO_3$	$\frac{1}{2}$	0·311 ±0·005	1·51 ±0·03	5×10^{-3}–4×10^{-2}		Magnetoelectric effect	Holmes *et al.* (1971)
FeF_2	2	0·325 ±0·010	1·36 ±0·03	5×10^{-4}–1×10^{-1}		Mössbauer effect	Wertheim and Buchanan (1967)
$FeCl_2$	2	0·29 ±0·01	1·47 ±0·02	1×10^{-3}–1×10^{-1}		Neutron scattering	Yelon and Birgeneau (1972)
Isotropic compounds							
$Cu(NH_4)_2Br_4 . 2H_2O$	$\frac{1}{2}$	0·38 ±0·04	1·33 ±0·15	1×10^{-2}–2×10^{-1}	3·9 ±0·2	Induction	Wielinga and Huiskamp (1969)
		0·34 ±0·02		1×10^{-2}–1×10^{-1}	6·75 ±0·5	Induction	Suzuki and Watanabe (1971)
		0·37	1·46	1×10^{-3}–5×10^{-2}		N.M.R.	Klaassen *et al.* (1972)
		0·40 ±0·01	1·56 ±0·04	1×10^{-3}–6×10^{-2}		N.M.R.	Renard and Velu (1970)
					4·3 ±0·2	Induction	Velu *et al.* (1972)
$CuRb_2Br_4 . 2H_2O$	$\frac{1}{2}$	0·37	1·44	1×10^{-3}–5×10^{-2}		N.M.R.	Klaassen *et al.* (1972)
$CrBr_3$	$\frac{3}{2}$	0·365 ±0·015	1·32 ±0·07	7×10^{-3}–5×10^{-2}		N.M.R.	Senturia and Benedek (1966)
		0·368 ±0·005		1×10^{-4}–8×10^{-2}	4·28 ±0·1	Faraday effect	Ho and Litster (1970)
MnF_2	$\frac{5}{2}$	0·335 ±0·005	1·193 ±0·01	7×10^{-5}–8×10^{-2}		N.M.R.	Heller (1966)
$RbMnF_3$	$\frac{5}{2}$	0·316 ±0·008		6×10^{-3}–5×10^{-2}		Neutron scattering	Corliss *et al.* (1969)
EuO	$\frac{7}{2}$	0·34 ±0·02	1·14 ±0·04	$2{\cdot}7\times10^{-2}$–7×10^{-2}		Mössbauer effect	Groll (1971)
		0·367 ±0·008	1·18 ±0·03	1×10^{-2}–1×10^{-1}		Neutron scattering	Als-Nielsen *et al.* (1971)
		0·368 ±0·005			4·46 ±0·1	Kink method	Menyuk *et al.* (1971)
EuS	$\frac{7}{2}$	0·33 ±0·015	1·145 ±0·02	1×10^{-2}–1×10^{-1}		N.M.R.	Heller and Benedek (1965)
		0·360 ±0·012	1·18 ±0·04	$1{\cdot}5\times10^{-2}$–1×10^{-1}		Neutron scattering	Als-Nielsen *et al.* (1971)

from ideality that may interfere with the critical behaviour under experimental conditions, it is surprising that the experimental findings are often so close to the theoretical predictions.

Only few δ's have been measured until now. The available values have been included in table 16. It may be seen that these are closer to a value of about 4·3 than to the theoretical prediction $\delta = 5$ for the 3-d magnets. In a recent analysis of experimental data on fluids and magnets, Vicentini-Missoni *et al.* (1970) also obtained $\delta \simeq 4{\cdot}4$ for both. Admittedly, the value of δ is very sensitive to the uncertainty in T_c. For instance, Ho and Litster (1970) have observed that the value of δ changed from 4·1 to 4·4 in going from $(T-T_c)/T_c = +6\times10^{-4}$ to -3×10^{-4}! This may also provide an explanation for the widely different δ values found for $Cu(NH_4)_2Br_4 . 2H_2O$. In any case, the scarce experimental information obtained thus far indicates that the theoretical prediction $\delta = 5$ is probably too large. It is of importance to remark that if the scaling relations are valid, the exponent δ will not be affected by the above mentioned renormalization by hidden variables (Fisher 1968).

In table 17 we have compiled the susceptibility parameters obtained on the same materials. Since the susceptibility singularity is again stronger than that of the spontaneous magnetization, the critical region extends still further away from T_c. Calculations (see, e.g., Wielinga 1971) show that the power-law behaviour may be expected to set in at $1-T_c/T \simeq 0{\cdot}2$ already. The susceptibility exponent is thus the easiest attainable critical index, since it may be obtained from measurements over the widest possible range of temperatures.

Most of the γ values in table 17 are seen to lie in between the Ising result 1·25 and the Heisenberg prediction 1·40, adopted in this paper. The latter is based upon the $S=\infty$ series (Bowers and Woolf 1969, Ferer *et al.* 1971). For $S=\frac{1}{2}$ indications for a higher γ ($=1{\cdot}43$) have been found (Baker *et al.* 1967 a, b). Bearing in mind the principle of universality (independence of S of the critical indices) and the fact that the estimates of γ for $S=\frac{1}{2}$ have ranged from 1·33 to 1·43 as more terms in the susceptibility series came available, we have adhered to a ' universal ' value of 1·40. For a discussion see Rushbrooke *et al.* (1973).

The fact that in many of the more or less isotropic salts the observed γ's are considerably lower than the Heisenberg value has been attributed by De Jongh *et al.* (1970) to the effect of anisotropy. Current theoretical research supports the hypothesis that further neighbour interactions will not change the exponent values. Moreover in the existing work on the Heisenberg model with anisotropic exchange (Dalton and Wood 1967, Obokata *et al.* 1967, Jasnow and Wortis 1968), indications are found that the exponent γ changes discontinuously from a Heisenberg to an Ising value upon the introduction of anisotropy, in agreement with the universality hypothesis. Coming back to the discussion of cross-over behaviour in the preceding pages, one may expect in the case of anisotropy to find a cross-over temperature or region in which the behaviour changes

Table 17. Experimental values of the critical exponents γ and γ' and the amplitudes C_0 and C_0' of the susceptibility (ferromagnetic or staggered antiferromagnetic).

	γ	γ'	C_0	C_0/C_0'	Range of relative temperature	Technique	Reference
Anisotropic compounds :							
$Tb(OH)_3$	$\simeq 1{\cdot}25$					Magnetization	Wolf *et al.* (1968)
FeF_2	$1{\cdot}38 \pm 0{\cdot}08$†	$1{\cdot}6 \pm 0{\cdot}2$		$6{\cdot}1 \pm 1{\cdot}0$	5×10^{-4}– 8×10^{-2}	Neutron scattering	Hutchings *et al.* (1972 a)
Isotropic compounds :							
$Cu(NH_4)_2Cl_4 \cdot 2H_2O$, $CuK_2Cl_4 \cdot 2H_2O$	$\simeq 1{\cdot}36$†				4×10^{-2}– 4×10^{-1}	Induction	Miedema *et al.* (1963)
$CuCs_2Cl_4 \cdot 2H_2O$	$1{\cdot}33 \pm 0{\cdot}02$				1×10^{-2}– 2×10^{-1}	Induction	Wielinga (1971)
$Cu(NH_4)_2Br_4 \cdot 2H_2O$	$1{\cdot}31 \pm 0{\cdot}02$		$1{\cdot}22 \pm 0{\cdot}15$		8×10^{-3}– 3×10^{-1}	Induction	De Jongh *et al.* (1970)
	$1{\cdot}25 \pm 0{\cdot}1$					Magnetization	Wielinga and Huiskamp (1969)
	$1{\cdot}30 \pm 0{\cdot}03$				2×10^{-3}– 1×10^{-1}	Induction	Suzuki and Watanabe (1971)
	$1{\cdot}33 \pm 0{\cdot}01$				$2{\cdot}5 \times 10^{-2}$– 2×10^{-1}	Magnetization	Cerdonio and Paroli (1970)
$CrBr_3$	$1{\cdot}215 \pm 0{\cdot}02$					Magnetization	Ho and Litster (1970)
MnF_2	$1{\cdot}27 \pm 0{\cdot}02$†	$1{\cdot}32 \pm 0{\cdot}06$	$1{\cdot}15 \pm 0{\cdot}03$	$4{\cdot}8 \pm 1{\cdot}0$	4×10^{-4}– 1×10^{-1}	Neutron scattering	Heller *et al.* (1971), Schulhof *et al.* (1970, 1971)
$RbMnF_3$	$1{\cdot}397 \pm 0{\cdot}034$†				4×10^{-4}– 4×10^{-2}	Neutron scattering	Lau *et al.* (1970)
EuO	$1{\cdot}396 \pm 0{\cdot}03$				1×10^{-2}–$3{\cdot}5 \times 10^{-1}$	Neutron scattering	Als-Nielsen *et al.* (1971)
	$1{\cdot}29 \pm 0{\cdot}01$					Magnetization	Menyuk *et al.* (1971)
EuS	$1{\cdot}390 \pm 0{\cdot}039$				2×10^{-2}–$1{\cdot}5 \times 10^{-1}$	Neutron scattering	Als-Nielsen *et al.* (1971)

† These γ's are the published values, which may be 0·01–0·02 too high (depending on the relative temperature range covered), because they were derived from plots of the susceptibility versus $T - T_c$, instead of $1 - T_c/T$.

its character from a Heisenberg to an Ising value. With a typical experimental anisotropy $H_A/H_E = 10^{-3}$–10^{-2} the cross-over may well occur in the region 10^{-3}–10^{-1}, accessible to the experiment. In that case log–log plots of experimental results will yield intermediate γ values, or, in case the anisotropy is large enough, the Ising value itself, since then the cross-over would occur outside the critical region. Indications for such a behaviour have already been met above in the discussion of the heat capacities and the β values. The same trend is observed in table 17, at least that is the explanation we offer for the variation in γ. Thus the highly isotropic compounds (EuO, EuS, $RbMnF_3$) have a γ near 1·40 (cross-over temperature too close to T_c to have an effect in the experimental region), whereas the fairly anisotropic compounds MnF_2 and $CrBr_3$ ($H_A/H_E \simeq 1{\cdot}6 \times 10^{-2}$) have a γ nearly equal to 1·25 (cross-over temperature outside the critical region already). The Cu salts are somewhere in between since they have a smaller anisotropy.

There remains then to discuss the high γ value in the anisotropic compound FeF_2. Re-examining the data of Hutchings *et al.* (1972 a), one finds that they are equally well represented by $\gamma = 1{\cdot}34$, in accordance with the high error margin of 0·08. Correcting for the fact that χ is plotted versus $T - T_c$, instead of χT versus $1 - T_c/T$, brings γ down to 1·32. A γ value lower than about 1·30 however does not seem to be consistent with the data, so that there remains a considerable discrepancy that is not readily explained.

The source of the low γ's reported by Wielinga and Huiskamp (1969) for $Cu(NH_4)_2Br_4 \, . \, 2H_2O$ and by Menyuk *et al.* (1971) for EuO is more easily traced. In these experiments the initial susceptibility was obtained from the isothermal magnetization as a function of field, by plotting the results in the form of M^2 versus H/M curves (Belov and Goryaga 1956, Kouvel and Fisher 1964), deducing the susceptibility by extrapolating the isotherms to $M^2 = 0$. This method is inadequate to obtain the initial susceptibility in the critical region and leads to too low values of γ, as witnessed by the other results for these compounds in table 17, found with better techniques. The same argument may explain why the γ value for $CrBr_3$ is lower than the Ising prediction, since also in this case the initial susceptibility was deduced from the magnetization.

Concerning the susceptibility index γ', on the low-temperature side, few results are as yet available. The values in table 17 for FeF_2 and MnF_2 do not seem to favour the scaling result $\gamma = \gamma'$, although this relationship is still fulfilled within the error margins. The ratios of the amplitudes C_0/C_0' may be compared with the predicted 5–5·5 for the cubic Ising magnets (e.g. Fisher 1967), the MF prediction of 2 and the value 5·5 found in beta-brass by Als-Nielsen (1969). For the 2-d Ising magnets the ratio is about 37.

Finally in table 18 are listed the results obtained so far for the exponents ν, ν' and η. Comparing these with the predicted Ising and Heisenberg values (table 15), there seems to be an agreement in that for the isotropic

Table 18. Experimental values of the critical exponents ν, ν' and η and of the amplitudes ratio N/N'. These exponents are associated with the inverse correlation length and the wave-vector dependent susceptibility. They have all been determined by neutron scattering.

	ν	ν'	N/N'	η	Range of relative temperature	Reference
Anisotropic compounds						
FeF_2	$0{\cdot}67 \pm 0{\cdot}04$				5×10^{-4}– 8×10^{-2}	Hutchings *et al.* (1972 a)
		$0{\cdot}7 \pm 0{\cdot}2$			$1{\cdot}2\times10^{-3}$– 3×10^{-2}	
Isotropic compounds						
MnF_2	$0{\cdot}634 \pm 0{\cdot}02$		$0{\cdot}6$		4×10^{-4}– 1×10^{-1}	Heller *et al.* (1971),
		$0{\cdot}56 \pm 0{\cdot}05$			1×10^{-3}– 1×10^{-1}	Schulhof *et al.* (1970, 1971)
				$0{\cdot}05$		
$RbMnF_3$	$0{\cdot}724 \pm 0{\cdot}008$				4×10^{-4}–$1{\cdot}8\times10^{-1}$	Lau *et al.* (1970)
		$0{\cdot}59 \pm 0{\cdot}03$			1×10^{-2}– 2×10^{-1}	
				$0{\cdot}067 \pm 0{\cdot}01$		
EuO	$0{\cdot}690 \pm 0{\cdot}023$				1×10^{-2}– 5×10^{-1}	Als-Nielsen *et al.* (1971)
EuS	$0{\cdot}702 \pm 0{\cdot}027$				2×10^{-2}–$1{\cdot}5\times10^{-1}$	Als-Nielsen *et al.* (1971)

salts EuO, EuS and $RbMnF_3$, the index ν is near to 0·71, whereas for the fairly anisotropic compound MnF_2 it is about equal to the Ising value 0·65. Again FeF_2 is the exception, since despite its high anisotropy the ν value is more Heisenberg-like. The reported uncertainty is fairly large, however. The experiments do not seem to favour the scaling relationship $\nu = \nu'$, although, in this case too, agreement can still be reached within the experimental uncertainties.

The results for η are even fewer in number and a comparison with theory is as yet not meaningful, taking also into account the large possible errors in the theoretical predictions. One may state, however, that indications for an $\eta > 0$ have indeed been found in the experiments.

With the aid of tables 16–18 sets of critical indices are derived for a number of compounds, with which some of the scaling relations may be tested. This has been done in table 19. Admittedly, an accurate test of these relations is as yet prohibited by the uncertainties in the listed numbers, caused by the possible errors in the individual exponent values, but a tentative comparison is certainly justified. Interestingly enough, one observes that within the experimental uncertainties the scaling relations can indeed be satisfied. Concerning the gap exponents, it is seen that the values for the highly isotropic Eu compounds and $RbMnF_3$ on the one hand, and for the less isotropic materials MnF_2, $CrBr_3$ and $Cu(NH_4)_2Br_4 \cdot 2H_2O$ on the other, are systematically closer to the Heisenberg and Ising predictions, respectively. Evidently, FeF_2 again forms the exception through its unexpectedly high γ value.

Another way of testing the scaling predictions is via the magnetic equation of state, i.e. the functional relationship among the variables M, H and T (Domb and Hunter 1965, Widom 1965, Kadanoff 1966, Griffiths 1967). Introducing the scaled variables

$$m = \sigma |\epsilon|^{-\beta}, \tag{4.39}$$

$$h = H |\epsilon|^{-\beta\delta}, \tag{4.40}$$

where $\sigma = M/M(T=0)$ and $\epsilon = (T - T_c)/T_c$, the scaling relations predict that h is a function of m only, so that the equation of state reads simply

$$h = h(m). \tag{4.41}$$

It remains, of course, to establish the mathematical form of the function $h(m)$. However, eqn. (4.41) implies that if, instead of the usual magnetization versus field isotherms, we plot m versus h, the different isotherms in the critical region should fall on a single curve, one for $T > T_c$ ($\epsilon > 0$) and one for $T < T_c$ ($\epsilon < 0$). Such analyses have recently been performed on a variety of materials (magnetic substances and fluids) with remarkable success. As an example that fits into the present context, we show in fig. 75 the h–m plot of $CrBr_3$, as reported by Ho and Litster (1969).

Obviously, two different approaches to the problem can be taken. Determining the critical parameters in the usual way from log–log plots, one may compute h and m and thus obtain an experimental prediction

Table 19. Test of a number of scaling relations, using sets of exponents derived from tables 16–18. The 3-d Ising and Heisenberg predictions for the gap exponents ($\Delta=\Delta'=\beta+\gamma=\beta\delta$) are $\simeq 1{\cdot}6$ and $\simeq 1{\cdot}8$, respectively.

	$2\beta+\gamma$ $(=2-\alpha)$	$\beta(\delta+1)$ $(=2-\alpha)$	$\gamma-\beta(\delta-1)$ $(=0)$	$\gamma-\nu(2-\eta)$ $(=0)$	$\beta+\gamma$ $(=\Delta=\Delta')$	$\beta\delta$ $(=\Delta=\Delta')$
FeF_2 ($\beta=0{\cdot}32$; $\gamma=1{\cdot}38$)	2·03				1·705	
$Cu(NH_4)_2Br_4 . 2H_2O$ ($\beta=0{\cdot}37$; $\gamma=1{\cdot}31$; $\delta=4{\cdot}3$)	2·05	1·96	0·09		1·68	1·59
$CrBr_3$ ($\gamma=1{\cdot}22$; $\beta=0{\cdot}37$; $\delta=4{\cdot}3$)	1·96	1·96	−0·001		1·59	1·59
MnF_2 ($\beta=0{\cdot}335$; $\gamma=1{\cdot}27$; $\nu=0{\cdot}63$; $\eta=0{\cdot}05$)	1·94			0·04	1·61	
$RbMnF_3$ ($\beta=0{\cdot}32$; $\gamma=1{\cdot}39$; $\nu=0{\cdot}72$; $\eta=0{\cdot}07$)	2·03			0·00	1·71	
EuO ($\beta=0{\cdot}37$; $\gamma=1{\cdot}40$; $\delta=4{\cdot}5$)	2·14	2·04	0·11		1·77	1·67
EuS ($\beta=0{\cdot}35$; $\gamma=1{\cdot}39$)	2·09				1·74	

Fig. 75

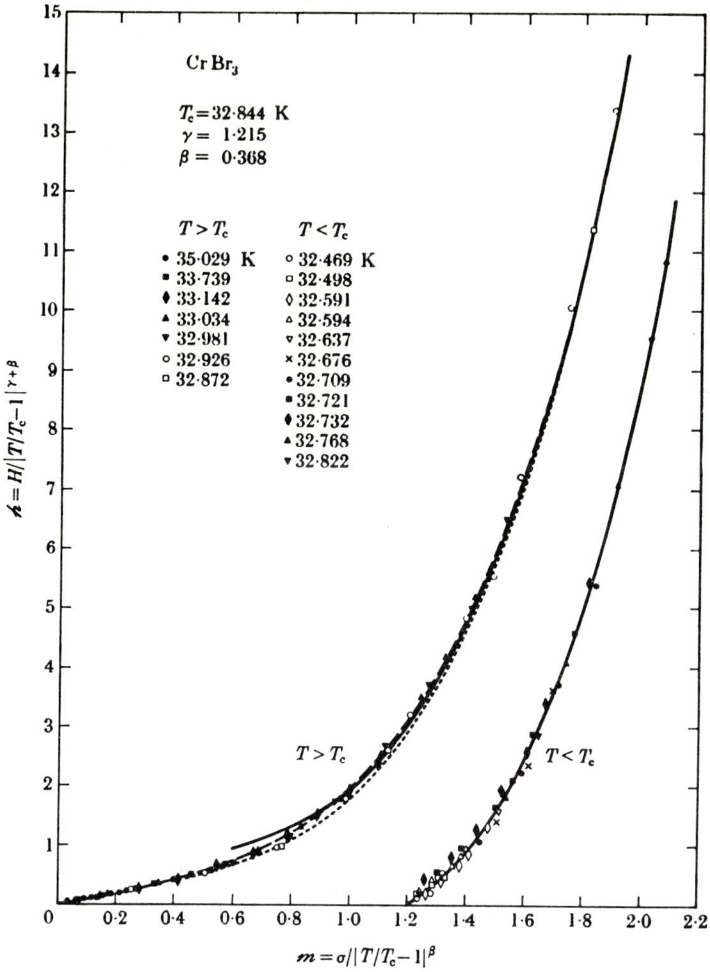

Scaled plot of the magnetization isotherms of $CrBr_3$. (Ho and Litster 1969.)

for the form of the equation of state that may be compared with theory. The analysis of Ho and Litster was performed in this spirit and the solid and dashed curves in fig. 75 represent, in fact, different assumed forms of the equation of state. Alternatively, a particular form may be assumed *a priori* and a set of critical parameters is derived in fitting the experimental isotherms to such a function. It would go too far to mention all the various forms of the equation of state that have been proposed, all the more since there exist review papers in which the interested reader can find extensive information on the subject (Vicentini-Missoni *et al.* 1969, 1970). We merely mention a recent development in this field, namely, the calculation of the equation of state from series expansions (Gaunt and Domb 1970, Milošević and Stanley 1972). Likewise we have refrained from going into the topic of dynamic scaling, a rapidly developing new trend that constitutes a generalization of the static scaling hypothesis to dynamic phenomena. The theory for isotropic magnetic systems has been discussed by Hohenberg and Halperin (1970), in which paper references to earlier work may be found, and has been extended to anisotropic systems by Riedel and Wegner (see, e.g., Riedel, 1971). Experimentally, neutron investigations have been performed on MnF_2, $RbMnF_3$ and FeF_2, which seem to be in good agreement with theory (see, e.g., Lau *et al.* 1970, Schulhof *et al.* 1970, 1971, Hutchings *et al.* 1972 a).

In this section we have thus far confined ourselves to the 3-d systems. In the remainder the critical behaviour observed in the lower-dimensional magnets, as discussed in §§ 3.1 and 3.2, will be briefly reviewed.

Concerning the 1-d systems we have seen that since the ideal 1-d system does not possess a finite transition point, any critical behaviour that is observed experimentally is of a 3-d character, the argument being that no matter how small the interchain interaction J' is in an assembly of magnetic chains, any finite J' will make such an assembly of a 3-d nature. According to the universality hypothesis the critical behaviour is therefore the same for all values of J'. This is nicely confirmed by the experiments since, e.g., the spontaneous magnetization that is observed experimentally below the (J'-induced) transition points, in all the investigated cases shows a 3-d behaviour ($\beta \simeq \frac{1}{3}$).

For the quasi 2-d systems the above argument must be slightly modified. Evidently, since an assembly of weakly coupled magnetic layers is likewise essentially a 3-d system, one expects, according to the same reasoning, that the critical behaviour will be 3-d if the transition point is approached closely enough. However, the fact that the ideal 2-d Ising model itself also possesses a transition to long-range order at a T_c differing from zero, alters the situation, in that for small enough J' there can be an intermediate critical region in which the 2-d character can manifest itself. Closer to T_c there occurs then a cross-over from 2-d to 3-d behaviour, through the effect of the finite J'. If the inter-layer coupling is not too small for the cross-over point to be reached experimentally, the cross-over may be spread out over a considerable portion of the critical region, so that log–log plots of the magnetization will appear to be rounded and β values derived from measurements in one or two decades in relative temperature may be in between the 3-d and the 2-d values.

This has been discussed at the end of § 3.2.3, where we have also pointed out that any quasi 2-d spontaneous magnetization observed experimentally must be anisotropy-induced, since in the isotropic Heisenberg limit there is no transition to long-range order. Quite convincingly the β values obtained on the more or less isotropic layer-type antiferromagnets with $|J'/J| \simeq 10^{-6}$, are close to the 2-d Ising prediction $\beta = \frac{1}{8}$. This culminates in the results found for the anisotropic salts K_2CoF_4 and Rb_2CoF_4, which have $\beta = \frac{1}{8}$ within the experimental uncertainties ! Thus below T_c a cross-over effect in the quasi 2-d salts can only be caused by the influence of J'. Until now no clear evidence for the expected kink in the log–log plot of the magnetization has been observed, presumably because either J' is too small or else because of the mentioned spreading of the cross-over temperature.

In the case of the susceptibility, above T_c, the situation is different because for the 2-d Heisenberg model there is likely also to be a transition point at which the susceptibility diverges. Therefore two cross-overs would in principle be found in the susceptibility of a quasi 2-d ferromagnet with small anisotropy (but with $H_A/H_E > |J'/J|$), namely, a first cross-over from 2-d isotropic to 2-d anisotropic behaviour, and, closer to T_c, a second from 2-d anisotropic to 3-d anisotropic behaviour. Such an argument could explain the features of the susceptibility behaviour found in the layer-type ferromagnets $(C_nH_{2n+1}NH_3)_2CuX_4$ (X = Cl or Br ; $n = 0, 1, 2, \ldots 10$). The properties of these salts have been discussed in § 3.2. Referring to that section for details, we mention that at the temperatures where the small λ-type anomalies are found in the heat capacity (cf. fig. 43), apparently the susceptibility is found to diverge, that is it reaches the limiting value expected for a ferromagnetic sample as estimated on the basis of the sample shape. The compound $(C_2H_5NH_3)_2CuCl_4$ forms the exception, since in that case the antiferromagnetic interlayer coupling is strong enough to lower the susceptibility from the diverging ferromagnetic curve long before the ferromagnetic limit is reached (cf. the discussion in § 3.2 and fig. 42). As examples the susceptibilities of $(CH_3NH_3)_2CuCl_4$ and $(C_{10}H_{21}NH_3)_2CuCl_4$ have been plotted in fig. 76 on a double logarithmic scale (De Jongh to be published) as $\chi T/C$ versus $1 - T_c/T$, where T_c is the experimentally observed transition point. For comparison the susceptibility of the 3-d ferromagnet $Cu(NH_4)_2Br_4 \cdot 2H_2O$ (De Jongh *et al.* 1970) has been included in the figure as the dashed line. Note that the 2-d susceptibilities tend to be one or two orders of magnitude larger than that of the 3-d salt over most of the critical region.

It is seen that in the high-temperature limit the susceptibility of both 2-d salts coincides with the high-temperature series expansion prediction (H.T.S.) for the $S = \frac{1}{2}$ Heisenberg quadratic ferromagnet (Baker *et al.* 1967 a, b). As T_c is approached they begin to differ, which is ascribed to a different degree of ideality. On the basis of the T_c/θ value (cf. table 7) one may conclude that $(CH_3NH_3)CuCl_4$ is a less ideal 2-d Heisenberg ferromagnet than $(C_{10}H_{21}NH_3)_2CuCl_4$, the latter having the lowest T_c/θ. It can further be seen from fig. 76 that we may interpret the data of both salts in such a way as to distinguish three different regions. Far away from T_c straight lines may be drawn through the data that have a slope $\gamma \simeq 2{\cdot}7$. In the intermediate region the apparent slope is about $\gamma = 1{\cdot}75$, which is the 2-d Ising value†, whereas nearest to T_c the points

† It is not obvious that the cross-over should be to 2-d Ising behaviour, since the planar part of the anisotropy (H_A^{II}) is larger than the Ising part (H_A^{I}) (cf. table 7). The behaviour in the intermediate region may thus be due to a mixture of the effects of both anisotropies.

Fig. 76

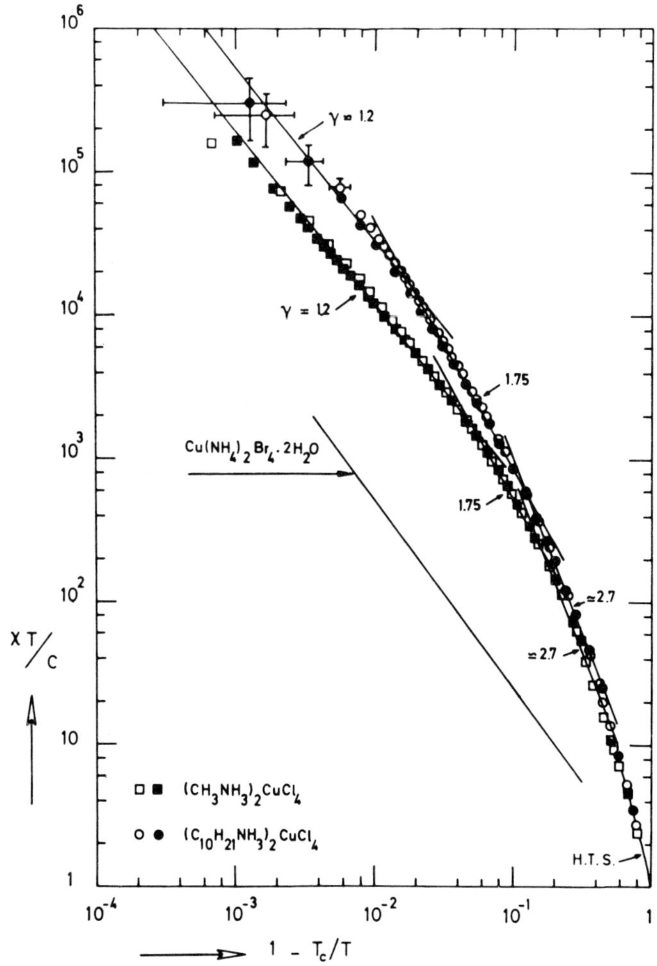

The susceptibilities of the ferromagnetic layer-type compounds $(CH_3NH_3)_2CuCl_4$ and $(C_{10}H_{21}NH_3)_2CuCl_4$. For comparison the results obtained in the 3-d copper compound $Cu(NH_4)_2Br_4 \cdot 2H_2O$ have been indicated by the solid curve. The susceptibilities are plotted as $\chi T/C$ versus $1-T_c/T$, where T_c is the experimentally observed transition temperature.

agree with a γ of roughly 1·2, close to the 3-d Ising value of 1·25. All susceptibilities have been corrected for demagnetizing effects to an infinitely long cylindrical sample shape. Note that the cross-over to 3-d behaviour in the less ideal $(CH_3NH_3)_2CuCl_4$ occurs farther away from T_c, as is to be expected.

Another way of interpreting the cross-over phenomenon is the following. We may conceive of an ideal transition temperature $T_c^{id.}$ at which the susceptibility of the 2-d Heisenberg ferromagnet diverges. The deviations from ideality in the experimental compounds will cause a shift of the experimental T_c with respect to $T_c^{id.}$ in the direction of higher temperature. For temperatures far away from T_c (large compared to the difference $T_c - T_c^{id.}$) the susceptibility will then behave two-dimensionally, as if it was going to diverge at $T_c^{id.}$. But as $T_c^{id.}$ is approached, the system 'remembers' its actual transition temperature and diverges at T_c, due to the fact that close enough to T_c the deviations from ideality start to have their influence. Accordingly if we would know the value of $T_c^{id.}$, and plot the susceptibility of the Cu compounds not as $\chi T/C$ versus $1 - T_c/T$, but as $\chi T/C$ versus $1 - T_c^{id.}/T$, the curves of the various compounds should coincide at high temperatures. As $T_c^{id.}$ is approached the curves of the individual compounds would start to diverge from this common curve, one by one, according to the degree of ideality reached in each compound.

In fig. 77 the susceptibilities of four of the Cu compounds have been plotted as $\chi T/C$ versus $1 - T_c^{id.}/T$ in order to demonstrate that such a picture can indeed be realized. Here $T_c^{id.}$ has been chosen such that the curves coincide over the largest possible region, since a theoretical prediction for $T_c^{id.}$ is not available. The percentage shifts $(T_c - T_c^{id.})/T_c$ of the individual compounds have been indicated. The resulting value for $kT_c^{id.}/J$ is 0·435, leading to $T_c^{id.}/\theta = 0{\cdot}22$, which gratifyingly is exactly the same prediction for the ideal transition temperature as obtained by Bloembergen from the extrapolation procedure in which he made use of the energy contents of the small λ-anomalies in the specific heat (cf. § 3.2.3 and table 9) ! Accordingly, the fractional shifts $(T_c - T_c^{id.})/T_c$ of the four compounds correspond quite well to the differences of their measured T_c/θ values (cf. table 7) from $T_c^{id.}/\theta$. We add, lastly, that a similar coincidence of the susceptibility curves of the various salts at high temperatures arises naturally by plotting $\chi T/C$ versus kT/J (cf. fig. 42). It is also remarked that here again the $T_c^{id.}/\theta$ obtained may still be affected slightly by the planar anisotropy (H_A^{II}), since the latter has about the same value ($\simeq 10^{-3}$) for all the compounds. Accordingly, possible shifts of the experimental T_c/θ values due to H_A^{II} (into the direction of the 2-d XY value $T_c/\theta \simeq 0{\cdot}45$, Betts *et al.* 1973) are probably not eliminated by the above extrapolation procedure.

Thus the picture sketched above is nicely confirmed. The least ideal salt $(C_3H_7NH_3)_2CuBr_4$ is the first to diverge from the common curve, at the point where its susceptibility diverges at the actual T_c. Last of course comes the nearest to the ideal salt $(C_{10}H_{21}NH_3)_2CuCl_4$. One could therefore regard this common curve, together with the extrapolation provided by the straight line drawn through the data, as being representative of the susceptibility of the ideal, $S = \frac{1}{2}$, 2-d Heisenberg ferromagnet. This straight line yields a slope $\gamma = 2{\cdot}7 \pm 0{\cdot}3$ and an amplitude $C \simeq 3{\cdot}3$. The γ value obtained agrees quite well with the estimate $\gamma = 3{\cdot}0 \pm 0{\cdot}5$ obtained by Ritchie and Fisher (1973) from their analysis of the high-temperature susceptibility series of quadratic Heisenberg ferromagnets with different S ($S = 1 - \infty$).

Fig. 77

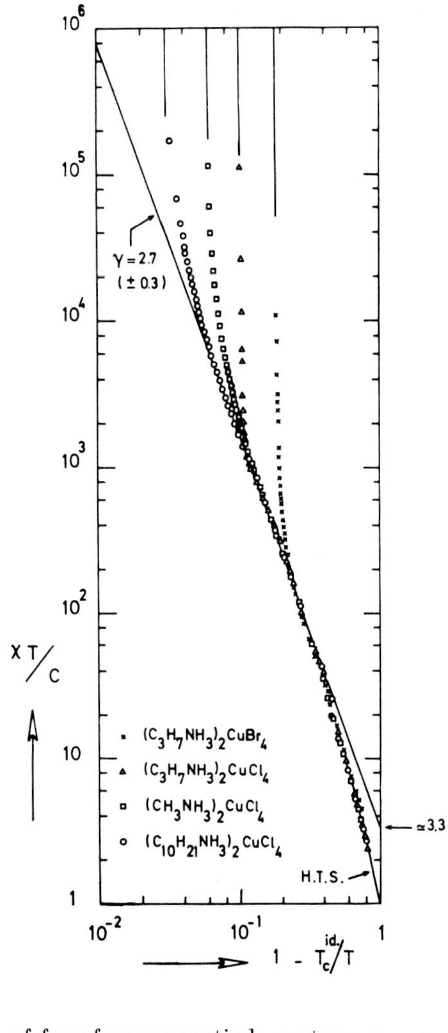

The susceptibilities of four ferromagnetic layer-type copper compounds plotted as $\chi T/C$ versus $1-T_c^{\text{id.}}/T$, where $T_c^{\text{id.}}$ is the assumed transition temperature of the ideal 2-d Heisenberg ferromagnet, that is in the absence of anisotropy and interlayer coupling. Since a theoretical prediction for $T_c^{\text{id.}}$ for the case $S=\frac{1}{2}$ is lacking we have determined $T_c^{\text{id.}}$ from the condition that the susceptibilities of the different compounds not only coincide in the high-temperature region but also on the same straight line over the largest possible temperature range. The fractional shifts $(T_c-T_c^{\text{id.}})/T_c$, where T_c is the observed transition temperature, are indicated by the vertical lines in the top of the figure, to which the susceptibilities of the individual compounds diverge. They are 3, 6, 10 and 18% for $(C_{10}H_{21}NH_3)_2CuCl_4$, $(CH_3NH_3)_2CuCl_4$, $(C_3H_7NH_3)_2CuCl_4$ and $(C_3H_7NH_3)_2CuBr_4$, respectively.

For the quadratic antiferromagnets few results for the critical behaviour above T_c have been reported thus far. Rather surprisingly, Birgeneau *et al.* (1971 b) found $\gamma \simeq 1{\cdot}0$ and $\nu \simeq 0{\cdot}57$ in K_2NiF_4, values that are closer to the MF predictions than to the 2-d Ising results $\gamma = 1{\cdot}75$ and $\nu = 1$. However, the data were obtained in one decade of relative temperature only (10^{-2}–10^{-1}) and in particular the susceptibility curve appears to be rounded. We point out that a similar analysis as given above, assuming a shifted experimental T_c with respect to $T_c^{\mathrm{id.}}$, can firstly straighten the log–log plots of χ and κ, and secondly bring the γ and ν values near to the 2-d Ising prediction. For instance, this could already be accomplished by taking a shift $(T_c - T_c^{\mathrm{id.}})/T_c$ of a few per cent only. Since in K_2NiF_4 the value of $|J'/J|$ is extremely small ($\simeq 10^{-6}$), the shift would be for the most part due to the anisotropy ($H_A/H_E = 2 \times 10^{-3}$).

4.5. *Field-dependent behaviour*

In recent years there has been a renewed interest in the study of magnetic systems as a function of field. This arises, amongst other things, from the analogues between the phase diagrams (H–T diagram) of certain antiferromagnetic systems and those of a quantum lattice gas (^{4}He), ^{3}He–^{4}He mixtures and systems undergoing structural phase transitions (NH_4Cl). Furthermore, one is interested to know whether or not the critical behaviour (exponents) is affected by the presence of a field and, if so, in what manner.

It is of importance in this respect to distinguish between an ‘ ordering ’ field (applied field for a ferromagnet, staggered field for an antiferromagnet) and a ‘ disordering ’ field, which is an applied or a staggered field in the case of an anti or a ferromagnet, respectively (Griffiths 1970 a). For instance, a finite external field applied to a ferromagnet will destroy the phase transition, whereas for an antiferromagnet the transition will remain sharp, although the transition point will (in general) be shifted to a lower temperature.

In this section we will be mainly concerned with the behaviour of antiferromagnets in an applied (disordering) field. Some results on ferromagnets will be mentioned too, but so far these have been scarce.

In the case of antiferromagnets, then, the behaviour in external fields depends strongly on the particular type of antiferromagnetic system considered. The degree and the type of anisotropy (exchange or single-ion) plays an important role. Moreover the presence of ferromagnetic interactions, in addition to the antiferromagnetic ones, may change the character of the field-dependent transitions in a fundamental way. This will become clear from the examples that we shall give below.

Let us first confine our attention to the Ising model, that is the case of fully anisotropic exchange, and assume only nearest-neighbour antiferromagnetic interactions to be present. In that case the antiferromagnetic phase diagram has the simple form already displayed in fig. 32 for the square lattice. At $T = 0$ the spins become ferromagnetically aligned at a critical field $H_c = H_{af}$, where H_{af} denotes the antiferromagnetic exchange field ($g\mu_B H_{af} = 2z|J|S$). By raising the temperature, the critical field $H_c(T)$ decreases continuously until it vanishes at the critical point. Alternatively, one may say that the critical temperature $T_c(H)$ is decreasing with increasing field. The variation of $T_c(H)$ with H has been studied by Bienenstock (1966), locating $T_c(H)$ from the singularity of the staggered susceptibility. He found that his results could be summarized by the (empirical) formula

$$T_c(H)/T_c(0) = [1-(H/H_c)^2]^{\xi}, \tag{4.42}$$

with $\xi = 0{\cdot}87$, $0{\cdot}35$ and $0{\cdot}36$ for the square, s.c. and b.c.c. Ising lattices, respectively. For $H \ll H_c$, this expression reduces to a quadratic dependence of $T_c(H)$ on H. The latter variation has also been found by Fisher (1960 b) from his (exactly soluble) model of a decorated quadratic lattice, already mentioned in § 3.2.1. He obtained the equation

$$\sinh 4|J|/kT_c(H) = (2+2\sqrt{2})^{1/2} \cosh g\mu_B H/kT_c(H), \tag{4.43}$$

which for small fields yields $T_c(H)/T_c(0) \simeq 1 - c_0(H/H_c)^2$, with c_0 a constant. At low temperatures ($H \simeq H_c$), on the other hand, eqn. (4.43) gives a linear dependence of $H_c(T)$ on temperature :

$$H_c(T)/H_c(0) \simeq 1 - c_1(T/T_c),$$

where c_1 is another constant. The Bienenstock formula (4.42) only yields a linear variation for $\xi \simeq 1$, but unfortunately his results for $H \simeq H_c$ were not conclusive, due to a decreasing rate of convergence with increasing H of the susceptibility series. From fig. 32 it is seen that the data on the quasi 2-d Ising antiferromagnet $CoCs_3Br_5$ agree rather well with Bienenstock's results.

Among the other relevant features of Fisher's decorated 2-d lattice model is the fact that the locus of transition points in the H–T diagram is a second-order transition curve, except at $T=0$, where the transition becomes of first order. Thus the magnetization (first derivative of the free energy) versus field isotherm for $T=0$ rises discontinuously to its saturation value at $H=H_c$, whereas for $T>0$ the behaviour of the magnetization is continuous, although for $T<T_c$ anomalies of the form $[H_c(T)-H] \ln |H_c(T)-H|$ occur at the transition fields $H_c(T)$. This implies that the susceptibility (second derivative) as a function of field for fixed $T<T_c$ will display a logarithmic singularity at $H_c(T)$. Likewise, if instead of a vertical path in the H–T diagram a horizontal path is followed, the susceptibility at fixed magnetic field as a function of temperature possesses a logarithmic singularity at $T_c(H)$. It is worth while to note that these susceptibility anomalies at the boundary separating the antiferromagnetic from the paramagnetic phase in fact reflect the specific heat singularity.

Next we will consider the field dependence of the Heisenberg antiferromagnet with small anisotropy, which is the most extensively studied example of field-dependent behaviour. In addition to a transition from an antiferromagnetic to the paramagnetic phase, this system displays the phenomenon of spin-flopping over a certain range of temperatures below T_c. This type of phase transition was predicted by Néel as far back as 1936, but was not discovered experimentally until 1952 (in $CuCl_2 . 2H_2O$) by the Leiden group. For references to the earlier theoretical and experimental papers see, e.g., Néel (1957) and Gorter (1957).

The spin-flop transition is most easily explained by considering a simple uniaxial two-sublattice MF model at $T=0$. If an external field is applied parallel to the preferred axis of antiferromagnetic alignment, the moments will have the tendency to orient themselves perpendicular to the field, since in so doing they gain a magnetic energy of $\frac{1}{2}(\chi_\perp - \chi_\parallel)H^2$. In small fields the anisotropy, that establishes the preferential direction, will exceed the field term, but at a certain critical field the spins will flip over to the perpendicular orientation. A further increase of H will gradually rotate the sublattice moments, until at the critical field H_c their mean direction is parallel to the easy axis and the paramagnetic phase is entered. At this point the antiferromagnetic interaction is balanced by the applied field and the anisotropy field.

The behaviour of the magnetization and the initial (differential) susceptibility as a function of field expected on basis of this model is sketched

Fig. 78

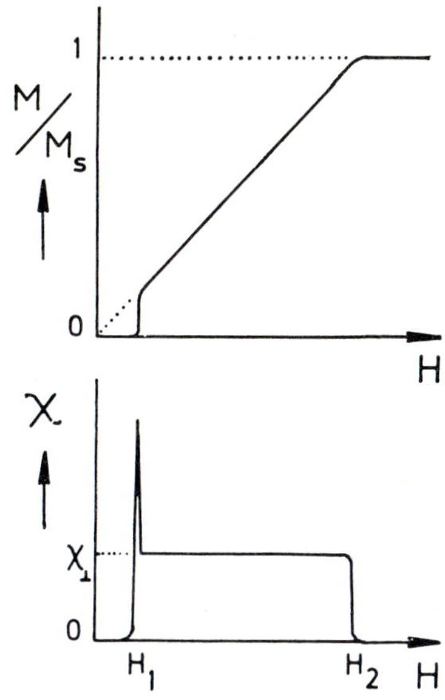

The behaviour as a function of field of the isothermal magnetization and differential susceptibility of a weakly anisotropic antiferromagnet, according to the MF theory for a temperature near $T=0$. The critical fields H_1 and H_2 correspond to the spin-flop transition field H_{SF} and the transition from the flopped to the paramagnetic phase (H_c), respectively, that are discussed in the text.

in fig. 78. In the MF treatment the spin-flop transition is of first order (discontinuity in M), whereas the transition from the spin-flopped to the paramagnetic phase (SF–P transition) is of second order (M continuous but χ discontinuous). In between the two transition fields the magnetization increases linearly with H according to

$$M/M_s = H/(2H_{af} - H_A). \tag{4.44}$$

For fields applied perpendicular to the easy axis the spin-flop transition obviously does not occur and the magnetization is given by

$$M/M_s = H/(2H_{af} + H_A), \tag{4.45}$$

until for $H = H_c$ the saturation value M_s is reached. The formulae for the critical fields are

$$H_{SF} = (2H_{af}H_A - H_A^2)^{1/2}, \tag{4.46}$$

$$H_c = 2H_{af} - H_A. \tag{4.47}$$

for fields parallel to the preferred direction and

$$H_c' = 2H_{af} + H_A \tag{4.48}$$

if the field is applied perpendicular to the easy axis. These critical fields are, of course, dependent on temperature and a theoretical phase diagram is shown in fig. 79. In this diagram the first-order antiferromagnetic to spin-flop transition (AF–SF) and the second-order SF–P transition curves are seen to meet in a triple point, together with the boundary separating the antiferromagnetic and the paramagnetic phase (AF–P), which is also thought to be of second order. A theoretical treatment of the antiferromagnetic phase diagram within the MF approximation can be found in the papers of Gorter and Van Peski-Tinbergen (1956) and of Van Wier *et al.* (1959). The analogy with the phase diagram of the quantum lattice gas has been pointed out by Fisher (see, e.g., Liu and Fisher 1973). Recently, spin-wave theoretical calculations on the Heisenberg antiferromagnet have been performed by Anderson and Callen (1964) and by Feder and Pytte (1968) (see also Keffer 1966). Since the spin-flop transition is of first order, hysteresis effects may be expected to occur and in fact spin-wave theory yields expressions for upper and lower spin-flop fields, quite similar to the case of supercooling and superheating in the liquid-gas transition (Anderson and Callen 1964). However, clear experimental evidence for this effect has not yet been obtained in the fairly isotropic antiferromagnets.

From eqns. (4.46) and (4.47) one observes that an increase in anisotropy increases H_{SF} while it lowers H_c. For $H_A = H_{af}$ the two critical fields become equal. It is then no longer energetically favourable to have an intermediate flopped phase; since the anisotropy is so large, the moments go over directly from an antiferromagnetic alignment (parallel to H) to a ferromagnetic alignment, at a field value $H_c = H_{af}$. Thus the situation resembles that of the Ising antiferromagnet in that the energy involved is just that needed to turn over one of the antiferromagnetically coupled

Fig. 79

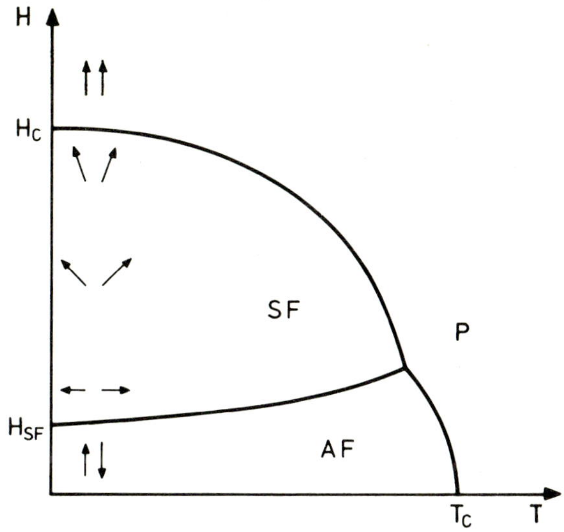

Theoretical magnetic phase diagram of a weakly anisotropic antiferromagnet.

moments. A subclass of substances that falls into this category are the so-called metamagnets that we will discuss next.

Originally this term was introduced for systems like $FeCl_2$, that consist of antiferromagnetically coupled ferromagnetic layers with $J_f \gg J_{af}$. Due to the large anisotropy, the transition from the antiferromagnetic to the paramagnetic phase indeed occurs in the above fashion, with for $T \ll T_c$ a discontinuous rise of the magnetization at the transition field to a value near to saturation. However, there also exist ferromagnetic layer-type systems with an antiferromagnetic inter-layer coupling which has a smaller anisotropy, like e.g. $(C_2H_5NH_3)_2CuCl_4$, that shows a behaviour similar to the 'normal' antiferromagnets with small anisotropy. A metamagnet is therefore best defined as an array of antiferromagnetically coupled ferromagnetic layers with an anisotropy that exceeds the antiferromagnetic exchange field.

The *AF–P* boundary of a metamagnet is, however, only of first order up to a certain temperature $T_t < T_c$, above which the transition changes into second order. The point (H_t, T_t) in the phase diagram is a tri-critical point, i.e. a point where three critical lines meet. This is best explained by considering the metamagnetic phase diagram in $\tilde{H}$, H, T space, as sketched in fig. 80, where H is the (staggered) ordering field (Griffiths 1970 a). The form of the phase diagram shown follows from MF calculations and the Landau phenomenological theory (Griffiths 1970 a). It is seen that there exist three surfaces that intersect along the dashed line in the H–T plane, which is the experimentally observed line of first-order transitions. The boundaries of these three surfaces at the high-temperature side meet in the tri-critical point.

Fig. 80

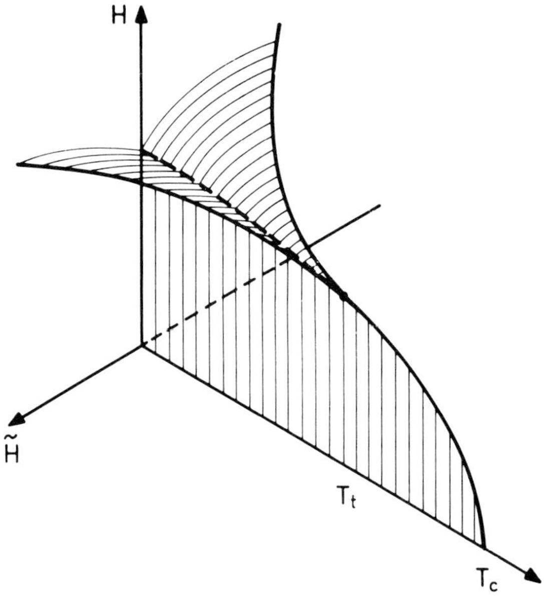

The theoretical phase diagram of a metamagnetic substance. H is the constant (non-ordering) field, $\tilde{H}$ the staggered (ordering) field. The dashed line is the intersection of three co-existence surfaces and is the experimentally observed line of first order transitions terminating in the tri-critical point (H_t, T_t). The phase boundary extending from the tri-critical point towards the temperature axis is presumably of second order.

From both MF theory and calculations on Ising chains (Nagle and Bonner 1971) it can be inferred that the ratio T_t/T_c depends on the ratio $|J_{af}/J_f|$ of antiferromagnetic inter-layer coupling and ferromagnetic intra-layer exchange. If J_f becomes large with respect to $|J_{af}|$, the tri-critical point approaches the critical point T_c ($H=0$). If there are only antiferromagnetic interactions, the tri-critical point recedes to $T=0$ ($H=H_c$) and the H–T diagram becomes that of an ordinary anisotropic antiferromagnet, as shown in fig. 81 (*a*). The same effects are found for the more general case of an antiferromagnet with both antiferro and ferromagnetic interactions (e.g. nearest and next-nearest neighbour interactions). The H–T diagram of the metamagnet is given in fig. 81 (*b*). For completeness the phase diagram of a ferromagnet in a normal field (or an antiferromagnet in a staggered field) is shown in fig. 81 (*c*). In that case the phase boundary is the $H=0$ axis up to $T=T_c$, where for $T<T_c$ this is a line of first-order transitions ending in the second-order transition point $T=T_c$.

One of the interesting features of the metamagnetic phase diagram is the field dependence of the critical exponents (β, γ, δ). Recent theoretical investigations (Harbus and Stanley 1972, Riedel 1972, Arora and Landau

Fig. 81

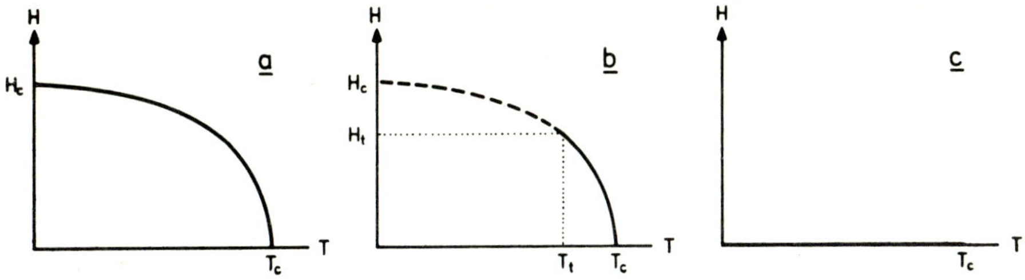

A comparison of the magnetic phase diagrams of (*a*) a 'normal' antiferromagnet, (*b*) a metamagnet, (*c*) a ferromagnet.

1972) suggests that the exponents should retain their $H=0$ values up to the tri-critical point, where they change discontinuously into tri-critical values. It should be remarked that the particular path followed in the H–T plane may be of importance. Until now no experimental investigations into this matter have, to our knowledge, been performed.

After having reviewed some of the principal aspects of the various magnetic phase diagrams, we discuss a few experimental examples. This will also give the opportunity to go a little deeper into the detailed features, some of which remain as yet unsolved. Since the only clear-cut example of a fully mapped phase diagram of an Ising antiferromagnet is that of the 2-d antiferromagnet $CoCs_3Br_5$, already treated above, we will first turn to the Heisenberg antiferromagnet with small anisotropy, which is at the same time the most extensively investigated category.

Fig. 82

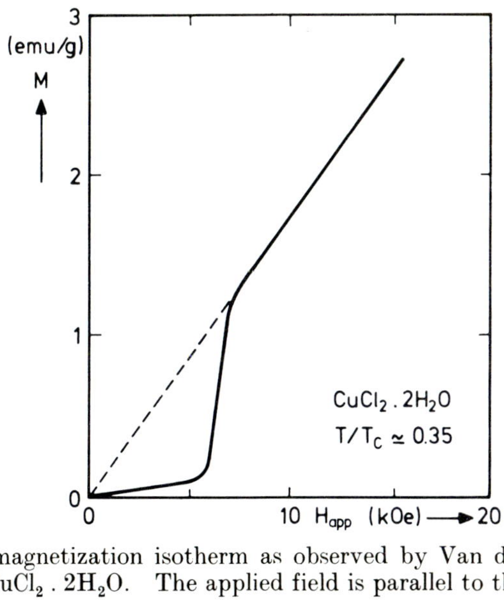

Experimental magnetization isotherm as observed by Van den Handel *et al.* (1952) in $CuCl_2 \cdot 2H_2O$. The applied field is parallel to the easy axis.

The classical example of a spin-flop transition is that observed in $CuCl_2 . 2H_2O$ by Van den Handel *et al.* (1952), shown in fig. 82. Since in this salt the exchange field and the anisotropy are both fairly small, the spin-flop transition occurs in a moderate field of a few kOe. More generally $2H_{af}$ will be of the order of 10^6 Oe, so that with an anisotropy of 1%, H_{SF} will be about 10^4–10^5 Oe (eqn. (4.46)). To observe the *SF–P* transition one would then need fields near to 10^6 Oe (eqn. (4.47)), which is clearly outside the limits of normal laboratory equipment. In order to measure a complete phase diagram, one has therefore to take recourse to the (hydrated) salts with small exchange fields. As mentioned in § 3.3 the *SF–P* transition field in $CuCl_2 . 2H_2O$ is about 150 kOe. An example of a non-hydrated salt that nevertheless has a low exchange field is $GdAlO_3$. Magnetization curves obtained by Cashion *et al.* (1970) at $T/T_c \simeq 0{\cdot}13$ are shown in fig. 83. In this salt (as in $CuCl_2 . 2H_2O$) the anisotropy is of orthorhombic symmetry, but within the easy plane a similar treatment, as in the uniaxial case, may be applied. The data shown are taken along the preferred and the next preferred axes and confirm the expectations based upon the MF approximations confined in eqns. (4.44)–(4.48).

Fig. 83

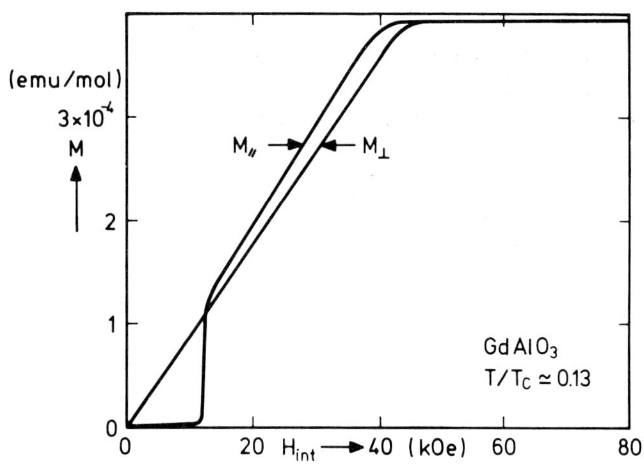

Experimental magnetization isotherms as observed in $GdAlO_3$ by Cashion *et al.* (1970). The internal field, that is the applied field corrected for demagnetizing effects, is parallel to the easy axis ($M_{\parallel}$) and to the next preferred axis ($M_{\perp}$).

It is of importance to note that in fig. 83 the magnetization is plotted against the internal field, that is the applied field corrected for demagnetizing effects, contrary to the $CuCl_2 . 2H_2O$ curves of fig. 82, where $M(H)$ is plotted versus the applied field. It is seen that the demagnetization correction is necessary in order to exhibit the first-order character of the transition (discontinuous jump in $M(H)$). In this respect one may recall the analogy between the spin-flop transition and the ferromagnetic transition, which is also of first order for $T < T_c$ (Anderson and Callen 1964). In the latter the discontinuity in $M(H)$, occurring at $H = 0$, is likewise masked by demagnetizing effects; plots of $M(H)$ versus the applied field yield a magnetization that increases linearly up to saturation, with a slope given by the reciprocal of the demagnetizing factor, as a consequence of the establishment of a domain structure. A similar division in domains of flopped and non-flopped spins will occur in the case of the spin-flop transition, since the increase in magnetization as a consequence of spin-flopping will lower the internal field, through the equation $H_{\text{int.}} = H_{\text{appl.}} - NM$, to a value that is below the critical field H_{SF} needed to overcome the anisotropy energy. A quite similar statement applies to the first-order transition in the metamagnetic substances, as we will see below.

As an example of a fully mapped antiferromagnetic phase diagram we show in fig. 84 the phase boundaries of $MnCl_2 \cdot 4H_2O$ as obtained by Giauque *et al.* (1970). The phase boundaries plotted here have been determined from maxima in the specific heat, measured as a function of temperature at constant fields, from maxima in the isentropic susceptibility and, additionally, from minima in the isentropic $(\partial T/\partial H)_S$ curves (making use of the magnetocaloric effect). This illustrates at the same time the variety of techniques that may be used in determining the phase

Fig. 84

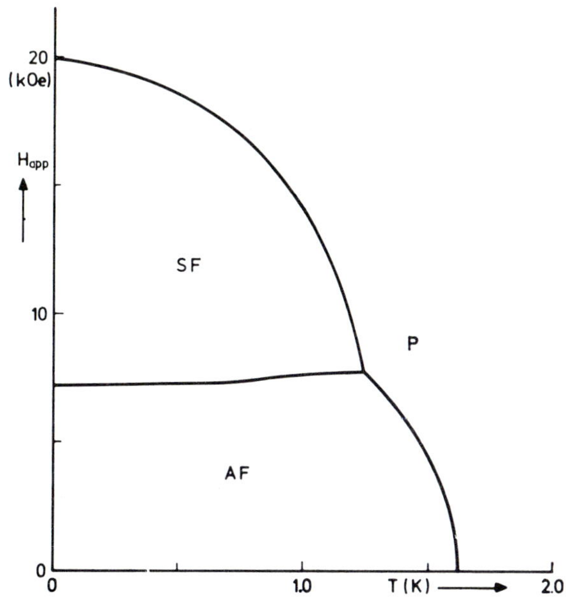

Experimental antiferromagnetic phase diagram as obtained by Giauque *et al.* (1970) for $MnCl_2 \cdot 4H_2O$. (H is the applied field.)

boundaries, in addition to the simple measurement of the M versus H curves already mentioned. It should be emphasized, however, that it may not be taken for granted that these different methods define exactly the same boundary (see, e.g., Giauque *et al.* 1970). For instance, the maximum in the susceptibility versus T curves need not coincide with the specific heat maximum, in case both remain finite at the transition point, as in any experiment. Still another method of locating the phase boundaries is from measurements of the ultrasonic attentuation, as has recently been applied by Shapira to a number of antiferromagnets (e.g. Shapira 1971). The phase transitions appear as anomalies in the ultrasonic attenuation, that have a different shape according to the order of the transition and to the mode of propagation.

In our opinion the most reliable criterion in determining the field-dependent transition points will be the specific heat anomaly. As an example the data of Reichert and Giauque (1969) and Giauque *et al.* (1970) on $MnCl_2 . 4H_2O$ are shown in fig. 85. Upon increasing the field the specific heat anomaly apparently remains as sharp as for the case $H=0$, although its height decreases. This has also been found in experiments on other compounds. Since the heat capacity was measured as a function of T at various constant H values, the anomaly traces the AF–P and SF–P boundaries. One would expect that a measurement of the specific heat as a function of field at constant temperatures below the

Fig. 85

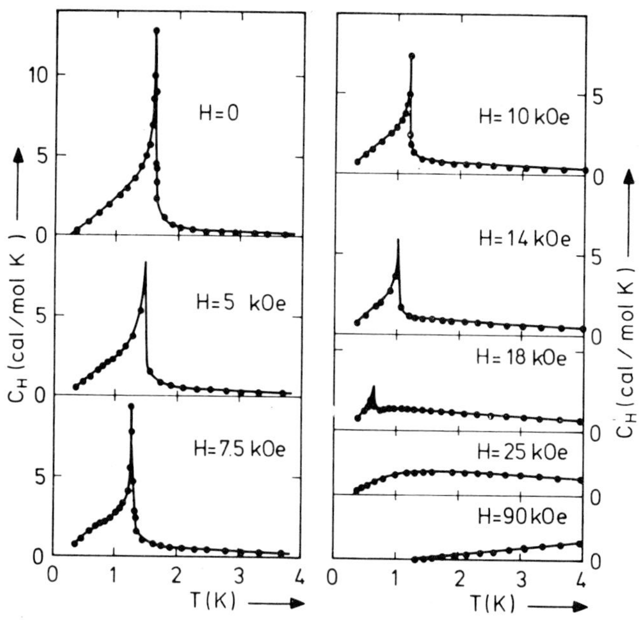

The specific heat of $MnCl_2 . 4H_2O$ in various (constant) applied fields. (After Reichert and Giauque (1969) and Giauque *et al.* (1970).

triple point will yield a sharp spike at the spin-flop transition, in addition to the λ-type anomaly at the transition to the paramagnetic state. These spikes should reflect the latent heat of transition associated with the first-order spin-flop transition.

At this point we would like to emphasize the advantage of the antiferromagnetic systems consisting of antiferromagnetically coupled ferromagnetic layers in the study of field-dependent behaviour. Since the antiferromagnetic interaction is in this case the (often extremely) weak coupling between the layers, saturation can be reached in moderate fields of a few kOe already. The metamagnetic substances will be discussed below. Here we want to concentrate on the quite isotropic layer-type compound $(C_2H_5NH_3)_2CuCl_4$ already discussed in the preceding pages. The phase diagram, as measured in the easy plane with the aid of differential susceptibility measurements, is given in fig. 86 (de Jongh *et al.* 1972 b, de Jongh 1972 a). It may be seen that the SF–P transition along the preferred and next-preferred axes are of the order of 1500 Oe, whereas the spin-flop transition is a mere 330 Oe. The critical field values are consistent with an antiferromagnetic exchange field $H_{af} \simeq 800$ Oe and an in-plane anisotropy $H_A{}^I \simeq 80$ Oe. Other interesting features are the

Fig. 86

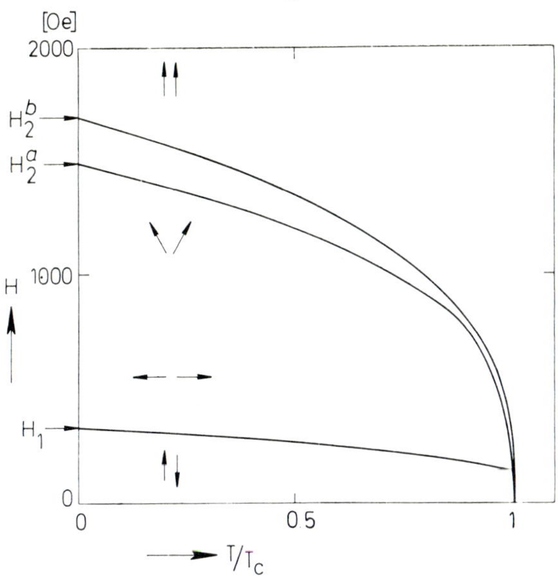

The antiferromagnetic phase diagram of $(C_2H_5NH_3)_2CuCl_4$ (De Jongh 1972 a). The critical fields H_1 and $H_2{}^a$ denote the spin-flop transition (H_{SF}) and the transition from the flopped to the paramagnetic phase (H_c), respectively, as obtained when the field is parallel to the easy axis. With H parallel to the next preferred direction, only the latter transition is observed ($H_2{}^b$, corresponding to $H_c{}'$ in eqn. (4.48)). The differences between $H_2{}^a$ and $H_2{}^b$ reflect the anisotropy within the easy phase (compare also with fig. 83). H is the internal field.

fact that H_{SF} decreases with temperature (compare fig. 84) and that the triple point is very close to T_c (at $T/T_c = 0{\cdot}997$, $H = 135$ Oe).

The low values of the critical fields admit of a careful study of the critical behaviour near the field-dependent transition points. In fig. 87 are plotted a number of isotherms of the differential susceptibility versus the applied field, from which the phase diagram shown in fig. 86 has been

Fig. 87

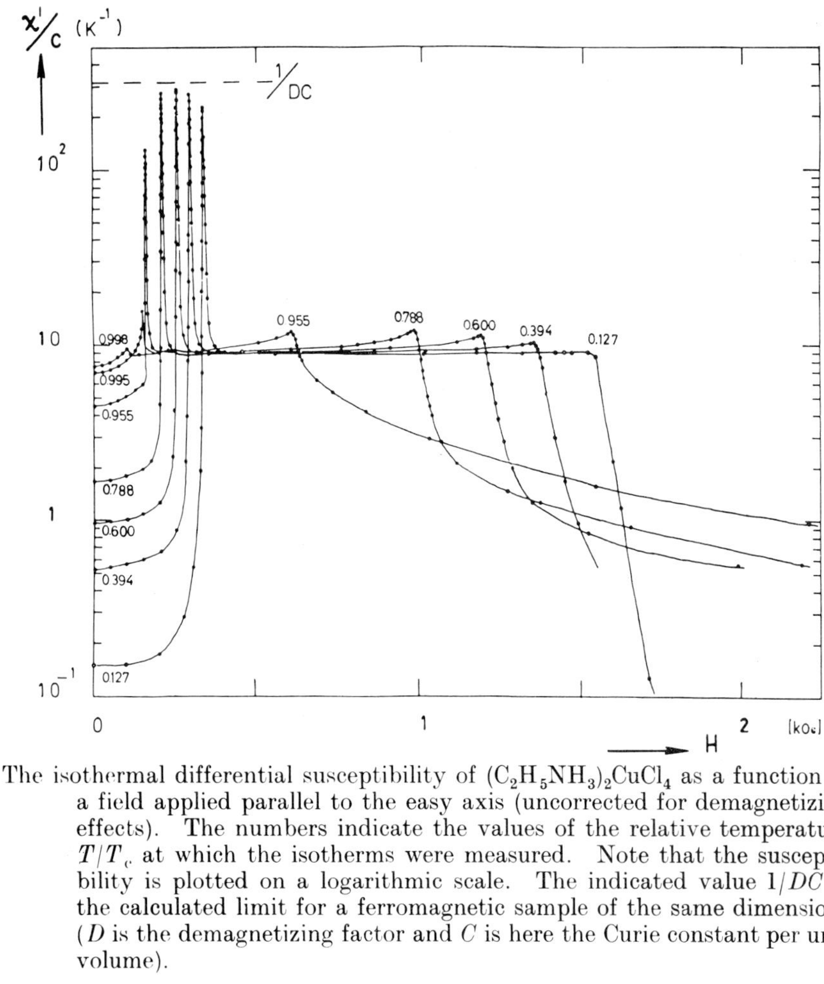

The isothermal differential susceptibility of $(C_2H_5NH_3)_2CuCl_4$ as a function of a field applied parallel to the easy axis (uncorrected for demagnetizing effects). The numbers indicate the values of the relative temperature T/T_c at which the isotherms were measured. Note that the susceptibility is plotted on a logarithmic scale. The indicated value $1/DC$ is the calculated limit for a ferromagnetic sample of the same dimensions (D is the demagnetizing factor and C is here the Curie constant per unit volume).

derived. The field is applied parallel to the easy axis. The general behaviour is seen to be in agreement with the simple MF model discussed above. The intercepts at the ordinate reflect the increase of the parallel susceptibility in zero field as the critical temperature is approached. At the spin-flop transition the susceptibility apparently diverges, since it reaches values nearly equal to the ferromagnetic limit, calculated on the basis of the sample shape. In between H_{SF} and H_c the susceptibility attains values that depend only weakly on temperature and are near to that of the perpendicular susceptibility. Moreover, the behaviour near H_c suggests, indeed, a discontinuity in the susceptibility in the limit $T \to 0$. What is plotted is the real part χ' of the a.c. susceptibility. Although the imaginary part is zero for most of the field range, sharp peaks in χ'' were observed at the spin-flop transitions, which is consistent with the notion of a domain formation at H_{SF}.

The peaks in the susceptibility at H_{SF} are seen to be very sharp, and correcting the applied field for demagnetizing effects would make them even sharper (this correction amounts to a few per cent of the applied field for $H > H_{SF}$). However, a diverging susceptibility merely implies an infinite derivative of the M versus H curve at $H = H_{SF}$ and not necessarily a discontinuity in M, in which case a plot of χ versus the applied field should give a susceptibility that retains the limiting (ferromagnetic) value over a certain finite field interval. Expanded plots of the experimental χ versus the applied field curves indicate indeed that χ has its maximum value over an interval of a few Oe, but it is hard to draw firm conclusions from such a small field range. That the interval is so small may of course be explained by the fact that the demagnetizing field just above the spin-flop transition is only about 9 Oe. Considering the other evidence obtained in, for example, $GdAlO_3$ and $MnCl_2 . 4H_2O$, however, where the magnetization jump can be made vertical within the uncertainties involved in correcting the applied field for demagnetizing effects, one would draw the conclusion that the SF transition is indeed of first order, in the sense that there occurs a discontinuity in the magnetization. The difference with the MF prediction is that the susceptibility is not constant above and below the SF transition. In fact the divergences shown in fig. 87 can be fitted to a power-law behaviour of the form

$$\chi/C \sim R_0 |(H - H_{SF})/H_{SF}|^{-\rho}, \tag{4.49}$$

with $\rho \simeq 2{\cdot}3$ over several decades in χ/C. This is exemplified in fig. 88 for the data on the high-field side of the SF transition. In order to carry out such an analysis, the value of the perpendicular susceptibility attained in between H_{SF} and H_c was subtracted from the measured χ values, and furthermore the data were corrected to an infinitely long cylindrical sample shape. For $H < H_{SF}$ a similar though less impressive fit to a power law with the same ρ value could be obtained.

Another difference with the MF theory that is observed in fig. 87 is the fact that in between H_{SF} and H_c the susceptibility does not remain constant but rises with H and even displays an anomaly at the SF–P transition, rather than a discontinuous decrease. This was observed in the flopped phase, with H parallel to the easy axis, as well as with the field in the perpendicular orientation. The latter results are shown in fig. 89, where χ is now plotted on a linear scale, so that the peaks are seen more clearly. It is observed that the amplitude of the singularity

Fig. 88

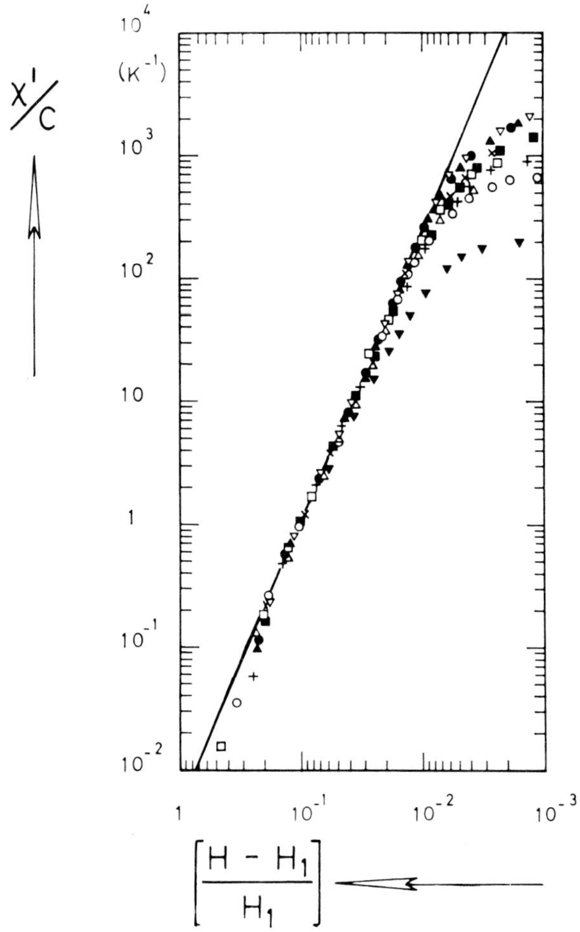

Critical behaviour of the susceptibility near to the spin-flop transition field (H_1, corresponding to H_{SF} in the text) in $(C_2H_5NH_3)_2CuCl_4$. The relative temperatures T/T_c at which the isotherms were taken are: ○ : 0·127 ; △ : 0·209 ; □ : 0·292 ; × : 0·394 ; ▽ : 0·495 ; ● : 0·600 ; ▲ : 0·691 ; ■ : 0·788 ; + : 0·870 ; ▼ : 0·955. The susceptibilities have been corrected to an infinitely long cylindrical sample shape, but this correction is negligible for $(H-H_1)/H_1 > 2\times10^{-2}$. The straight line represents a power-law divergence with an exponent of 2·3. (eqn. (4.49)).

depends on temperature ; for $T\rightarrow 0$ the MF prediction is the more closely approximated.

Deviations of the perpendicular magnetization curves from linear behaviour (constant $\chi_{\perp}$) have been found in many other materials. As examples we mention here EuTe (Jacobs and Silverstein 1964, Oliveira *et al.* 1972), $FeCl_2$ (Carrara *et al.* 1969), and $CoCl_2 . 6H_2O$ and $CoBr_2 . 6H_2O$

Fig. 89

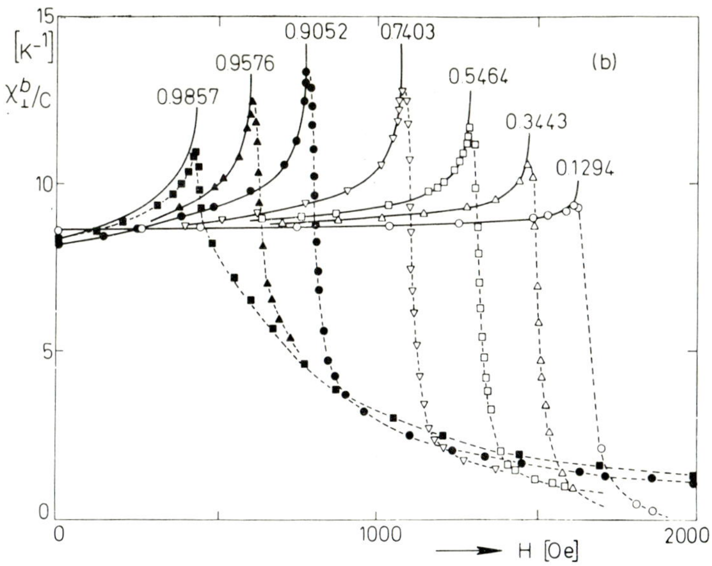

Critical behaviour of the perpendicular susceptibility of $(C_2H_5NH_3)_2CuCl_4$ near to the transition to the paramagnetic phase. The (internal) field is parallel to the next preferred axis. The numbers again indicate the relative temperatures T/T_c. The solid curves drawn through the data on the low-field side of H_c have been calculated from eqn. (4.50) with $A = 1{\cdot}52 \times 10^{-2}\ K^{-1}$ and $\tau = 0{\cdot}32$.

(Metselaar and De Klerk 1973 a, b). In EuTe the temperature dependence of the amplitude of the singularity in $\chi_\perp$ at H_c is similar to that observed in $(C_2H_5NH_3)_2CuCl_4$ (Oliveira *et al.* 1972). Quite contrarily, the data on $CoBr_2 . 6H_2O$ (Metselaar and De Klerk 1973 b), reproduced in fig. 90, show an exactly opposite behaviour, in that the amplitude increases with decreasing temperature. The magnetization curve of $FeCl_2$ reported by Carrara *et al.* (1969) is again different, since even at a temperature as low as $T \simeq 0{\cdot}2\ T_c$ the magnetization was found to increase linearly for $H > H_c$ up to $H \simeq 6H_c$, instead of saturating. These various phenomena may be explained by considering the different mechanisms that can cause these deviations from the simple MF prediction. The following four mechanisms are mentioned in the literature.

(i) Zero-point spin deviations. A gradual suppression of the effects of zero-point spin deviations upon $\chi_\perp$ will occur when the sublattice moments are rotated by the field from an antiferromagnetic to a ferromagnetic configuration (Kanamori and Yosida 1955, Jacobs and Silverstein 1964). This result in an extra increase of the magnetization, in addition to the MF term $M = \chi_\perp{}^0 H$(cf. eqn. (3.6)).

Fig. 90

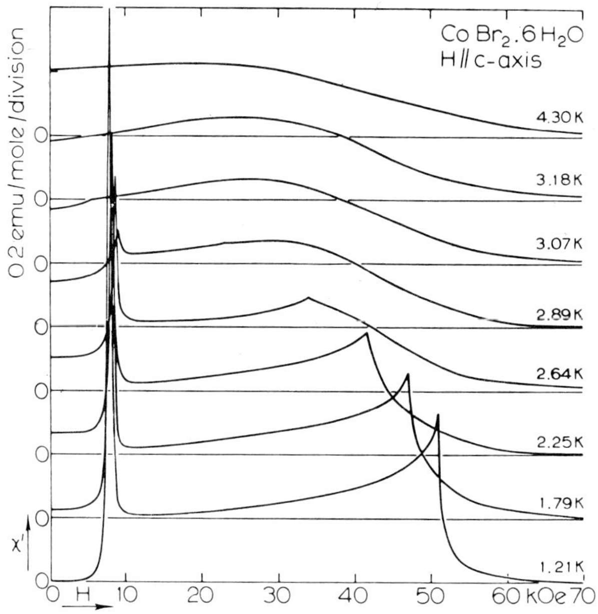

The isothermal differential susceptibility of $CoBr_2 . 6H_2O$ as a function of a (applied) field parallel to the easy axis, for different temperatures. The transition temperature is $T_c = 3{\cdot}14$ K. (After Metselaar and De Klerk (1972 b.)

(ii) Besides the above effect, spin-wave theory predicts the occurrence of instabilities at the SF–P transition (Feder and Pytte 1968), arising from the fact that at H_c the magnon dispersion relation changes suddenly from a linear (antiferromagnetic) to a quadratic (ferromagnetic) wave-vector dependence. This leads to a divergent term in the susceptibility at H_c (on both sides of H_c) which, to first order in the $1/2S$ expansion, is of the form $\chi_{\perp} \sim HT(H_c - H)^{-1/2}$ for $H \rightarrow H_c^-$ and similarly for $H \rightarrow H_c^+$.

(iii) Anisotropy effects. Carrara *et al.* (1969) have shown that for a metamagnet (large anisotropy and $H_f \gg H_{af}$) the susceptibility increases for $H \rightarrow H_c^-$, where the transition field is lowered from the MF prediction. For $H > H_c$ the magnetization rises slowly with H towards saturation.

(iv) Biquadratic exchange, arising from the strain dependence of the exchange energy (Kittel 1960, Jacobs and Silverstein 1964). This also introduces an extra field-dependent term in the susceptibility that reaches a maximum at $H = H_c$.

The last effect is difficult to estimate quantitatively, since for most salts knowledge of the amount of magnetostriction is lacking. We feel however that it will be small in most cases ; in any case it cannot explain the temperature dependences of the amplitude of the singularity in $\chi_{\perp}$ shown in figs. 89 and 90. Discarding mechanism (iv) therefore for the present discussion, we remark that the behaviour of $FeCl_2$ is adequately described by the mechanism (iii), as expected from its metamagnetic properties (see below). For $FeCl_2$ and $(C_2H_5NH_3)_2CuCl_4$ the mechanism (i) can be disregarded as a possible source, since, by the fact that $J_f \gg J_{af}$, the zero-point spin reduction will be very small (De Jongh 1972 a). Calculations of Lebesque of our laboratory pertinent to the Cu compound yield a reduction $\Delta S \simeq 10^{-4}$ only. On the other hand, the anisotropy mechanism (iii) does not apply for the fairly isotropic compounds EuTe, $(C_2H_5NH_3)_2CuCl_4$ and $CoBr_2 . 6H_2O$. (Note that for the latter two the anisotropy within the easy plane is considered.)

Thus, through the properties of the individual salts, we are in a position to choose for one, in some cases two mechanisms. As mentioned, for $FeCl_2$ the anisotropy effect will be predominant, although (ii) will perhaps also contribute. For $(C_2H_5NH_3)_2CuCl_4$ we can definitely choose for mechanism (ii) and in fact the spin-wave calculations yield the proper temperature dependence of the amplitude of the singularity. Guided by Feder and Pytte's result, De Jongh (1972 a) fitted the data for $H < H_c$ to the formula

$$[\chi_\perp(H) - \chi_\perp(0)]/\chi_\perp(0) = AT(H/H_c)[1 - H/H_c]^{-\tau}, \qquad (4.50)$$

where A and τ are (positive) constants. It turned out that all the measured isotherms between $0{\cdot}12 < T/T_c < 0{\cdot}96$, for the field in the perpendicular direction, as well as parallel to the easy axis for $H > H_{SF}$, could be described within 1 to 2% by this formula up to fields $H/H_c < 0{\cdot}98$, with the same values for A and τ ($\tau \simeq 0{\cdot}32$). The discrepancy with the spin-wave prediction $\tau = 0{\cdot}5$ should not be taken too serious, amongst other things because it is only the leading term in the $1/2S$ expansion. The heavy curves drawn through the data for $H < H_c$ in fig. 89 are calculated from eqn. (4.50) with the experimental values for A and τ. The data for $H > H_c$ are not easily analysed in a similar way because the ‘ paramagnetic ’ contribution to the susceptibility is difficult to subtract. This contribution arises, of course, from the fact that for $T > 0$, after the rotation of the mean directions of the sublattice moments towards the easy axis is completed, there still remains the saturation of these sublattice momenta to be accomplished.

The compound $CoBr_2 . 6H_2O$, on the other hand, is a clear candidate for mechanism (i), since through its 2-d XY character it will no doubt have substantial zero-point spin deviations (cf. the discussion in § 3.2.2). The effect of (i) should become the more apparent the lower the temperature, firstly because the contribution of (ii) decreases with temperature, and secondly because the effects of the zero-point spin deviations will be the larger the lower the temperature. The data of fig. 90 are seen to be in good agreement with these simple arguments.

In EuTe one may expect a mixture of both effects (i) and (ii), but since this is a 3-d compound with a high spin value the zero-point effects are very small and thus mechanism (ii) will likely dominate. The data of

Oliveira *et al.* (1972) show a temperature dependence of the amplitude that is similar to that observed in the Cu compound. Unfortunately these particular measurements were performed on a conducting sample. In addition we may add that Jacobs and Silverstein (1964) analysed an EuTe isotherm measured at $T/T_c \simeq 0{\cdot}22$ in terms of mechanism (i), showing that even at this (low) temperature, zero-point reduction alone cannot account for the observed behaviour.

In conclusion we may say that the experiments on the isotropic antiferromagnets strongly support the first-order character of the spin-flop transition, although the nature of the susceptibility divergence at H_{SF} is certainly different from the MF theory (as yet there is no theoretical treatment available to explain the data in fig. 88). In the case of the SF–P transition spin-wave theory predicts divergences in the susceptibility at H_c, which would make the transition to be still of second order, as it is in the MF approximation although this predicts discontinuities in χ. Indeed, anomalies in the experimental susceptibilities are found at H_c, the dependence on temperature being in apparent agreement with spin-wave theory. However, the possibility of finite cusps in the susceptibility cannot be excluded by the experiments, so that the SF–P transition could still be of an order higher than 2. This question must be solved by further theoretical and experimental work.

The same problem arises when one considers the nature of the AF–P transition (for temperatures above the triple point) which is believed to be of second order. The field-dependent heat capacity (e.g. fig. 85) seems to support this, since for the experimental examples available the specific heat anomaly apparently remains sharp, as for $H=0$, although the height of the experimental maxima decreases with H. Also the peaks in the susceptibility observed at the AF–P boundary in e.g. $MnCl_2 . 4H_2O$ (Giauque *et al.* 1970), $GdAlO_3$ (Blazey *et al.* 1971) and $(C_2H_5NH_3)_2CuCl_4$ (De Jongh *et al.* 1972 b) have more the character of finite cusps than divergences, although the peaks in the latter two examples are more pronounced than in the first. It has yet to be investigated whether the small size of the observed anomalies arises from the experimental conditions (naturally the experimental specific heat singularities are also always finite even for $H=0$). In any case, if the AF–P transition be of order higher than 2 this would mean either that the transition at $H=0$ ($T=T_c$) is a special point, since the transition at T_c is generally accepted to be of second order, or that this latter assumption is false and that also for $H=0$ the transition is of higher order. Neither of these possibilities is in accordance with current theoretical ideas on the subject, which expect the order of the transition not to be affected by a magnetic field and thus equal to 2.

Lastly we turn to the experimental examples of metamagnetic behaviour. We have already mentioned $FeCl_2$; other materials are $FeBr_2$, $CoCl_2$, DAG (with H parallel to the [111] axis) and $Ni(NO_3)_2 . 2H_2O$. The magnetization isotherms of $FeCl_2$ measured by Jacobs and Lawrence

Fig. 91

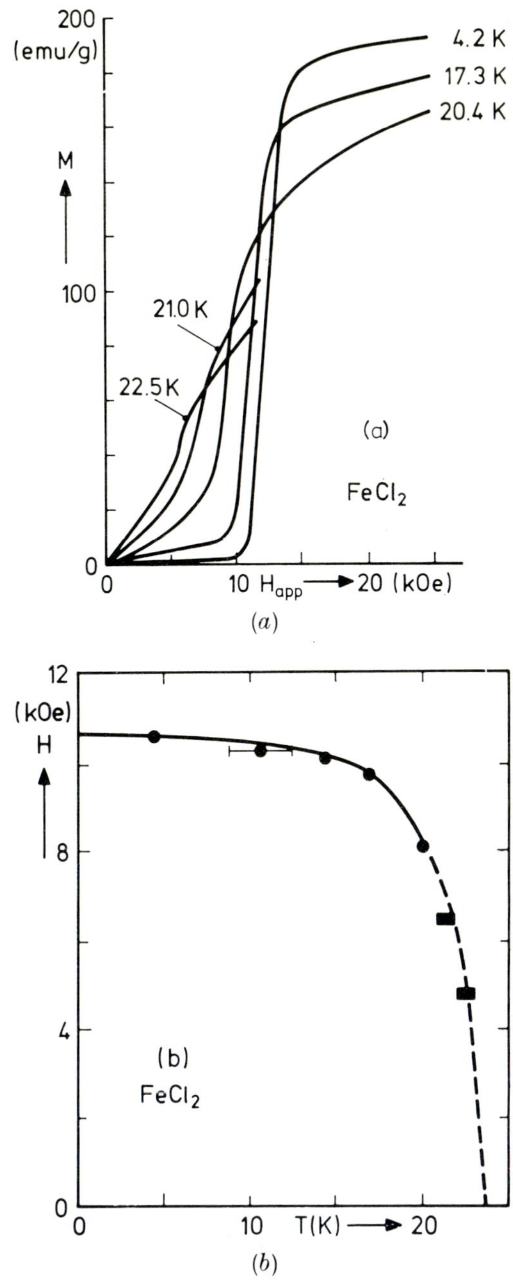

(*a*) Magnetization isotherms of the metamagnet $FeCl_2$ as measured by Jacobs and Lawrence (1967). The transition temperature is $T_c = 23{\cdot}5$ K. (*b*) The metamagnetic phase diagram of $FeCl_2$, showing the first-order transition line (solid curve) and the higher order line (dashed curve). The tri-critical point is estimated to be at about 20·4 K.

(1967) are shown in fig. 91 (*a*). One observes that up to $T=20{\cdot}4$ K the curves of the magnetization versus the applied field show a linear portion, with a slope that is independent of temperature. Within the (unfortunately large) experimental errors, this slope equals the reciprocal of the demagnetizing factor of the sample, in agreement with the expected first-order character of the transition. For higher temperatures the linear region apparently is absent, instead the curves display an inflexion point, indicating a transition of higher order. Correcting for demagnetizing effects, a phase diagram is obtained (fig. 91 (*b*)) that is in qualitative agreement with the theoretical expectation outlined above. Considerable hysteresis in the metamagnetic transition was reported.

Additional evidence for the first-order character of the transition is provided, for example, by the work on $Ni(NO_3)_2 \cdot 2H_2O$ (Schmidt and Friedberg 1970) and DAG (Landau *et al.* 1971), in which case the demagnetizing factor of the sample was known to much higher accuracy (about 1% for DAG), and the discontinuities in the magnetization versus the internal field curves were indeed found to be vertical within the error involved in the demagnetizing correction. In between the value for the applied field at which the magnetization starts to rise and that at which it levels of, the system is in a mixed state of co-existing antiferromagnetic and paramagnetic regions. By making a plot of applied field versus temperature one obtains a phase diagram as shown in fig. 92 for DAG (Landau *et al.* 1971). The different symbols refer to the various thermal and magnetic measurements used in locating the phase boundaries. From such a graph one may determine accurately the value of the tri-critical point. Correction for the demagnetizing factor ($N=5{\cdot}35$) of the sample

Fig. 92

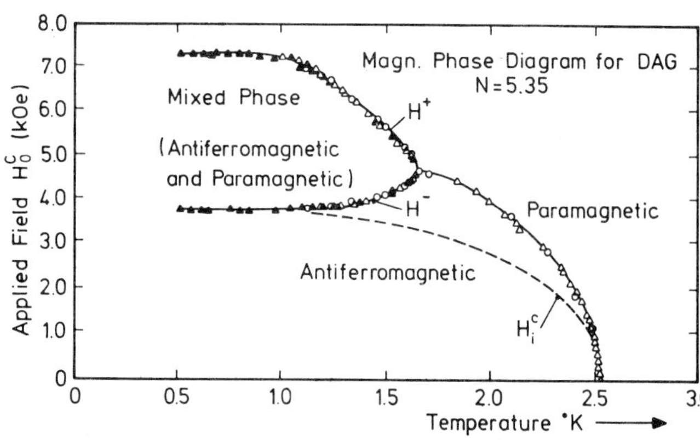

The metamagnetic phase diagram of dysprosium aluminium garnet plotted as $H_{\text{appl.}}$ versus T (solid curve) and $H_{\text{int.}}$ versus T (dashed curve). N denotes the demagnetizing factor of the sample. (After Landau *et al.* 1971.)

yields the broken curve that is similar to the one shown for $FeCl_2$. It is interesting to observe that within the experimental accuracy there occurs no kink at the tri-critical point.

Instead of the $H_{app.}$ versus T plot one can also draw a magnetization versus T diagram, given in fig. 93. Below the tri-critical point there are two branches, which are the loci of the ends of the vertical discontinuities in the magnetization isotherms. Between the tri-critical point and T_c the AF–P phase boundary was identified by the maxima in the isothermal susceptibility measured as a function of field. A similar diagram for $Ni(NO_3)_2 \cdot 2H_2O$ was constructed by Schmidt and Friedberg. It is this M–T diagram that is the analogue of the composition-temperature diagram for ^3He–^4He mixtures (Griffiths 1970 a). In the latter the tri-critical temperature corresponds to that below which spontaneous phase separation takes place. The upper and lower branches below the tri-critical point give the temperature variation of the ^3He-rich and the ^4He-rich phases, respectively. The boundary in between the tri-critical point and T_c is the line of λ transitions of the single phase.

Fig. 93

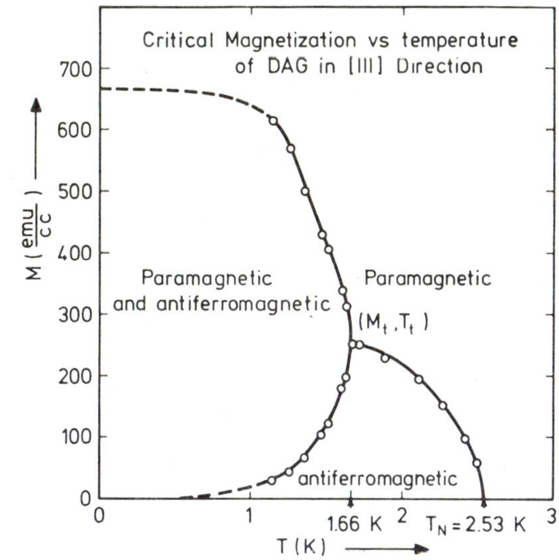

The M–T diagram of DAG, which can be considered to be a magnetic analogue of the composition-temperature diagram for ^3He–^4He mixtures.

Among the other interesting features emerging from the study of Landau *et al.* on DAG is the fact that the specific heat, measured as a function of temperature in constant internal fields, displays sharp, possibly infinite, peaks at the first-order transitions below the tri-critical point. These peaks reflect the latent heat of transition accompanying the first-order transition. About the nature of the phase boundary in between T_c and the tri-critical point, the same questions pertain as to the AF–P transition in the isotropic antiferromagnets. Although sharp peaks are observed in the heat capacity and the susceptibility, they appear to be finite so that it is still doubtful what the order of this transition will be. We remark in conclusion that in the experiments on antiferromagnetic substances performed thus far, the temperature dependence of the AF–P phase boundary has always been found to be of parabolic form, in agreement with theory.

In concluding this section we briefly comment on the field-dependent behaviour in ferromagnets. To our knowledge the only experiments performed so far are the specific heat measurements of Miedema *et al.* (1963) on $CuK_2Cl_4 \cdot 2H_2O$ and those on EuS by Teaney *et al.* (1968). In both cases the application of a constant field changes the appearance of the specific heat curve from a sharp peak into an apparently non-anomalous curve with a rounded maximum, the position of which was, in the case of $CuK_2Cl_4 \cdot 2H_2O$, found to decrease firstly with respect to T_c in small fields, and then moving upwards again as the field increases to higher values. A decrease of the temperature of the maximum as a function of field was also observed in EuS. These features seem to be consistent with the expectation quoted above that the transition in an isotropic ferromagnet will be destroyed on application of a non-zero external field.

However, Teaney *et al.*, analysing their data in terms of a complex critical temperature, claim that the apparent logarithmic divergence, found for $H=0$ as $T \to T_c^+$ (see above), is preserved in non-zero fields. They attribute the observed rounding to inhomogeneous magnetization and report some evidence for this by comparing the rounding found in a spherical sample with that in a cylindrically shaped specimen, for which it was found to be considerably larger.

Leaving it to the reader to judge the merits of the approach of Teaney *et al.* (1968), we mention the paper of Wojtowicz and Rayl (1968) that was inspired by the work of Griffiths and Arrott (see Arrott 1968). Within the MF approximation, Wojtowicz and Rayl show the possible existence of transitions from a non-uniformly to a uniformly magnetized state in an isotropic ferromagnet with dipolar interactions included. In low fields these transitions are reflected as sharp anomalies in the heat capacity, which occur at a temperature that decreases roughly quadratically with field and broaden as the field is increased. Furthermore, these sharp peaks are superimposed upon much broader maxima associated with the development of long-range order in the uniform state by the field. Although the existence of the sharp anomalies is as yet not verified by the experiments, the shape and temperature dependence of the broad maxima are qualitatively in good accord with the observations in $CuK_2Cl_4 \cdot 2H_2O$. Yet another mechanism for the occurrence of field-induced phase transitions in the experimental ferromagnets may be the presence of anisotropy, as shown by calculations bearing on this problem (see, e.g., Pfeifer 1971, Durczewski 1970). In this case the transition temperature also decreases with field.

Lastly we mention that in the work on EuS (Van der Hoeven *et al.* 1968) the specific heat at the critical isotherm was found to vary logarithmically with field. Moreover the authors analysed their data in different fields in terms of an equation of state, rather similarly to the example given above in § 4.4.

§ 5. Concluding remarks

The first aim in writing this paper has been to present a catalogue of those magnetic crystals that from experimental investigations have proven to be approximate representatives of one of the simple models used in theoretical descriptions of magnetic ordering phenomena. The present review clearly lacks completeness, i.e. it covers a limited variation in the type of magnetic interaction only and furthermore has been concentrated on insulators. We have considered three different dimensionalities for the spin (Ising, X–Y or Heisenberg), three magnetic lattice dimensionalities and two signs for the exchange interaction. However, we did not include magnetic crystals with predominantly dipolar interactions, magnetic metals in which the magnetic moments are not localized (itinerant magnets) and metals in which the interactions between localized moments are of the long-range, oscillating type (Rudermann–Kittel interaction).

The choice made has been influenced by the authors' own interests and activities. One may say that the selection comes down to a preference for those magnetic systems that on the one hand are the relatively most simple ones and on the other hand the most extensively investigated substances.

In addition to the systematic presentation of a collection of simple magnetic systems, i.e. how they can be conceived and where they have been found in Nature, we have tried to make clear why this type of research in magnetism has drawn so much attention. We have collected a number of experiments which present convincing experimental verification of theoretical predictions derived in fields as the theory of spin waves, series expansion methods and theories which treat the critical behaviour that accompanies phase transitions in general. Concerning the latter, we restricted ourselves to magnetic phase transitions, which means a severe limitation in view of the extensive literature on, for instance, gas–liquid phase transitions or ordering phenomena in alloys.

Having viewed the present collection of crystals that approximate simple magnetic models, one may ask which types are already sufficiently covered experimentally and for which types there is still a need for more complete experimental information. Firstly, considering the one-dimensional systems, we conclude that in fact only for Heisenberg antiferromagnetic chains is the situation completely satisfactory. Apparently, ferromagnetic Heisenberg chains and both ferro and antiferromagnetic Ising chains are more difficult to realize. A reason for this may be the large reduction of the spin moment that is inherent in an isotropic low-dimensional antiferromagnetic system. In spite of the fact that there are no good examples of Ising chains, however, one cannot positively conclude that it would be worth while to put considerable experimental effort into finding better ones. The theory for one-dimensional Ising systems is exact, while for Heisenberg ferromagnetic chains, too, the theoretical predictions derived by extrapolation from properties of rings with a finite, increasing number of spins are apparently sufficiently accurate to warrant the assumption that no new surprising phenomena are to be discovered by doing more experiments. Undoubtedly all of the thermodynamic behaviour can be calculated accurately enough theoretically, while the specific properties that have to do with an exceptionally large degree of short-range magnetic order can be (and have been) investigated equally well in the one-dimensional Heisenberg antiferromagnets, of which so many representatives are known at present. Generally speaking, we would say that a search for still more examples of simple magnetic model systems would be justified only in the case where the additional experimental work is expected to provide new contributions to the mutual stimulation of theory and experiment.

An interest in one-dimensional systems that remains is connected with studies of the magnetic interaction in insulators (its quantitative value). Low-dimensional magnetic systems offer the possibility of a relatively accurate and easy determination of the value of the exchange constant; also, the much smaller interaction between magnetic atoms of neighbouring chains, which gives information concerning complicated exchange paths, will be reflected in the experimental T_c values.

Turning next to the two-dimensional magnetic systems, we can conclude that there is a large number of Heisenberg substances, ferromagnetic as well as antiferromagnetic. The number of crystals in which the two-dimensional Ising model is approximated is again much smaller but likewise this does not necessarily imply a real need for further experimental research. The zero-field two-dimensional Ising problem is exactly soluble and also the field-dependent properties of the 'ideal system' can be adequately studied theoretically so that there seems to be no need of additional experimental verification. Note that the situation for Heisenberg magnets in two dimensions is fully different. Here the theory is far from complete, whereas experimentally it has been possible for instance to derive with good accuracy the specific heat of the quadratic Heisenberg ferromagnet with $S=\frac{1}{2}$, theoretical predictions being restricted to the low and high-temperature limiting cases only. Surprisingly, the series expansion analyses have led to the prediction of a phase transition in the magnetic susceptibility (Stanley and Kaplan 1966). Experiments have confirmed that this should be a phase transition to a state of infinite susceptibility but no spontaneous magnetization.

We suggest that a search for new examples of two-dimensional magnetic crystals is not necessary. Any forthcoming fundamental question, which is in principle open for experimental studies, can likely be answered from an investigation of the series of compounds known.

Similar to one-dimensional magnetic crystals, two-dimensional compounds will remain attractive for investigations of magnetic interaction constants. Apart from an accurate determination of the value of the main exchange constant, the quasi two-dimensional crystals in addition often offer the possibility of studying in detail the interaction in the third dimension and the effects of magnetic anisotropy.

Surprisingly enough, the number of fair approximations of simple models in three dimensions is quite limited. The few good examples do agree with theoretical predictions, but a conclusion of how important it would be to have more or better experimental realizations of three-dimensional Ising and Heisenberg magnets cannot be drawn straightforwardly, amongst other things because of the approximate nature of the theoretical results obtained so far.

In this section we did not mention the XY model until now. In this case both theoretical and experimental information are far from complete. For instance, the spin-wave theory for the planar models has received little attention. Quite generally, one expects an XY system to behave as intermediate between Ising and Heisenberg (critical exponents); in the special case of two-dimensional lattices the XY model is similar to the Heisenberg model, in that it will also possess the new type of phase transition to a state of infinite susceptibility without long-range order.

Further experiments on XY type crystals would therefore be quite interesting. However, the XY models is relatively difficult to realize. In Fe^{2+} and Ni^{2+} compounds, in which the crystal field would produce a singlet state lying lowest in the absence of magnetic interactions, the XY model can only be a relatively poor approximation. At the higher temperatures these substances will behave as anisotropic Heisenberg rather than XY systems, and for this reason the effective spin value is no simple constant. We suggest that compounds similar to $CoCl_2 . 6H_2O$ are more attractive in studying the XY model. One would like to have an isolated crystal-field doublet to be the only populated level at temperatures of the order of the exchange constant (effective $S=\frac{1}{2}$), while the XY character then arises from the anisotropy in the g tensor ($g_\perp > g_\parallel$). There are possibilities in finding $g_\perp \gg g_\parallel$ among rare-earth ions, but in that case it is difficult to produce magnetic interactions that are predominantly of the exchange type.

A conclusion that can be drawn from § 4.2 is that the applicability of the spin-wave approximation in describing the low-temperature properties of magnetically ordered crystals is beyond doubt, including the case of antiferromagnetism with the inherent phenomenon of zero-point spin deviations. For chains with only short-range order the situation is not fully clear. Apparently it makes sense to speak about spin-wave-like excitations also in systems which have no long-range order. It is possible to study the dispersion relations in neutron scattering experiments, but in particular the static properties of one-dimensional systems (specific heat and suceptibility) only qualitatively agree with simple spin-wave theory. The key to the solution of this problem may lie in the study of those spin-wave-like excitations which have wavelengths nearly equal to the correlation length, i.e. neutron diffraction experiments in which for a specific k value or energy the effect of a varying degree of short-range order is studied. In this respect neutron scattering experiments have a clear advantage over magnetic resonance experiments in which the temperature dependence of the linewidth (N.M.R., E.S.R.) is used as a tool to study the effects of the anomalously large degree of short-range order in low-dimensional magnetic crystals (Nagata and Date 1964, Bucci and Guidi 1970, Maarschall 1970, de Wijn, Walker, Davis and Guggenheim 1973 b). A resonance linewidth represents an integral property over the full spectrum of excitations, whereas in a neutron diffraction study an excitation of a particular frequency may be selected.

Closely connected to the problem of how to describe spin waves at temperatures above T_c is the question of how to renormalize the dispersion relation as a function of temperature. The energies of long wavelength spin waves become renormalized according to the spontaneous magnetization, short wavelength excitations on the other hand apparently have their frequencies renormalized according to the magnetic energy. For intermediate frequencies one has no clear idea yet as to how renormalization effects have to be taken into account.

Returning to the correlation length, we want to stipulate the prime importance of this quantity in understanding the behaviour of quasi one or two-dimensional systems. Differentiating between the inter-planar (or inter-chain) correlations and those within the magnetic layers (or chains), we have seen in the preceding pages that the former quite often play a very minor role, becoming manifest only at the lowest temperatures. In addition to the difference in strength between the inter and intra-layer interaction, this may be attributed to the difference in the dependence on temperature and on distance of the correlations in two and in three dimensions. Theoretical work (Jasnow and Fisher 1967, Lines 1970) has shown that the correlations in two-dimensional systems fall of considerably more slowly with distance than in three dimensions. The fact that at the same kT/J the correlation length is much larger in e.g. two dimensions, explains the large degree of short-range intra-planar order, long before the inter-planar correlations come into play. Thus the behaviour as a function of temperature of an array of nearly isolated layers is two-dimensional over a very wide temperature range, until, sometimes extremely close to T_c, the effect of the correlations in the third dimension is felt. This contrasts with the situation in thin magnetic films, for which one may argue that they will also approximate a two-dimensional system when the film thickness becomes of the same order as the lattice constant. In the case of thin films the high-temperature properties will be that of the bulk material, since only as the temperature is lowered to T_c will the correlation length become comparable with the film thickness, so that a two-dimensional character may manifest itself. We remark that thin films in general present a more difficult problem than crystals consisting of nearly isolated layers, since in the former the boundary effects are of considerably more importance than in the latter.

In § 4.3 we have stipulated the usefulness of predictions based on series expansion methods. It is surprising that a truncated series can provide accurate quantitative results for thermodynamic properties of interest at temperatures as close to T_c as 1%. Also the results derived for critical exponents appear to be quite satisfactory.

In § 4.4 we have witnessed that unusual values for the critical exponents β and γ (i.e. values different from those of three-dimensional lattices) have indeed been observed experimentally. The experimental analysis in these cases has to be performed in a temperature region slightly away from T_c rather than as near to T_c as possible. For instance, studying two-dimensional Heisenberg compounds a relatively weak interaction in the third dimension or a small anisotropy will lead to a three-dimensional or Ising-type critical behaviour, respectively, at temperatures which are sufficiently near to T_c. (In practical cases this may amount to $(T-T_c)/T_c \simeq 10^{-3}$–$10^{-2}$ already.)

A conclusion of § 4.4 is that the experimental information on critical behaviour agrees with theoretical expectations. Further experiments, in particular on systems with unusual exponent values, would be of much interest, but within the group of systems covered in the present paper this will be difficult. Approximately one-dimensional lattices will show three-dimensional critical behaviour (if any), quasi two-dimensional crystals too will show three-dimensional behaviour at temperatures sufficiently close to T_c, which region may in the highly anisotropic cases be preceded by one where it is two-dimensional Ising-like. Likewise an approximate two-dimensional Heisenberg compound may, away from T_c, show the critical behaviour characteristic for the ideal system. For this model two cross-overs may be observed as T_c is approached, namely, firstly from two-dimensional Heisenberg to two-dimensional Ising (or XY) and thereafter to three-dimensional Ising behaviour. Obviously a critical exponent for the heat capacity may in this case only be observed in the three-dimensional (or in the two-dimensional Ising) region.

We suggest that in addition to giving much attention to the dependence of critical exponents on lattice dimensionality, one should like to have more information on sets of critical exponents for magnetic systems unusual in other ways, viz. magnetic systems having long-range interactions, magnetic crystals with well-defined temperature-dependent interaction constants or dilute magnetic systems (crystals in which a fraction of the magnetic sites is occupied by diamagnetic atoms).

One may conclude from § 4.5 that there is still a lot of interesting work to be done concerning field-dependent properties and field-induced phase transitions in general. Since one is restricted to long-range ordered systems, one basically deals with either 3-d or 2-d-Ising systems. Nevertheless, approximately low-dimensional crystals have an advantage that, when considering their three-dimensional magnetic properties at temperatures below T_c, the magnetic field region of interest when studying transitions in the H–T phase diagram is often restricted to easily accessible low field values.

Fig. 94

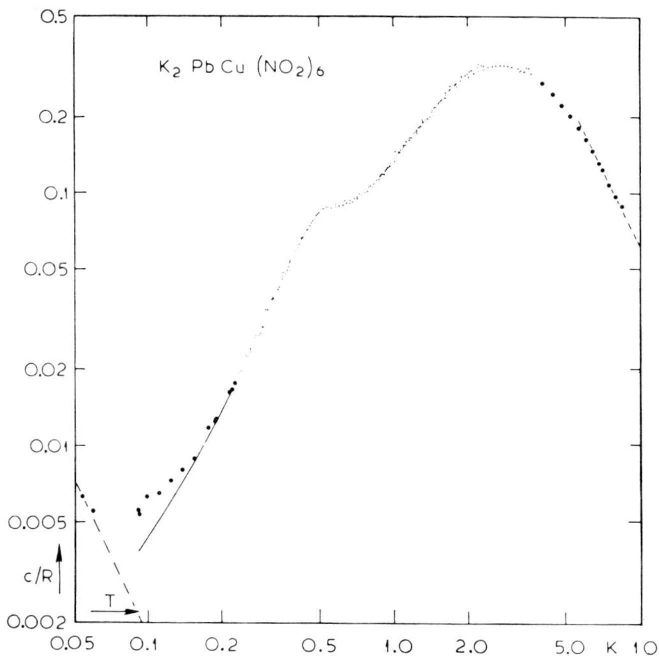

The magnetic heat capacity of $K_2PbCu(NO_2)_6$, an example of an f.c.c. antiferromagnet with $S=\frac{1}{2}$. The curve displays the broad maximum (height about $C/R=0{\cdot}3$) characteristic for magnetic systems that (ideally) would have no transition to long-range order. The temperature of the maximum is about 3 K. The small anomaly found at $T\simeq 0{\cdot}5$ K could be interpreted as a transition to long-range order, caused by next-nearest neighbour interactions. This temperature should be compared to the molecular field prediction for the transition temperature, which is of the order of $\theta\simeq 10$ K, as deduced from estimates of the exchange constant. (After Blöte, private communication ; see also Huiskamp (1966).)

Concentrating a little more on possibilities for future work it will come as no surprise that we suggest that one should pay more attention to some of the magnetic systems that have not been treated in the present review. For instance, we did not consider the f.c.c. or triangular antiferromagnets that have transition temperatures strongly dependent on the value of the second nearest-neighbour interaction, since they will not show long-range order in the case of only nearest-neighbour interactions (see fig. 94). Also one may study magnetic ordering in the case of only dipolar interactions, in helical spin systems, weak ferromagnets, metallic itinerant magnets, alloys of rare-earth metals with localized moments coupled by the mechanism of magnetically polarized conduction electrons, diluted magnetic insulators and the magnetic substances whose properties are affected by the size of the crystal. Here a lot of interesting work is still to be done, theoretically as well as experimentally.

Fig. 95

(*a*) (*b*)

'Staring crowd' analogue of phase transition. (*a*) 'normal' phase. (*b*) 'condensed' phase.

The 'staring crowd' phenomenon (Mattuck and Johansson 1968) is an illustration of the fact that phase transitions are certainly not confined to physics but is a more general phenomenon that can be found everywhere in Nature whenever one deals with a system consisting of elements between which there exists some sort of feed-back mechanism (exchange). If one of the elements (spins, human beings) gets conditioned in a certain fashion—in the above example the attention of one of the persons is attracted by something at the window—the neighbouring elements become conditioned in a similar way, even though there is no external force present that compels them to do so (they may see nothing at all at the window in question). Another example from daily life is the 'spontaneous buying' of luxury goods as colour-television sets, new cars, etc., which occurs when a given person has enough neighbours around him that possess such an item.

The main point at this moment for doing such research may be to get additional evidence supporting the ideas of a universal description of phase transitions in terms of a few parameters. Taking this argument as a start, it is only a small step to fields outside magnetism, phase transitions being of importance for molecular physics and metallurgy as well as for biology and as illustrated by fig. 95 also, for example, for the social sciences.

A final conclusion to be drawn from the present paper concerns materials science in magnetism. In the above we have learned that it has been possible to find experimental approximants of various highly artificial theoretical models which theoreticians are forced to use. In many cases the compounds were not discovered by accident but have been searched for systematically. This leads us to the optimistic conclusion that, if it is made sufficiently clear which combination of properties has to be looked for, even the apparently most unlikely combinations may be realized in practice.

Acknowledgments

The preparation of the manuscript has evidently involved a large amount of typing, drawing and reproduction work. We want therefore to thank firstly the members of the administrative and technical staff of our laboratory who took care of these matters, in particular Miss M. Bos, Mr. B. E. Leonards and Mr. P. de Vries. They managed to deal with our never-ending demands.

Furthermore, we have profited from discussions, correspondence, comments, etc., with workers in the field. In alphabetical order we mention G. A. Baker, Jr., D. D. Betts, R. J. Birgeneau, H. W. J. Blöte, J. C. Bonner, D. E. Cox, W. Duffy Jr., M. Eibschütz, M. E. Fisher, S. A. Friedberg, R. B. Griffiths, W. J. Huiskamp, E. Lagendijk, M. Matsuura, J. F. Nagle, E. J. Samuelsen, I. F. Silvera, J. Skalyo Jr., H. E. Stanley and coworkers, G. de Vries, L. R. Walker and R. E. Walstedt. In particular, one of the authors (D.J.) expresses his gratitude to his American colleagues for the warm hospitality received during a trip to the U.S.A.

Last but not least we want to thank the many members of our laboratory who have contributed, especially those of the group working on two-dimensional magnetism : A. C. Botterman, D. J. Breed, P. Bloembergen, J. H. P. Colpa, F. W. Gorter, J. V. Lebesque, E. P. Maarschall, A. H. M. Schrama, and the students that took part.

This paper will serve as part of a Doctor's Thesis of one of the authors (D.J., to be promoted by M.), submitted to the University of Amsterdam. He wishes to thank the 'Stichting voor Fundamenteel Onderzoek der Materie' (F.O.M.), which is supported by the 'Nederlandse Organisatie voor Zuiver Wetenschappelijk Onderzoek' (Z.W.O.), for financial support during the time of preparation of the manuscript, and last but not least his family for their patience.

REFERENCES

ABE, H., and MATSUURA, M., 1964, *J. phys. Soc. Japan*, **19,** 1867.

ABE, H., and TORII, K., 1965, *J. phys. Soc. Japan*, **20,** 183.

ABRAHAMS, S. C., and PRINCE, E., 1962, *J. chem. Phys.*, **36,** 50.

ACHIWA, N., 1969, *J. phys. Soc. Japan*, **27,** 561.

AJIRO, Y., 1969, *J. phys. Soc. Japan*, **27,** 829.

AJIRO, Y., and TERATA N., 1970, *Proc. 12th Int. Conf. Low Temp. Phys.*, Kyoto.

AJIRO, Y., TERATA, N., MATSUURA, M., and HASEDA, T., 1970, *J. phys. Soc. Japan*, **28,** 1587.

AKIMITSU, J., ISHIKAWA, Y., and ENDOH, Y., 1970, *Solid St. Commun.*, **8,** 87.

ALS-NIELSEN, J., 1969, *Phys. Rev.*, **185,** 664.

ALS-NIELSEN, J., DIETRICH, O. W., KUNNMANN, W., and PASSELL, L., 1971, *Phys. Rev. Lett.*, **27,** 741.

ANDERSON, F. B., and CALLEN, H. B., 1964, *Phys. Rev.* A, **136,** 1068.

ANDERSON, P. W., 1952, *Phys. Rev.*, **86,** 694.

ARGYLE, B. E., MIYATA, N., and SCHULTZ, T. D., 1967, *Phys. Rev.*, **160,** 413.

ARORA, B. L., and LANDAU, D. P., 1972, *Proc. 18th Ann. Conf. Magn. Magn. Materials*, Denver (*AIP Conf. Proc.*, **10,** 870).

ARROTT, A., 1968, *Phys. Rev. Lett.*, **20,** 1029.

BAKER, Jr., G. A., 1963, *Phys. Rev.*, **129,** 99.

BAKER, Jr., G. A., EVE, J., and RUSHBROOKE, G. S., 1970, *Phys. Rev.* B, **2,** 706.

BAKER, Jr., G. A., and GAUNT, D. S., 1967, *Phys. Rev.*, **155,** 545.

BAKER, Jr., G. A., GILBERT, H. E., EVE, J., and RUSHBROOKE, G. S., 1967 a, *Phys. Lett.* A, **25,** 207 ; 1967 b, *Phys. Rev.*, **164,** 800.

BAKER, Jr., G. A., RUSHBROOKE, G. S., and GILBERT, H. E., 1964, *Phys. Rev.* A, **135,** 1272.

BAKER, J. M., 1971 a, *Rep. Prog. Phys.*, **34,** 109 ; 1971 b, *J. Phys.* C, **4,** 1631.

BALL M., HUTCHINGS, M. T., LEASK, M. J. M., and WOLF, W. P., 1962, *Proc. 8th Int. Conf. Low Temp. Phys.*, London, p. 248.

BALL, M., LEASK, M. J. M., WOLF, W. P., and WYATT, A. F. G., 1963, *J. appl. Phys.*, **34,** 1104.

BARENDREGT, F., and SCHENK, H., 1970, *Physica*, **49,** 465.

BARRACLOUGH, C. G., and NG, C. F., 1964, *Trans. Faraday Soc.*, **60,** 836.

BELJERS, H. G., BONGERS, P. F., VAN STAPELE, R. P., and ZIJLSTRA, H., 1964, *Physics Lett.*, **12,** 81 ; 1966, *J. chem. Phys.*, **44,** 3719.

BELOV, K. P., and GORYAGA, A. N., 1956, *Fiz. Met. Metallov.*, **2,** 3.

BEREZINSKII, 1970, *Zh. eksp. Teor. Fiz.*, **59,** 907 ; 1971, *Soviet Phys. J.E.T.P.*, **32,** 493.

BERLIN, T. H., and KAC, M., 1952, *Phys. Rev.*, **86,** 821.

BERTAUT, E. F., BURLET, P., and BURLET, P., 1969, *Solid St. Commun.*, **7,** 343.

BETTS, D. D., 1973, *Phase Transitions and Critical Phenomena*, Vol. 3, edited by C. Domb and M. S. Green (Academic Press).

BETTS, D. D., DITZIAN, R. V., ELLIOTT, C. J., and LEE, M. H., 1970, *Proc. Int. Conf. Magn.*, Grenoble (*J. Phys.*, **32,** suppl. C–1, p. 356).

BETTS, D. D., ELLIOTT, C. J., and DITZIAN, R. V., 1971, *Can. J. Phys.*, **49,** 1327.

BETTS, D. D., TSAI, J.-T., and ELLIOTT, C. J., 1973, *Proceed. Int. Conf. Magn.*, Moscow.

BIDAUX, R., and MÉRIEL, P., 1968, *J. Phys.*, **29,** 220.

BIENENSTOCK, A., 1966, *J. appl. Phys.*, **37,** 1459.

BIRGENEAU, R. J., DEROSA, F., and GUGGENHEIM, H. J., 1970 a, *Solid St. Commun.*, **8,** 13.

BIRGENEAU, R. J., DINGLE, R., HUTCHINGS, M. T., SHIRANE, G., and HOLT, S. L., 1971 a, *Phys. Rev. Lett.*, **26,** 718.

Birgeneau, R. J., Guggenheim, H. J., and Shirane, G., 1969, *Phys. Rev. Lett.*, **22,** 720 ; 1970 b, *Phys. Rev.* B, **1,** 2211 ; 1973, *Ibid.* B, **8,** 304.
Birgeneau, R. J. Shirane, G., and Kitchens, T. A., 1972 a, *Proc. 14th Low Temp. Conf.*, Boulder, Colorado.
Birgeneau, R. J., Skalyo, Jr., J., and Shirane, G., 1970 c, *J. appl. Phys.*, **41,** 1303 ; 1971 b, *Phys. Rev.* B, **3,** 1736.
Birgeneau, R. J., Yelon, W. B., Cohen, E., and Makovsky, J., 1972 b, *Phys. Rev.* B, **5,** 2607.
Bizette, H., Terrier, C., and Tsaï, B., 1956, *C.r. Séanc. Acad. Sci., Paris*, **243,** 1295 ; 1965 *Ibid.*, **261,** 653.
Blazey, K. W., and Rohrer, H., 1969, *Phys. Rev.*, **185,** 712.
Blazey, K. W., Rohrer, H., and Webster, R., 1971, *Phys. Rev.* B, **4,** 2287.
Bloch, D., 1966, *J. Phys. Chem. Solids*, **27,** 881.
Bloembergen, P., Tan, K. G., Lefèvre, F. H. J., and Bleyendaal, A. H. M., 1970, *Proc. Int. Conf. Magn.*, Grenoble (*J. Phys.*, **32,** suppl. C–1, p. 879).
Bloembergen, P., and Franse, J. J. M., 1972, *Solid St. Commun.*, **10,** 325.
Bloembergen, P., Berkhout, P. J., and Franse, J. J. M., 1972, *Proc. 18th Ann. Conf. Magn. Magn. Materials*, Denver (*AIP Conf. Proc.*, **10,** 1598) ; 1973 (in print).
Blöte, H. W. J., and Huiskamp, W. J., 1969, *Phys. Lett.* A, **29,** 304.
Blöte, H. W. J., 1972, Thesis, University of Leiden.
Boesch, H. R., Schmocker, U., Waldner, F., Emerson, K., and Drumheller, J. E., 1971, *Phys. Lett.* A, **36,** 461.
Bongers, P. F., Van Bruggen, C. F., Koopstra, J., Omloo, W. P. F. A. M., Wiegers, G. A., and Jellinek, F., 1968, *J. Phys. Chem. Solids*, **29,** 977
Bonner, J. C., and Fisher, M. E., 1964, *Phys. Rev.* A, **135,** 640.
Bonner, J. C., Friedberg, S. A., Kobayashi, H., and Myers, B. E., 1970, *Proc. 12th Low Temp. Conf.*, Kyoto.
Botterman, A. C., De Jonge, W. J. M., and De Leeuw, P., 1969. *Physics Lett.* A **30,** 150.
Bowers, R. G., and Woolf, M. E., 1969, *Phys. Rev.*, **177,** 917.
Boyd, E. L., 1966, *Phys. Rev.*, **145,** 174.
Bozorth, R. M., and Nielsen, J. W., 1958, *Phys. Rev.*, **110,** 879.
Breed, D. J., 1966, *Phys. Lett.*, **23,** 181 ; 1967, *Physica*, **37,** 35 ; 1969, Thesis, University of Amsterdam.
Breed, D. J., Gilijamse, K., and Miedema, A. R., 1969, *Physica*, **45,** 205.
Brout, R., 1965, *Phase Transitions* (New York : Benjamin).
Brown, M. R., Coles, B. A., Owen, J., and Stevenson, R. W. H., 1961, *Phys. Rev. Lett.*, **7,** 246.
Bucci, C., and Guidi, G., 1970, *Proc. Int. Conf. Magn.*, Grenoble (*J. Phys.*, **32,** suppl. C–1, p. 887) ; 1974, *Phys. Rev.* B (to be published).
Burlet, P., Burlet, P., Bertaut, E. F., Roult, G., and De Combarieu, A., 1969, *Solid St. Commun.*, **7,** 1403.
Butterworth, G. J., and Zidell, V. S., 1969, *J. appl. Phys.*, **40,** 1033.
Butterworth, G. J., and Woollam, J. A., 1969, *Physics Lett.* A, **29,** 259.
Callaway, J., and McCollum, D. C., 1963, *Phys. Rev.*, **130,** 1741.
Carrara, P., De Gunzbourg, J., and Allain, Y., 1969, *J. appl. Phys.*, **40,** 1035.
Cashion, J. D., Cooke, A. H., Thorp, T. L., and Wells, M. R., 1968, *J. Phys* C, **1,** 539 ; 1970, *Proc. R. Soc.* A, **318,** 473.
Cerdonio, M., and Paroli, P., 1970, *Physics Lett.* A, **33,** 217.
Chang, C. H., 1952, *Phys. Rev.*, **88,** 1422.
Charap, S. H., and Boyd, E. L., 1964, *Phys. Rev.* A, **133,** 811.
Child, H. R., Koehler, W. C., Lau, H. Y., and Stout, J. W., 1970, *J. appl. Phys.*, **41,** 1274.

CHINN, S. R., ZEIGER, H. J., and O'CONNOR, J. R., 1971, *Phys. Rev.* B, **3,** 1709.
CHISHOLM, R. C., and STOUT, J. W., 1962, *J. chem. Phys.*, **36,** 972.
CITTEUR, C. A. W., 1973, Thesis, University of Leiden. See also CITTEUR, C. A. W., and KASTELEYN, P. W., 1972, *Physica*, **62,** 17, and to be published.
CLARK, R. H., and MOULTON, W. G., 1972, *Phys. Rev.* B, **5,** 788.
CLAY, R. M., and STAVELEY, L. A. K., 1966, *Proc. Low Temp. Calorimetry Conf.*, Helsinki (*Ann. Acad. Sci. Fennicae* A, VI, 210, p. 194).
COHEN, A. F., FRIEDBERG, S. A., and WAGNER, G. R., 1964, *Physics Lett.*, **11,** 198.
COLPA, J. H. P., 1972 a, *Physica*, **56,** 185 and 205 ; 1972 b, *Ibid.*, **57,** 347.
COLPA, J. H. P., SIEVERTS, E. G., and VAN DER LINDE, R. H., 1971, *Physica*. **51,** 573.
COOKE, A. H., EDMONDS, D. T., FINN, C. B. P., and WOLF, W. P., 1968, *Proc. R. Soc.* A, **306,** 313 and 335.
COOKE, A. H., EDMONDS, D. T., MCKIM, F. R., and WOLF, W. P., 1959, *Proc. R. Soc.* A, **252,** 246.
COOKE, J. F., and GERSCH, H. A., 1967, *Phys. Rev.* **153,** 641.
COOPER, M. J., and NATHANS, R., 1966, *J. appl. Phys.*, **37,** 1041.
CORLISS, L. M., DELAPALME, A., HASTINGS, J. M., LAU, H. Y., and NATHANS, R., 1969, *J. appl. Phys.*, **40,** 1278.
COX, D. E., EIBSCHÜTZ, M., GUGGENHEIM, H. J., and HOLMES, L., 1970, *J. appl. Phys.*, **41,** 943.
COX, D. E., and MINKIEWICZ, V. J., 1971, *Phys. Rev.* B, **4,** 2209.
COX, D. E., SHIRANE, G., BIRGENEAU, R. J., and MACCHESNEY, J. B., 1969, *Phys. Rev.*, **188,** 930.
DALTON, N. W., and WOOD, D. W., 1967, *Proc. phys. Soc.*, **90,** 459 ; 1969, *J. Math. Phys.*, **10,** 1271.
DATE, M., 1961, *J. phys. Soc. Japan*, **16,** 1337.
DAVIDSON, G. E., EIBSCHÜTZ, M., COX, D. E., and MINKIEWICZ, V. J., 1971, *17th Ann. Conf. Magn. Magn. Materials*, Chicago (*AIP Conf. Proc.* **5,** 436).
DAVIS, H. L., 1960, *Phys. Rev.*, **120,** 789.
DAVIS, H. L., and NARATH, A., 1964, *Phys. Rev.* A, **134,** 433.
DE HAAS, W. J., and GORTER, C. J., 1931, *Commun. Leiden*, 215a.
DE JONGH, L. J., BOTTERMAN, A. C., DE BOER, F. R., and MIEDEMA, A. R., 1969, *J. appl. Phys.*, **40,** 1363.
DE JONGH, L. J., 1972 a, *Solid St. Commun.*, **10,** 537 ; 1972 b, *Physics Lett.* A, **40,** 33 ; 1972 c, *Proc. 18th Ann. Conf. Magn. Magn. Materials*, Denver (*AIP Conf. Proc.*, **10,** 561).
DE JONGH, L. J., BLOEMBERGEN, P., and COLPA, J. H. P., 1972 a, *Physica*, **58,** 305.
DE JONGH, L. J., MIEDEMA, A. R., and WIELINGA, R. F., 1970, *Physica*, **46,** 44.
DE JONGH, L. J., and VAN AMSTEL, W. D., 1970, *Proceed. Int. Conf. Magn.*, Grenoble (*J. Phys.*, **32,** suppl. C–1, p. 880).
DE JONGH, L. J., VAN AMSTEL, W. D., and MIEDEMA, A. R., 1972 b, *Physica*, **58,** 277.
DE WIJN, H. W., WALKER, L. R., GESCHWIND, S., and GUGGENHEIM, H.J., 1973 a, *Phys. Rev.* B, **8,** 299.
DE WIJN, H. W., WALKER, L. R., and WALSTEDT, R. E., 1973 b, *Phys. Rev.* B, **8,** 285.
DE WIJN, H. W., WALSTEDT, R. E., WALKER, L. R., and GUGGENHEIM, H. J., 1971, *J. appl. Phys.*, **42,** 1595.
DIETZ, R. E., MERRITT, F. R., DINGLE, R., HONE, D., SILBERNAGEL, B. G., and RICHARDS, P. M., 1971, *Phys. Rev. Lett.*, **26,** 1186.

DILLON, Jr., J. F., 1964, *J. phys. Soc. Japan*, **19,** 1662.
DILLON, Jr., J. F., and OLSON, C. E., 1965, *J. appl. Phys.*, **36,** 1259.
DINGLE, R., LINES, M. E., and HOLT, S. L., 1969, *Phys. Rev.*, **187,** 643.
DOBRUSHIN, R. L., 1965, *Teor. Veroyat. Ee Primen.*, **10,** 209 ; *Theory Probab. Its. Appl.*, **10,** 193.
DOMB, C., 1960, *Adv. Phys.*, **9,** 149 ; 1968, *Phys. Rev. Lett.*, **20,** 1425.
DOMB, C., and BOWERS, R. G., 1969, *J. Phys.* C, **2,** 755.
DOMB, C., and DALTON, N. W., 1966, *Proc. phys. Soc.*, **89,** 859.
DOMB, C., and HUNTER, D. L., 1965, *Proc. phys. Soc.*, **86,** 1147.
DOMB, C., and MIEDEMA, A. R., 1964, *Progress in Low Temp. Phys.*, Vol. IV, edited by C. J. Gorter : (Amsterdam : North-Holland), Chap. 4.
DOMB, C., and WYLES, J. A., 1969, *J. Phys.* C, **2,** 2435.
DUFFY, Jr., W., DUBACH, J. F., PIANETTA, P. A., DECK, J. F., STRANDBURG, D. L., and MIEDEMA, A. R., 1972, *J. chem. Phys.*, **56,** 2555.
DUFFY, Jr., W., LUBBERS, J., VAN KEMPEN, H., HASEDA, T., and MIEDEMA, A. R., 1962, *Proc. 8th Int. Conf. Low Temp. Phys.*, London.
DUFFY, Jr., W., and STRANDBURG, D. L., 1967, *J. chem. Phys.*, **46,** 456.
DUFFY, Jr., W., STRANDBURG, D. L., and DECK, J. F., 1969, *Phys. Rev.*, **183,** 567.
DUPAS, A., and RENARD, J.-P., 1970 a, *C.r. Séanc. Acad. Sci., Paris* B, **271,** 154 ; 1970 b, *Physics Lett.* A, **33,** 470.
DURCZEWSKI, K., 1970, *Physics Lett.* A, **31,** 56.
DYSON, F. J., 1956, *Phys. Rev.*, **102,** 1217 and 1230.
EDELSTEIN, A. S., 1964, *J. chem. Phys.*, **40,** 488.
EHRENFREUND, E., RYBACZEWSKY, E. F., GARITO, A. F., HEEGER, A. J., and PINCUS, P., 1973, *Phys. Rev.*, B, **7,** 421.
EIBSCHÜTZ, M., DAVIDSON, G. R., GUGGENHEIM, H. J., and COFF, D. E., 1971 a, *17th Ann. Conf. Magn. Materials*, Chicago (*AIP Conf. Proc.*, **5,** 670).
EIBSCHÜTZ, M., HOLMES, L., and GUGGENHEIM, H. J., 1970, *Proc. Int. Conf. Magn.*, Grenoble (*J. Phys.*, **32,** suppl. C–1, p. 759).
EIBSCHÜTZ, M., GUGGENHEIM, H. J., and HOLMES, L., 1971 b, *J. appl. Phys.*, **42,** 1485.
EIBSCHÜTZ, M., GUGGENHEIM, H. J., HOLMES, L., and BERNSTEIN, J. L., 1972 a, *Solid St. Commun.*, **11,** 457.
EIBSCHÜTZ, M. ,HOLMES, L., GUGGENHEIM, H. J., and COX, D. E., 1972 b, *Phys. Rev.* B, **6,** 2677.
EPSTEIN, A., GUREWITZ, E., MAKOVSKY, J., and SHAKED, H., 1970 b, *Phys. Rev.* B, **2,** 3703.
EPSTEIN, A., MAKOVSKY, J., and SHAKED, H., 1971, *Solid St. Commun.*, **9,** 249.
ESSAM, J. W., and FISHER, M. E., 1963, *J. chem. Phys.*, **38,** 802.
ESSAM, J. W., and HUNTER, D. L., 1968, *J. Phys.* C, **1,** 392.
ESSAM, J. W., and SYKES, M. F., 1963, *Physica*, **29,** 378.
FAULHABER, R., and HUFNER, S., 1969, *Z. Phys.*, **228,** 235.
FEDER, J., and PYTTE, E., 1968, *Phys. Rev.*, **168,** 640.
FERER, M., MOORE, M. A., and WORTIS, M., 1971, *Phys. Rev.* B, **4,** 3954.
FIGGIS, B. N., GERLOCH, M., and MASON, R., 1964, *Acta. crystallogr.*, **17,** 506.
FIGGIS, B. N., and ROBERTSON, G. B., 1965, *Nature, Lond.*, **205,** 694.
FISHER, M. E., 1960 a, *Physica*, **26,** 618 ; 1960 b, *Proc. R. Soc.* A, **254,** 66 ; *Ibid.*, **256,** 502 ; 1962, *Phil. Mag.*, **7,** 1731 ; 1963, *J. Math. Phys.*, **4** 124 ; 1964, *Am. J. Phys.*, **32,** 343 ; 1965 a, *Phys. Rev.*, **176,** 257 ; 1965 b, *Lectures in Theoretical Physics*, Vol. VIII C (Boulder : University of Colorado Press), p. 61 ; 1967, *Rep. Prog. Phys.*, **30,** p. 615 ; 1968, *Phys. Rev.*, **176,** 257 ; 1973, *Essays in Physics* (Academic Press).
FISHER, M. E., and BURFORD, R. J., 1967, *Phys. Rev.*, **156,** 583.

FISHER, M. E., and JASNOW, D., 1971, *Phys. Rev.* B, **3,** 907.
FLIPPEN, R. B., and FRIEDBERG, S. A., 1960, *J. appl. Phys.*, **31,** 338 S ; 1963, *J. chem. Phys.*, **38,** 2652.
FOLEN, V. J., 1972, *Proc. 18th Ann. Conf. Magn. Magn. Materials, Denver* (*AIP Conf. Proc.*, **10**).
FOLEN, V. J., KREBS, J. J., and RUBINSTEIN, M., 1968, *Solid St. Commun.*, **6,** 865.
FONER, S., 1963, *Magnetism*, edited by G. T. Rado and H. Suhl ; 1964, *Proc. Int. Conf. Magn.*, Nottingham (London : The Institute of Physics and the Physical Society).
FORD, Jr., N. C., and JEFFRIES, C. D., 1966, *Phys. Rev.*, **141,** 381.
FORSTAT, H., TAYLOR, G., and SPENCE, R. D., 1959, *Phys. Rev.*, **116,** 897.
FRANZBLAU, M. C., EVERETT, G. E., and LAWSON, A. W., 1967, *Phys. Rev.*, **164,** 716.
FREISER, M. J., SEIDEN, P. E., and TEANEY, D. T., 1963, *Phys. Rev. Lett.*, **10,** 293.
FRIEDBERG, S. A., 1952, *Physica*, **18,** 714.
FRIEDBERG, S. A., COHEN, A. F., and SCHELLENG, J. H., 1961, *Proc. Int. Conf, Magn. Crystallography*, Kyoto (*J. phys. Soc. Japan*, **17,** suppl., B-1. p. 515).
FRIEDBERG, S. A., and RAQUET, C. A., 1968, *J. appl. Phys.*, **39,** 1132.
FRITZ, J. J., and PINCH, H. L., 1957, *J. Am. chem. Soc.*, **79,** 3644.
FUESS, H., KALLEL, A., and TCHÉOU, F., 1971, *Solid St. Commun.*, **9,** 1949.
GARBER, M., 1960, *J. Phys. Soc. Japan*, **15,** 734.
GAUNT, D. S., and DOMB, C., 1970, *J. Phys.* C, **3,** 1442.
GEBALLE, T. H., and GIAUQUE, W. F., 1952, *J. Am. chem. Soc.*, **74,** 3513.
GELLER, S., and BOND, W. L., 1958, *J. chem. Phys.*, **29,** 925.
GERRITSEN, H. J., OKKES, R., BÖLGER, B., and GORTER, C. J., 1955, *Physica*, **21,** 629.
GERSTEIN, B. C., GEHRING, F. D., and WILLETT, R. D., 1972 a, *J. appl. Phys.*, **43,** 1932 ; 1972 b (to be published).
GIAUQUE, W. F., FISHER, R. A., HORNUNG, E. W., and BRODALE, G. E., 1970, *J. chem. Phys.*, **53,** 3733.
GIBBONS, D. F., 1959, *Phys. Rev.*, **115,** 1194.
GOLDING, B., 1971, *Phys. Rev. Lett.*, **27,** 1142.
GORTER, C. J., 1957, *Nuov. Cim.*, suppl., **6,** 923.
GORTER, C. J., and VAN PESKI-TINBERGEN, T., 1956, *Physica*, **22,** 273.
GORTER, F. W., NOORDERMEER, L. J., KOP, A. R., and MIEDEMA, A. R., 1969, *Physics Lett.* A, **29,** 331.
GOTTLIEB, A. M., and HELLER, P., 1971, *Phys. Rev.* B, **3,** 3615.
GRIFFITHS, R. B., 1962, Thesis, Stanford University ; 1964 a, *Phys. Rev.*, A, **133,** 768 ; 1964 b, *Ibid.*, A, **135,** 659 ; 1964 c, *Ibid.* A, **136,** 437 ; 1965, *Phys. Rev. Lett.*, **14,** 623 ; 1967, *Phys. Rev.*, **158,** 176 ; 1970 a, *Phys. Rev. Lett.*, **24,** 715 ; 1970 b, *Ibid.*, **24,** 1479.
GROLL, G., 1971, *Z. Phys.*, **243,** 60.
GRÜNBERG, P., HÜFNER, S., ORLICH, E., and SCHMITT, J., 1969, *Phys. Rev.*, **184,** 285.
GUREWITZ, E., EPSTEIN, A., MAKOVSKY, J., and SHAKED, H., 1970, *Phys. Rev. Lett.*, **25,** 1713.
GUTTMANN, A. J., DOMB, C., and FOX, P. F., 1970, *Proc. Int. Conf. Magn.*, Grenoble (*J. Phys.*, **32,** suppl. C-1, p. 354).
HAMILTON, W. O., and PAKE, G. E., 1963, *J. chem. Phys.*, **39,** 2694.
HANKEY, A., and STANLEY, H. E., 1972, *Phys. Rev.* B, **6,** 3515.
HANSEN, W. N., and GRIFFEL, M., 1958, *J. chem. Phys.*, **28,** 902 ; 1959, *Ibid.*, **30,** 913.

HARBUS, F., and STANLEY, H. E., 1972, *Phys. Rev. Lett.*, **29,** 58.
HARDEMAN, G. E. G., and POULIS, N. J., 1955, *Physica*, **21,** 728.
HASEDA, T., 1960, *J. phys. Soc. Japan*, **15,** 483.
HASEDA, T., and KOBAYASHI, H., 1964, *J. phys. Soc. Japan*, **19,** 1260.
HASEDA, T., KOBAYASHI, H., and DATE, M., 1959, *J. phys. Soc. Japan*, **14,** 1724.
HASEDA, T., and MIEDEMA, A. R., 1961, *Physica*, **27,** 1102.
HEGER, G., and GELLER, R., 1972, *Phys. Stat. Sol.*, **53,** 227.
HEGER, G., GELLER, R., and BABEL, D., 1971, *Solid St. Commun.*, **9,** 335.
HELLER, G. S., 1964, *Proc. Int. Conf. Magn.*, Nottingham (London : The Institute of Physics and the Physical Society).
HELLER, P., 1966, *Phys. Rev.*, **146,** 403 ; 1967, *Rep. Prog. Phys.*, **30,** 731.
HELLER, P., and BENEDEK, G., 1965, *Phys. Rev. Lett.*, **14,** 71.
HELLER, P., SCHULHOF, M. P., NATHANS, R., and LINZ, A., 1971, *J. appl. Phys.*, **42,** 1258.
HENDERSON, Jr., A. J., BROWN, G. R., REED, T. B., and MEYER, H., 1970, *J. appl. Phys.*, **41,** 946.
HERWEYER, A., DE JONGE, W. J. M., BOTTERMAN, A. C., BONGAARTS, A. L. M., and COWEN, J. A. W., 1972, *Phys. Rev.* B, **5,** 4618.
HEWSON, A. C., TER HAAR, D., and LINES, M. E., 1965, *Phys. Rev.*, A, **137,** 1465.
HIRAKAWA, K., HIRAKAWA, K., and HASHIMOTO, T., 1960, *J. phys. Soc. Japan*, **15,** 2063.
HIRAKAWA, K., and KADOTA, S., 1967, *J. phys. Soc. Japan*, **23,** 756.
HIRAKAWA, K., YAMADA, I., and KUROGI, Y., 1970, *Proc. Int. Conf. Magn.*, Grenoble (*J. Phys.*, **32,** suppl. C–1, p. 890).
HO, J. T., and LITSTER, J. D., 1969, *Phys. Rev. Lett.*, **22,** 603 ; 1970, *Phys. Rev.* B, **2,** 4523.
HOHENBERG, P. C., 1967, *Phys. Rev.*, **158,** 383.
HOHENBERG, P. C., and HALPERIN, B. I., 1970, *J. appl. Phys.*, **41,** 1390.
HOLMES, L., EIBSCHÜTZ, M., and GUGGENHEIM, H. J., 1969, *Solid St. Commun.*, **7,** 973.
HOLMES, L. M., VAN UITERT, L. G., and HULL, G. W., 1971, *Solid St. Commun.*, **9,** 1373.
HONE, D., 1971, *Proc. 17th Ann. Conf. Magn. Magn. Materials*, Chicago (*AIP Conf. Proc.*, **5,** 413).
HOY, G. R., and DE S. BARROS, F., 1965, *Phys. Rev.*, **139,** A 929.
HOY, G. R., DE S. BARROS, S., DE S. BARROS, F., and FRIEDBERG, S. A., 1965, *J. appl. Phys.*, **36,** 936.
HUANG, N. L., ORBACH, R., and SIMANEK, E., 1966, *Phys. Rev. Lett.*, **17,** 134.
HUANG, N. L., ORBACH, R., SIMANEK, E., OWEN, J., and TAYLOR, D. R., 1967, *Phys. Rev.*, **156,** 383.
HUISKAMP, W. J., 1966, *Proc. Low Temp. Calorimetry Conf.*, Helsinki (*Ann. Acad. Sci. Fennicae*, A VI, p. 210).
HUTCHINGS, M. T., BIRGENEAU, R. J., and WOLF, W. P., 1968, *Phys. Rev.*, **168,** 1026.
HUTCHINGS, M. T., and GUGGENHEIM, H. J., 1970, *J. Phys.* C, **3,** 1303.
HUTCHINGS, M. T., RAINFORD, B. D., and GUGGENHEIM, H. J., 1970, *J. Phys.*, C, **3,** 307.
HUTCHINGS, M. T., SAMUELSEN, E. J., SHIRANE, G., and HIRAKAWA, K., 1969, *Phys. Rev.*, **188,** 919.
HUTCHINGS, M. T., SCHULHOF, M. P., and GUGGENHEIM, H. J., 1972 a, *Phys. Rev.* B, **5,** 154.
HUTCHINGS, M. T., SHIRANE, G., BIRGENEAU, R. J., and HOLT, S. L., 1972 b, *Phys. Rev.* B, **5,** 1999.
IKEBE, M., and DATE, M., 1971, *J. phys. Soc. Japan*, **30,** 93.

Ikeda, H., and Hirakawa, K., 1972, *J. phys. Soc. Japan*, **33,** 393 ; 1973, *Ibid.* (to be published).
Inawashiro, S., and Katsura, S., 1965, *Phys. Rev.*, **140,** A 892.
Ishikawa, T., and Oguchi, T., 1971, *J. Phys. Soc. Japan*, **31,** 1021.
Ising, E., 1925, *Z. Phys.*, **31,** 253.
Jacobs, I. S., 1961, *J. appl. Phys.*, **32,** 61 S.
Jacobs, I. S., and Lawrence, P. E., 1967, *Phys. Rev.*, **164,** 866.
Jacobs, I. S., Roberts, S., and Silverstein, S. D., 1968, *J. appl. Phys.*, **39,** 816.
Jacobs, I. S., and Silverstein, S. D., 1964, *Phys. Rev. Lett.*, **13,** 272.
Jasnow, D., and Wortis, M., 1968, *Phys. Rev.*, **176,** 739.
Jennings, L. D., and Hansen, W. N., 1965, *Phys. Rev.*, **139,** A 1694.
Joenk, R. J., 1962, *Phys. Rev.*, **126,** 565.
Johnson, F. M., and Nethercot, Jr., A. H., 1959, *Phys. Rev.*, **114,** 705.
Joshua, S. J., 1970, *Phys. Stat. Sol.*, **38,** 643.
Kadanoff, L. P., 1966, *Physics*, **2,** 263 ; 1970, *Proc. Fermi Summer School of Physics*, Varenna, edited by M. S. Green (New York : Academic Press).
Kadanoff, L. P., Götze, W., Hamblen, D., Hecht, R., Lewis, E. A. S., Palciauskas, V. V., Rayl, M., and Swift, J., 1967, *Rev. mod. Phys.*, **39,** 395.
Kadota, S., Yamada, I., Yoneyama, S., and Hirakawa, K., 1967, *J. phys. Soc. Japan*, **23,** 751.
Kanamori, J., 1963, *Magnetism*, Vol. I, edited by G. T. Rado and H. Suhl,
Kanamori, J., and Itoh, Y., 1968, *J. appl. Phys.*, **39,** 1358.
Kanamori, J., and Yosida, K., 1955, *Prog. theor. Phys.*, **14,** 423.
Karimov, Y. S., 1969, *Zh. eksp. teor. Fiz.*, **57,** 1962 ; *Soviet Phys. J.E.T.P.*, **30,** 1062.
Kats, E. I., 1969, *Zh. eksp. teor. Fiz.*, **56,** 2043 ; *Soviet Phys. J.E.T.P.*, **29,** 1098.
Katsura, S., 1962, *Phys. Rev.*, **127,** 1508 ; 1963, *Ibid.*, **129,** 2835.
Kaufman, B., and Onsager, L., 1949, *Phys. Rev.*, **76,** 1244.
Keen, B. E., Landau, D. P., and Wolf, W. P., 1967, *J. appl. Phys.*, **38,** 967.
Keffer, F., 1952, *Phys. Rev.*, **87,** 608 ; 1966, *Encyclopedia of Physics*, Vol. XVIII, edited by H. P. J. Wijn (New York : Springer Verlag), **2,** p. 109.
Keve, E. T., Abrahams, S. C., and Bernstein, J. L., 1969, *J. chem. Phys.*, **51,** 4928.
Kittel, C., 1958, *Phys. Rev.*, **110,** 836 ; 1960, *Ibid.*, **120,** 335.
Klaassen, T. O., Gevers, A., and Poulis, N. J., 1972, *Physica*, **61,** 95.
Kleinberg, R., 1970, *J. chem. Phys.*, **53,** 2660.
Kobayashi, H., and Haseda, T., 1963, *J. Phys. Soc. Japan*, **18,** 541.
Kobayashi, H., Tsujikawa, I., and Friedberg, S. A., 1972 (to be published).
Kouvel, J. S., and Fisher, M. E., 1964, *Phys. Rev.*, **136,** A 1626.
Kubo, R., 1952, *Phys. Rev.*, **87,** 568.
Landau, D. P., 1970, *Proc. Int. Conf. Magn.*, Grenoble (*J. Phys.*, **32,** suppl., p. 1012).
Landau, D. P., Keen, B. E., Schneider, B., and Wolf, W. P., 1971, *Phys. Rev.* B, **3,** 2310.
Landau, L., 1933, *Phys. Z. SowjetUn*, **4,** 675.
Landau, L. D., and Lifshitz, E. M., 1958, *Statistical Physics* (London : Pergamon Press), p. 482.
Lau, H. Y., Corliss, L. M., Delapalme, A., Hastings, J. M., Nathans, R., and Tucciarone, A., 1970, *J. appl. Phys.*, **41,** 1384.
Lau, H. Y., Stout, J. W., Koehler, W. C., and Child, H. R., 1969, *J. appl. Phys.*, **40,** 1136.
Lebesque, J. V., Snel, J., and Smit, J. J., 1973, *Solid St. Commun.*, **13,** 371.

Lécuyer, B., Renard, J. P., and Herpe, A., 1972, *C.r. Séanc. Acad. Sci., Paris*, B, **275,** 73.
Legrand, E., and Plumier, R., 1962 a, *Phys. Stat. Sol.*, **2,** 317 ; 1962 b, *J. Phys.*, **23,** 474 ; 1963, *Ibid.*, **24,** 741 ; **25,** 578.
Legrand, E., and Van den Bosch, A., 1969, *Solid St. Commun.*, **7,** 1191.
Lemaire, H., Rey, P., Rassat, A., De Combarieu, A., and Michel, J. C., 1968, *Molec. Phys.*, **14,** 201.
Lines, M. E., 1963, *Phys. Rev.*, **131,** 546 ; 1964, *Ibid.*, **133,** A 841 ; 1967 a, *Ibid.*, **156,** 543 ; 1967 b, *Ibid.*, **164,** 736 ; 1970, *J. Phys. Chem. Solids*, **31,** 101 ; 1971, *Phys. Rev.* B, **3,** 1749.
Liu, K. S., and Fisher, M. E., 1973, to be published.
Loly, P. D., 1968, *J. appl. Phys.*, **39,** 1109.
Loopstra, B. O., Van Laar, B., and Breed, D. J., 1968, *Physics Lett.* A, **26,** 526.
Lovesey, S. W., 1972, *J. Phys.* C, **5,** 2769.
Low, G. G., 1965, *Inelastic scattering of Neutrons in Solids and Liquids* (Vienna : Int. Atomic Energy Agency).
Lowndes, Jr., D. H., Finegold, L., and Rogers, R. N., 1969, *Phys. Rev.*, **186,** 515.
Maarschall, E. P., 1970, *Proc. XVIth Colloq. A.M.P.E.R.E.*, Bucharest (Publish. House Acad. Soc. Rep. Romania).
Maarschall, E. P., Botterman, A. C., Vega, S., and Miedema, A. R. 1969, *Physica*, **41,** 473.
MacChesney, J. B., Williams, H. J., Potter, J. F., and Sherwood, R. C., 1967, *Phys. Rev.*, **164,** 779.
McElearney, J. N., Forstat, H., and Bailey, P. T., 1969, *Phys. Rev.*, **181,** 887.
McGuire, T. R., Torrance, J. B., and Shafer, M. W., 1972, *Proc. 18th Ann. Conf. Magn. Magn. Materials*, Denver (*AIP Conf. Proc.*, **10,** 1289).
McLean, F. B., and Blume, M., 1973, *Phys. Rev.* B, **7,** 1149.
Marshall, W., 1958, *J. Phys. Chem. Solids*, **7,** 159.
Marshall, W., 1965, *Proc. Conf. Critical Phenomena*, Washington (NBS Misc. Publ. 273).
Marshall, W., and Lowde, R. D., 1968, *Rep. Prog. Phys.*, **31,** 705.
Marshall, W., and Lovesey, S. W., 1971, *Theory of Thermal Neutron Scattering* (Oxford : Clarendon Press).
Martin, R. L., and Waterman, H., 1959, *J. chem. Soc.*, p. 1359.
Matsuura, M., 1971, *Physics Lett.* A, **34,** 274.
Matsuura, M., Ajiro, Y., and Haseda, T., 1969, *J. phys. Soc. Japan*, **26,** 665.
Matsuura, M., Blöte, H. W. J., and Huiskamp, W. J., 1970 a, *Physica*, **50,** 444.
Matsuura, M., Gilijamse, K., Sterkenburg, J. E. W., and Breed, D. J., 1970 b, *Physics Lett.* A, **33,** 363.
Matthias, B. T., Bozorth, R. M., and Van Vleck, J. H., 1961, *Phys. Rev. Lett.*, **7,** 160.
Mattuck, R. D., and Johansson, B., 1968, *Adv. Phys.*, **17,** 509.
Mazzi, F., 1955, *Acta crystallogr.*, **8,** 137.
Meissner, H. E., and Wolf, W. P., 1969, *J. appl. Phys.*, **40,** 1038.
Mekata, M., Adachi, K., Takadi, H., and Achiwa, N., 1970, *Proc. 12th Int. Conf. Low Temp. Phys.*, Kyoto.
Menyuk, N., Dwight, K., and Reed, T. B., 1971, *Phys. Rev.*, B, **3,** 1689.
Mermin, N. D., and Wagner, H., 1966, *Phys. Rev. Lett.*, **17,** 1133.
Mess, K. W., Lagendijk, E., Curtis, D. A., and Huiskamp, W. J., 1967, *Physica*, **34,** 126.
Metselaar, J. W., and De Klerk, D., 1973 a, *Physica*, **63,** 191 ; 1973 b, *Ibid.*, **65,** 208.

MIEDEMA, A. R., VAN KEMPEN, H., HASEDA, T., and HUISKAMP, W. J., 1962, *Physica*, **28,** 119.
MIEDEMA, A. R., VAN KEMPEN, H., and HUISKAMP, W. J., 1963, *Physica*, **29,** 1266.
MIEDEMA, A. R., WIELINGA, R. F., and HUISKAMP, W. J., 1965, *Physica*, **31,** 1585.
MILOŠEVIĆ, S., and STANLEY, H. E., 1972, *Phys. Rev.* B, **6,** 986 and 1002.
MINKIEWICZ, V. J., COX, D. E., and SHIRANE, G., 1970, *Solid St. Commun.*, **8,** 1001.
MIYATA, N., and ARGYLE, B. E., 1967, *Phys. Rev.*, **157,** 448.
MIZUNO, J., 1960, *J. phys. Soc. Japan*, **15,** 1412 ; 1961, *Ibid.*, **16,** 1574.
MONTGOMERY, H., 1966, *Proc. Low Temp. Calorimetry Conf.*, Helsinki (*Ann. Acad. Sci. Fennicae* A, VI **210,** p. 214).
MOORE, M. A., 1969, *Phys. Rev. Lett.*, **23,** 861.
MORIGAKI, H., and ABE, H., 1967, *J. phys. Soc. Japan*, **23,** 462.
MORIGAKI, K., 1961, *J. phys. Soc. Japan*, **16,** 1639.
MOSTAFA, M. F., and WILLETT, R. D., 1971, *Phys. Rev.* B, **4,** 2213.
MURRAY, T. E., and WESSEL, G. K., 1968, *J. phys. Soc. Japan*, **24,** 738.
NAGAI, O., 1963, *J. phys. Soc. Japan*, **18,** 510.
NAGATA, K., and DATE, M., 1964, *J. phys. Soc. Japan*, **19,** 1823.
NAGATA, K., and TAZUKE, Y., 1970, *Physics Lett.* A, **31,** 293.
NAGLE, J. F., and BONNER, J. C., 1971, *J. chem. Phys.*, **54,** 729.
NARATH, A., 1965, *Phys. Rev.*, A, **140,** 854.
NARATH, A., and DAVIS, H. L., 1965, *Phys. Rev.* A, **137,** 163.
NATHANS, R., MENZINGER, F., and PICKART, S. J., 1968, *J. appl. Phys.*, **39,** 1237.
NÉEL, L., 1936, *Ann. Phys.*, **5,** 232 ; 1957, *Nuovo Cim.*, suppl. **6,** 942.
NORVELL, J. C., WOLF, W. P., CORLISS, L. M., HASTINGS, J. M., and NATHANS, R., 1969, *Phys. Rev.*, **186,** 557 and 567.
NOUET, J., ZAREMBOWITCH, A., PISAREV, R. V., FERRE, J., and LECOMTE, M., 1972, *Appl. Phys. Lett.*, **21,** 161.
OBOKATA, T., and OGUCHI, T., 1968, *J. phys. Soc. Japan*, **25,** 322.
OBOKATA, T., ONO, I., and OGUCHI, T., 1967, *J. phys. Soc. Japan*, **23,** 516.
OGUCHI, T., 1955, *Prog. theor. Phys.*, **13,** 148 ; 1964, *Phys. Rev.*, **133,** A 1098.
OHTSUKA, T., 1961 a, *J. Phys. Soc. Japan*, **16,** 1549 ; 1961 b, *Proc. Int. Conf. Magn. and Crystallography*, Kyoto, (*J. phys. Soc. Japan*, **17,** suppl. B-1, p. 472).
OKAZAKI, A., 1969 a, *J. phys. Soc. Japan*, **26,** 870 ; 1969 b, *Ibid.*, **27,** 267.
OKAZAKI, A., and SUEMUNE, Y., 1961 a, *J. phys. Soc. Japan*, **16,** 176 ; 1961 b, *Ibid.*, **16,** 671.
OKAZAKI, A., TURBERFIELD, K. C., and STEVENSON, R. W. H., 1964, *Physics Lett.*, **8,** 9.
OLIVEIRA, Jr., N. F., FONER, S., SHAPIRA, Y., and REED, T. B., 1972, *Phys. Rev.* B, **5,** 2634.
OLLIVIER, G., and BUISSON, G., 1971, *Solid St. Commun.*, **9,** 235.
ONN, D. G., MEYER, H., and REMEIKA, J. P., 1967, *Phys. Rev.*, **156,** 663.
ÔNO, K., ITO, A., and FUJITA, T., 1964, *J. phys. Soc. Japan*, **19,** 2119.
ÔNO, K., and OHTSUKA, M., 1958, *J. phys. Soc. Japan*, **13,** 206.
ÔNO, K., SHINOHARA, M., ITO, A., SAKAI, N., and SUENAGA, M., 1970, *Phys. Rev. Lett.*, **24,** 770.
ONSAGER, L., 1944, *Phys. Rev.*, **65,** 117 ; 1949, *Nuovo Cim.*, Suppl., **6,** 261.
OWEN, J., and TAYLOR, D. R., 1966, *Phys. Rev. Lett.*, **16,** 1164 ; 1968, *J. appl. Phys.*, **39,** 791.
PASSELL, L., DIETRICH, O. W., and ALS-NIELSEN, J., 1971, *17th Ann. Conf. Magn. Magn. Materials*, Chicago (*AIP Conf. Proc.*, **5,** 1251).

Passenheim, B. C., McCollum, Jr., D. C., and Callaway, J., 1966, *Physics Lett.*, **23,** 634.

Paul, G., and Stanley, H. E., 1971 a, *Physics Lett.* A, **37,** 328 ; 1971 b, *Ibid.*, 347.

Peierls, R., 1936, *Proc. Camb. phil. Soc.*, **32,** 477. See also H. G. Wannier, *Elements of Solid State Theory* (Cambridge University Press), p. 105.

Peter, M., and Moriya, T., 1962, *J. appl. Phys.*, **33,** 1304.

Pfeiffer, H., 1971, *Acta phys. polon.* A, **39,** 213.

Pickart, S. J., Collins, M. F., and Windsor, C. G., 1966, *J. appl. Phys.*, **37,** 1054.

Pierce, R. D., and Friedberg, S. A., 1968, *Phys. Rev.*, **165,** 680 ; 1971, *Ibid.* B, **3,** 934.

Poulis, N. J., and Hardeman, G. E. G., 1952, *Physica*, **18,** 201.

Racquet, C. A., and Friedberg, S. A., 1973 (to be published).

Rado, G. T., 1969, *Phys. Rev. Lett.*, **23,** 644.

Rainford, B. D., Houmann, J. G., and Guggenheim, H. J., 1972, *Proc. 5th Symp. Neutron Inelast. Scatt.*, Grenoble (Vienna : IAEA).

Rao, D. V. G. L. N., and Narasimhamurty, A., 1963, *Phys. Rev.*, **132,** 961.

Rayl, M., Vilches, O. E., and Wheatley, J. C., 1968, *Phys. Rev.*, **165,** 698.

Reichert, T. A., and Giauque, W. F., 1969, *J. chem. Phys.*, **50,** 4205.

Renard, J.-P., and Velu, E., 1970, *Proc. Int. Conf. Magn.*, Grenoble (*J. Phys.*, **32,** Suppl., p. 1154).

Riedel, E. K., 1971, *J. appl. Phys.*, **42,** 1383 ; 1972, *Proc. 18th Ann. Conf. Magn. Magn. Materials*, Denver (*AIP Conf. Proc.*, **10,** 865).

Rioux, F. J., and Gerstein, B. C., 1970, *J. chem. Phys.*, **53,** 1789.

Ritchie, D. S., and Fisher, M. E., 1972, *Phys. Rev.* B, **5,** 2668 ; 1973 (to be published).

Rives, J. E., Dixon, G. S., and Walton, D., 1969, *J. appl. Phys.*, **40,** 1555.

Robinson, W. K., and Friedberg, S. A., 1960, *Phys. Rev.*, **117,** 402.

Rogers, R. N., and Dempesey, C. W., 1967, *Phys. Rev.*, **162,** 333.

Rubinstein, M., and Folen, V. J., 1968, *Physics Lett.* A, **28,** 108.

Rushbrooke, G. S., Baker, Jr., G. A., and Wood, P. J., 1973, *Phase Transitions and Critical Phenomena*, Vol. 3, edited by C. Domb and M. S. Green

Rushbrooke, G. S., and Wood, P. J., 1958, *Molec. Phys.*, **1,** 257 ; 1963, *Ibid.*, **6,** 409.

Saito, S., Kumano, M., and Kanda, E., 1970, *Proc. 12th Int. Conf. Low Temp. Phys.*, Kyoto, p. 809.

Salamon, M. B., and Hatta, I., 1971, *Physics Lett.* A, **36,** 85.

Salamon, M. B., and Ikeda, H., 1973, *Phys. Rev.* B, **7,** 2017.

Salamon, M. B., and Ikushima, A., 1971, *Proc. 17th Ann. Conf. Magn. Magn. Materials*, Chicago (*AIP Conf. Proc.*, **5,** 1269).

Samuelsen, E. J., 1973, *Proc. Int. Conf. Magn.*, Moscow.

Samuelsen, E. J., Silberglitt, R., Shirane, G., and Remeika, J. P., 1971, *Phys. Rev.* B, **3,** 157.

Sawaoka, A., Miyahara, S., Akimoto, S., and Fujisawa, H., 1966, *J. phys. Soc. Japan*, **21,** 185.

Sawatzky, E., and Bloom, M., 1962, *Physics Lett.*, **2,** 28.

Scatturin, V., Corliss, L., Elliott, N., and Hastings, J., 1961, *Acta Cryst.*, **14,** 19.

Scharenberg, W., and Will, G., 1971, *Int. J. Magn.*, **1,** 277.

Schmidt, V. A., and Friedberg, S. A., 1967, *J. appl. Phys.*, **38,** 5319 ; 1970, *Phys. Rev.* B, **1,** 2250.

Schrama, A. H. M., 1972, *J. magn. Res.*, **6,** 432 ; 1973 a, *Physica*, **66,** 131 ; 1973 b, *Ibid.* (in the press).

Schuchert, H., Hüfner, S., and Faulhaber, R., 1969, *Z. Phys.*, **222,** 105.

SCHULHOF, M. P., HELLER, P., NATHANS, R., and LINZ, A., 1970, *Phys. Rev.* B, **1,** 2304.

SCHULHOF, M. P., NATHANS, R., HELLER, P., and LINZ, A., 1971, *Phys. Rev.* B **4,** 2254.

SCHUTTER, J. W., METSELAAR, J. W., and DE KLERK, D., 1972, *Physica*, **61,** 250.

SCOTT, P. D., MEISSNER, H. E., and CROSSWHITE, H. M., 1969, *Physics Lett.* A, **28,** 489.

SCOTT, P. D., and WOLF, W. P., 1969, *J. appl. Phys.*, **40,** 1031.

SEEHRA, M. S., 1969, *Physics Lett.* A, **28,** 754.

SEEHRA, M. S., and CASTNER, Jr., T. G., 1970, *Phys. Rev.* B, **1,** 2289.

SEMURA, J. S., and HUBER, D. L., 1971, *J. appl. Phys.*, **42,** 1415.

SENTURIA, S. D., and BENEDEK, G. B., 1966, *Phys. Rev. Lett.*, **17,** 475.

SHAPIRA, Y., 1970, *Phys. Rev.* B, **2,** 2725 ; 1971, *J. appl. Phys.*, **42,** 1588.

SHINOHARA, M., ITO, A., SUENAGA, M., and ÔNO, K., 1972, *J. phys. Soc. Japan*, **34,** 77.

SHIRANE, G., PICKART, S. J., and ISHIKAWA, Y., 1959, *J. phys. Soc. Japan*, **14,** 1352.

SHIRANE, G., FRAZER, B. C., and FRIEDBERG, S. A., 1965, *Physics Lett.*, **17,** 95.

SHULMAN, R. G., and WYLUDA, B. J., 1961, *J. chem. Phys.*, **35,** 1498.

SKALYO, Jr., J., COHEN, A. F., FRIEDBERG, S. A., and GRIFFITHS, R. B., 1967, *Phys. Rev.*, **164,** 705.

SKALYO, Jr., J., and FRIEDBERG, S. A., 1964, *Phys. Rev. Lett.*, **13,** 133.

SKALYO, Jr., J., SHIRANE, G., BIRGENEAU, R. J., and GUGGENHEIM, H. J., 1969 a, *Phys. Rev. Lett.*, **23,** 1394.

SKALYO, Jr., J., SHIRANE, G., and FRIEDBERG, S. A., 1969 b, *Phys. Rev.*, **188,** 1037.

SKALYO, Jr., J., SHIRANE, G., FRIEDBERG, S. A., and KOBAYASHI, H., 1970, *Phys. Rev.* B, **2,** 1310 and 4632.

SKJELTORP, A. T., and WOLF, W. P., 1971, *J. appl. Phys.*, **42,** 1487.

SMART, J. S., 1965, *Magnetism*, Vol. 3, edited by G. T. Rado and H. Suhl.

SMITH, J., GERSTEIN, B. C., LIU, S. H., and STUCKY, G., 1970, *J. chem. Phys.*, **53,** 418.

SMITH, T., and FRIEDBERG, S. A., 1968, *Phys. Rev.*, **176,** 660.

SPENCE, R. D., MIDDENTS, P., ELSAFFAR, Z., and KLEINBERG, R., 1964, *J. appl. Phys.*, **35,** 854.

SRIVASTAVA, K. G., 1963, *Physics Lett.*, **4,** 55.

STANLEY, H. E., 1967, *Phys. Rev.*, **158,** 537 and 546 ; 1968 a, *Phys. Rev. Lett.*, **20,** 150 ; 1968 b, *Ibid.*, **20,** 589 ; 1968 c, *Phys. Rev.*, **176,** 718 ; 1969 a, *J. appl. Phys.*, **40,** 1546 ; 1969 b, *Phys. Rev.*, **179,** 570 ; 1971, *Introduction to Phase Transitions and Critical Phenomena* (Oxford : Clarendon Press).

STANLEY, H. E., and BETTS, D. D., 1972 (to be published).

STANLEY, H. E., and KAPLAN, T. A., 1966, *Phys. Rev. Lett.*, **17,** 913.

STARR, C., BITTER, F., and KAUFMAN, A. R., 1940, *Phys. Rev.*, **58,** 977.

STEINER, M., 1971, *Z. angew. Phys.*, **32,** 116.

STEINER, M., and DACHS, H., 1971, *Solid St. Commun.*, **9,** 1603.

STEINER, M., KRÜGER, W., and BABEL, D., 1971, *Solid St. Commun.*, **9,** 227.

STELL, G., 1968, *Phys. Rev. Lett.*, **20,** 533.

STEPHENSON, J., 1971, *Physical Chemistry*, Vol. 8B, edited by Henderson and Eyring, p. 717.

STEPHENSON, R. L., PIRNIE, K., WOOD, P. J., and EVE, J., 1968, *Physics Lett.* A, **27,** 2.

STEPHENSON, R. L., and WOOD, P. J., 1968, *Phys. Rev.*, **173,** 475.

STICKLER, J. J., KERN, S., WOLD, A., and HELLER, G. S., 1967, *Phys. Rev.*, **164,** 765.

STOUT, J. W., and BOO, W. O. J., 1966, *J. appl. Phys.*, **37,** 966.

STOUT, J. W., and CATALANO, E., 1955, *J. chem. Phys.*, **23,** 2013.

STOUT, J. W., and CHISHOLM, R. C., 1962, *J. chem. Phys.*, **36,** 972 and 979.

STOUT, J. W., and LAU, H. Y., 1967, *J. appl. Phys.*, **38,** 1472.

SUZUKI, M., TSUJIYAMA, B., and KATSURA, S., 1967, *J. Math. Phys.*, **8,** 124.

SUZUKI, H., and WATANABE, T., 1967, *Physics Lett.* A, **26,** 103 ; 1971, *J. phys. Soc. Japan*, **30,** 367.

SYKES, M. F., ESSAM, J. W., and GAUNT, D. S., 1965, *J. Math. Phys.*, **6,** 283.

SYKES, M. F., and FISHER, M. E., 1962, *Physica*, **28,** 919 and 939.

SYKES, M. F., HUNTER, D. L., MCKENZIE, D. S., and HEAP, B. R., 1972, *J. Phys.* A, **5,** 667.

SYKES, M. F., MARTIN, J. L., and HUNTER, D. L., 1967, *Proc. phys. Soc.*, **91,** 671.

TAKAYANAGI, S., and WATANABE, T., 1970, *J. Phys. Soc. Japan*, **28,** 296.

TAKEDA, K., HASEDA, T., and MATSUURA, M., 1971 a, *Physica*, **52,** 225.

TAKEDA, K., and KAWASAKI, K., 1971, *J. phys. Soc. Japan*, **31,** 1026.

TAKEDA, K., and MATSUKAWA, S., 1971, *J. phys. Soc. Japan*, **30,** 887.

TAKEDA, K., MATSUKAWA, S., and HASEDA, T., 1971 b, *J. phys. Soc. Japan*, **30,** 1330.

TAKEDA, K., MATSURA, M., MATSUKAWA, S., AJIRO, Y., and HASEDA, T., 1970, *Proc. 12th Int. Conf. Low Temp. Phys.*, Kyoto, p. 803.

TAZAWA, S., NAGATA, K., and DATE, M., 1965, *J. phys. Soc. Japan*, **20,** 181.

TEANEY, D. T., 1965, *Phys. Rev. Lett.*, **14,** 898.

TEANEY, D. T., FREISER, M. J., and STEVENSON, R. W. H., 1962, *Phys. Rev. Lett.*, **9,** 212.

TEANEY, D. T., MORUZZI, V. L., and ARGYLE, B. E., 1966, *J. appl. Phys.*, **37,** 1122.

TEANEY, D. T., VAN DER HOEVEN, Jr., B. J. C., and MORUZZI, V. L., 1968, *Phys. Rev. Lett.*, **20,** 722.

THOULESS, D. J., 1969, *Phys. Rev.*, **187,** 732.

TRAPP, C., and STOUT, J. W., 1963, *Phys. Rev. Lett.*, **10,** 157.

TURBERFIELD, K. C., OKAZAKI, A., and STEVENSON, R. W. H., 1965, *Proc. phys. Soc.*, **85,** 743.

UMEBAYASHI, H., FRAZER, B. C., COX, D. E., and SHIRANE, G., 1968, *Phys. Rev.*, **167,** 519.

URYÛ, N., 1965, *J. chem. Phys.*, **42,** 234.

URYÛ, N., and FRIEDBERG, S. A., 1965, *Phys. Rev.*, **140,** A 1803.

VAN AMSTEL, W. D., and DE JONGH, L. J., 1972, *Solid St. Commun.*, **11,** 1423.

VAN AMSTEL, W. D., MATSUURA, M., and DE JONGH, L. J., 1974, *Solid St. Commun.* (in the press).

VAN DEN HANDEL, J., GIJSMAN, H. M., and POULIS, N. J., 1952, *Physica*, **18,** 862.

VAN DER HOEVEN, Jr., B. J. C., TEANEY, D. T., and MORUZZI, V. L., 1968, *Phys. Rev. Lett.*, **20,** 719.

VAN DER LEEDEN, P., VAN DALEN, P. A., and DE JONGE, W. J. M., 1967, *Physica*, **33,** 202.

VAN DER LUGT, W., and POULIS, N. J., 1960, *Physica*, **26,** 917.

VAN DER MAREL, L. C., VAN DER BROEK, J., WASSCHER, J. D., and GORTER, C. J., 1955, *Physica*, **21,** 685.

VAN NIEKERK, J. N., and SCHOENING, F. R. L., 1953, *Acta crystallogr.*, **6,** 227 ; *Nature, Lond.*, **171,** 36.

VAN DER SLUYS, J. C. A., ZWEERS, B. A., and DE KLERK, D., 1967, *Physics Lett.* A, **24,** 637.

Van Stapele, R. P., Henning, J. C. M., Hardeman, G. E. G., and Bongers, P. F., 1966, *Phys. Rev.*, **150,** 310.

Van Wier, O. P., Van Peski-Tinbergen, T., and Gorter, C. J., 1959, *Physica*, **25,** 116.

Velu, E., Cadoul, D., Lécuyer, B., and Renard, J. P., 1971, *Physics Lett.* A, **36,** 443.

Velu, E., Renard, J.-P., and Dupas, C., 1972, *Solid St. Commun.*, **11,** 1.

Vicentini-Missoni, M., Joseph, R. I., Green, M. S., and Levelt Sengers, J. M. H., 1970, *Phys. Rev.* B, **1,** 2312.

Vicentini-Missoni, M., Levelt Sengers, J. M. H., and Green, M. S., 1969, *J. Res. Nat. Bur. Stand.* A, **73,** 563.

Von Schnering, H. G., 1967, *Z. anorg. allg. Chem.*, **353,** 13.

Walker, L. R., Dietz, R. E., Andres, K., and Darack, S., 1972, *Solid St. Commun.*, **11,** 593.

Walstedt, R. E., De Wijn, H. W., and Guggenheim, H. J., 1970, *Phys. Rev. Lett.*, **25,** 1119.

Watanabe, T., and Haseda, T., 1958, *J. chem. Phys.*, **29,** 1429.

Watson, R. E., Blume, M., and Vineyard, G. H., 1970, *Phys. Rev.* B, **2,** 684.

Weng, C. Y., 1969, Thesis, Carnegie-Mellon University.

Wertheim, G. K., and Buchanan, D. N. E., 1967, *Phys. Rev.*, **161,** 478.

Wertheim, G. K., Guggenheim, H. J., Levinstein, H. J., Buchanan, D. N. E., and Sherwood, R. C., 1968, *Phys. Rev.*, **173,** 614.

Widom, B., 1965, *J. chem. Phys.*, **43,** 3892.

Wielinga, R. F., 1968, Thesis, University of Leiden ; 1971, *Prog. Low Temp. Phys.*, Vol. VI, edited by C. J. Gorter (Amsterdam : North-Holland).

Wielinga, R. F., Blöte, H. W. J., Roest, J. A., and Huiskamp, W. J., 1967, *Physica*, **34,** 223.

Wielinga, R. F., and Huiskamp, W. J., 1969, *Physica*, **40,** 602.

Wilkinson, M. K., Cable, J. W., Wollan, E. O., and Koehler, W. C., 1959, *Phys. Rev.*, **113,** 497.

Windsor, C. G., and Stevenson, R. W. H., 1966, *Proc. phys. Soc.*, **87,** 501.

Wittekoek, S., Klaassen, T. O., and Poulis, N. J., 1968, *Physica*, **39,** 293.

Witteveen, H. T., 1973, Thesis, University of Leiden.

Wojtowicz, P. J., and Rayl, M., 1968, *Phys. Rev. Lett.*, **20,** 1489.

Wolf, W. P., 1970, *Proc. Int. Conf. Magn.*, Grenoble (*J. Phys.*, **32,** Suppl. C–1, p. 26).

Wolf, W. P., Ball, M., Hutchings, M. T., Leask, M. J. M., and Wyatt, A. F. G., 1962, *Proc. Int. Conf. Magn. Crystallography*, Kyoto (*J. phys. Soc. Japan*, **17,** Suppl. B, p. 443).

Wolf, W. P., Meissner, H. E., and Catanese, C. A., 1968, *J. Appl. Phys.*, **39,** 1134.

Wolf, W. P., Schneider, B., Landau, D. P., and Keen, B. E., 1972, *Phys. Rev.* B, **5,** 4472.

Wolf, W. P., and Wyatt, A. F. G., 1964, *Phys. Rev. Lett.*, **13,** 368.

Wood, D. W., and Dalton, N. W., 1966, *Proc. phys. Soc.*, **87,** 755 ; 1972, *J. Phys.* C, **5,** 1675.

Wrege, D. E., Spooner, S., and Gersch, H. A., 1971, *Proc. 17th Ann. Conf. Magn. Magn. Materials*, Chicago (*AIP Conf. Proc.*, **5,** 1334).

Wright, J. C., Moos, H. W., Colwell, J. H., Magnum, B. W., and Thornton, D. D., 1971, *Phys. Rev.* B, **3,** 843.

Yamada, I., 1970, *J. phys. Soc. Japan*, **28,** 1585 ; 1972, *Ibid.*, **33,** 979.

Yamada, I., Kubo, H., and Shimohigashi, K., 1971, *J. phys. Soc. Japan*, **30,** 896.

Yamagata, K., 1967, *J. phys. Soc. Japan*, **22,** 582.

YAMAGUCHI, Y., FUJITO, T., NISHIGUCHI, H., and DEGUCHI, Y., 1970, *Proc, 12th Int. Conf. Low Temp. Phys.*, Kyoto, p. 805.
YAMAGUCHI, Y., and SAKAMOTO, N., 1969, *J. phys. Soc. Japan*, **27,** 1444.
YAMAZAKI, H., 1973, *Proceed. Int. Conf. Magn.*, Moscow.
YAMAZAKI, H., WATANABE, K., and ABE, H., 1972, *J. phys. Soc. Japan*, **32,** 862.
YANG, C. N., 1952, *Phys. Rev.*, **85,** 808.
YANG, C. N., and YANG, C. P., 1966, *Phys. Rev.*, **150,** 321.
YELON, W. B., and BIRGENEAU, R. J., 1972, *Phys. Rev.* B, **5,** 2615.

Author Index

Benedek, G. B., 121, 122, 159, 204

Index to Substances

Bold figures refer to sections specifically dealing with the substance

YSICS

Technology, London

sity of Cambridge

EL SYSTEMS